Universitext

Universitext is a series of textbooks that presents material from a wide variety of mathematical disciplines at master's level and beyond. The books, often well class-tested by their author, may have an informal, personal, or even experimental approach to their subject matter. Some of the most successful and established books in the series have evolved through several editions, always following the evolution of teaching curricula, into very polished texts.

Thus as research topics trickle down into graduate-level teaching, first textbooks written for new, cutting-edge courses may find their way into *Universitext*.

Junjiro Noguchi

Basic Oka Theory in Several Complex Variables

Springer

Junjiro Noguchi
Graduate School of Mathematical Sciences
The University of Tokyo
Tokyo, Japan

ISSN 0172-5939 ISSN 2191-6675 (electronic)
Universitext
ISBN 978-981-97-2055-2 ISBN 978-981-97-2056-9 (eBook)
https://doi.org/10.1007/978-981-97-2056-9

This Springer imprint is published by the registered company Springer Nature Singapore Pte Ltd.
The registered company address is: 152 Beach Road, #21-01/04 Gateway East, Singapore 189721, Singapore

Paper in this product is recyclable.

Preface

This book provides a comprehensive self-contained account of Oka theory, mainly concerned with the proofs of the Three Big Problems of approximation, Cousin and pseudoconvexity (Hartogs, Levi) stated below, which were solved by Kiyoshi Oka and form the basics of complex analysis in several variables. It is the purpose to present a textbook of course lectures that follow on from complex function theory of one variable. The presentation is intended to be readable, enjoyable and self-contained for those from beginners in mathematics to researchers interested in complex analysis in several variables and complex geometry.

The nature of the present book should be evinced by the following two points:

- We develop the theory by the method of the Oka Extension of holomorphic functions from a complex submanifold of a polydisk to the whole polydisk (Oka's *Joku-Iko Principle*[1]);
- We represent Oka's original proofs, following his unpublished five papers of 1943 and Oka IX (1953).

In those unpublished five papers (in Japanese), historically, the pseudoconvexity problem (Hartogs' Inverse Problem, Levi's Problem) was first solved for domains of $\mathbf{C}^n$ ($n \geq 2$), and furthermore for unramified domains over $\mathbf{C}^n$ as well (see [50], [44] which contains the English translation of the last one of the five papers).

We derive the Oka Extension of the Joku-Iko Principle from the coherence of the sheaf $\mathscr{O}_{\mathbf{C}^n}$ of holomorphic functions on $\mathbf{C}^n$ (Oka's First Coherence Theorem), which is proved by Weierstrass' Preparation Theorem; Weierstrass' Preparation Theorem is shown by the residue theorem in one variable. In this way we use only elementary techniques, yet reach the core of the theory.

The basis of analytic function theory of several variables or complex analysis in several variables was founded till the middle of the 1950s. Just afterwards new theories and generalizations were developed. Also the simplification of the theory

[1] "Joku-Iko" is pronounced "dzóuku ikou" (not aiko). The Joku-Iko Principle is a term due to K. Oka himself, and the guiding methodological principle of Oka theory all through his works (cf. footnote 2).

has been done, but the difficulties of the introductory part from the theory of one variable to that of several variables has remained considerably for beginners. The present book aims to provide a smooth introduction for that part (cf. [39], [43], [2]).

The Three Big Problems which were summarized by Behnke–Thullen [4] in 1934 are stated as follows:

(P1) Approximation Problem (problem of developments) (Runge's Theorem in one variable).

(P2) Cousin Problem (I and II) (Mittag-Leffler and Weierstrass Theorems in one variable).

(P3) Pseudoconvexity Problem (Hartogs' Inverse Problem, Levi's Problem) (the natural boundary problem of analytic continuation).

K. Oka solved all of these problems in Oka I–IX ([48], [49]); they are roughly classified into two groups:

(G1) Oka I–VI + IX.

(G2) Oka VII, VIII, IX.

Oka IX contains works belonging to both of the groups, and the Three Big Problems were solved by the first group (G1); in the second group (G2) he proved his Three Coherence Theorems, aiming for development beyond the original problems.

In the present book we restrict ourselves to the results of group (G1); this is the reason for the use of "Basic" in the title. Here we do not use:

- Cohomology theory with coefficients in sheaves;
- L^2-$\bar{\partial}$ method.

The solution of the Pseudoconvexity Problem (P3) is the culmination of the works (G1); Oka's methods consist of:

(i) (In VI, 1942; *univalent domains of* $\dim n = 2$) Cousin Problem & Weil's integral formula & Fredholm integral equation of the second kind;

(ii) (In VII–XI, 1943, unpublished (cf. [44]); *unramified multivalent domains over* $\mathbf{C}^n$ *of general* $n \geq 2$) Cousin Problem & "Primitive Coherence Theorem" & the *Joku-Iko Principle* [2] & Fredholm integral equation of the second kind type (a variant of the equation of the second kind combined with the Joku-Iko Principle);

---

[2] A direct translation might be "a transfer (=Iko) to an upper space (=Joku)". He found the principle in the study of Oka I (1936) and II (1937), and used it all through his works, till Oka IX (1953). It is an idea to solve the problem caused by the increased number of variables by increasing the number of variables more; one embeds the initial domains into simply shaped polydisks of higher dimensions, extends the problems over the polydisks, and then solves them by making use of the simplicity of polydisks.

In T. Nishino [37] the term was translated to the "Lifting Principle". As a matter of fact, the statement itself holds more generally for any subvarieties of Stein spaces, and so may be called an analytic extension or interpolation; then, however, the spirit of the wording will be lost, since the general case is proved through embeddings of analytic polyhedra into polydisks. So, here we prefer to use the original term as in [39].

(iii) (In IX, 1953, published; *unramified multivalent domains over* $\mathbf{C}^n$ *of general* $n \geq 2$) Cousin Problem & "Coherence Theorem" & the *Joku-Iko Principle* & Fredholm integral equation of the second kind type.

After all, K. Oka proved the Pseudoconvexity Problem (P3) three times. It is noticed that in (ii) and (iii) above, Oka proved the Approximation Theorem and Cousin Problem for unramified domains, multivalent in general, over $\mathbf{C}^n$ by a new method, which had been proved for univalent domains in his former papers (I–III). The content of Oka IX ((iii) above) is essentially the same as that of (ii) except for the part of the "Coherence Theorem" obtained in (G2).

In Oka VI ((i) above), he mentioned the validity of the result for general dimension $n \geq 2$ in a modest phrase at the end of the paper:

> L'auteur pense que cette conclusion sera aussi indépendante des nombres de variables complexes.

In fact, the method of (i) was later generalized to univalent domains of general dimension n by S. Hitotsumatsu [28] 1949, H.J. Bremermann [8] 1954, and F. Norguet [45] 1954, independently. The method of (iii) was generalized for abstract complex spaces by T. Nishino [36] 1962 (cf. [37]), and A. Andreotti–R. Narasimhan [3] 1964.

In (ii) above, Oka formulated and proved a kind of "Primitive Coherence Theorem" with a certain condition, yet sufficient for the purpose, and he used some type of the Fredholm integral equation of the second kind. In Oka IX, he replaced the "Primitive Coherence Theorem" with his Coherence Theorems. The present book, hopefully, presents an easy comprehensive account of that theory.

H. Cartan has written ([49], p. XII):

> ....
>
> Mais il faut avouer que les aspects techniques de ses démonstrations et le mode de présentation de ses résultats rendent difficile la tâche du lecteur, et que ce n'est qu'au prix d'un réel effort que l'on parvient à saisir la portée de ses résultats, qui est considérable. C'est pourquoi il est peut-être encore utile aujourd'hui, en hommage au grand créateur que fut Kiyoshi OKA, de présenter l'ensemble de son œuvre.
>
> ....

In English (by Noguchi),

> ....
>
> But we must admit that the technical aspects of his proofs and the mode of presentation of his results make it difficult to read, and that it is possible only at the cost of a real effort to grasp the scope of its results, which is considerable. This is why it is perhaps still useful today, for the homage of the great creator that was Kiyoshi OKA, to present the collection of his work.
>
> ....

It is interesting that, looking for an easier introduction of analytic function theory of several variables, we came back to Oka's original method.

To the best of the author's knowledge, there is no book or monograph presenting Oka's original method except for Nishino [37], while there are many for the developments or other proofs obtained after Oka's works. The author hopes that the present book fills the gap even a little, and is useful to recognize Oka's original ideas.

It should be worthwhile for students and researchers to look into the original work of K. Oka, which may still contain some new ideas. Therefore the prerequisites of the present book are minimized to the contents from standard complex analysis in one variable (cf., e.g., [38]), which range from Cauchy's integral formula to Riemann's mapping theorem. We explain the necessary contents of topology, rings and modules; if they are not sufficient, it may suffice to confer any nearby books on those elementary materials. We avoided the general notion of manifolds.

Now we briefly describe the contents of the present book. In Chapter 1 we begin with the definition of holomorphic or analytic functions of several variables, and convergent power series. We then explain Hartogs' phenomenon, which was the starting point of analytic function theory of several variables. For the preparation of the chapters in the sequel we show the Runge approximation theorem on convex cylinder domains, and explain the Cousin integral and analytic subsets.

Chapter 2 describes the notion of analytic sheaves and the coherence. Analytic sheaves will be defined just as sets or as collections of rings or modules without topology. We then show Weierstrass' Preparation Theorem by making use of the residue theorem in one variable. We then prove Oka's First Coherence Theorem, Cartan's Matrix Lemma and then, Oka's Syzygy Lemma, with which we finally derive the Oka Extension of the Joku-Iko Principle.

Chapter 3 is devoted to the theory of domains of holomorphy and holomorphically convex domains. We prove the foundational Cartan–Thullen Theorem, which asserts the equivalence of those two domains in the univalent (schlicht) case. Then an analytic polyhedron is introduced, and the Oka–Weil Approximation Theorem is proved as a solution of the First Big Problem (P1) above by means of the Joku-Iko Principle. As a special case of one variable, we show Runge's Approximation Theorem.

Subsequently the Cousin Problem (the Second Big Problem (P2)) is dealt with. Here we formulate the *Continuous Cousin Problem* by which we unify the treatment of the Cousin I, II Problems and the problem of $\bar{\partial}$-equation for functions, where the *Oka Principle* is included. We then solve the Continuous Cousin Problem on holomorphically convex domains, equivalently on domains of holomorphy in the univalent case. We discuss the applications to the case of one variable, proving the Mittag-Leffler and Weierstrass Theorems.

As an application of the Continuous Cousin Problem we prove the Hartogs extension of holomorphic functions over a compact subset of a domain of $\mathbf{C}^n$. By a similar method of the proof of the Continuous Cousin Problem, we solve the *interpolation problem* for complex submanifolds of univalent domains of holomorphy.

At the end of Chapter 3 we introduce the notion of multivalent domains over $\mathbf{C}^n$, which are here assumed to be unramified. We define the envelope of holomorphy of such a domain, and the notion of domains of holomorphy in the multivalent case. We introduce Stein domains in the multivalent case, and see that the results obtained for univalent domains of holomorphy remain valid for multivalent Stein domains.

In Chapters 4 and 5 we deal with the Pseudoconvexity Problem (P3) for domains over $\mathbf{C}^n$, where domains are assumed to be multivalent in general. Chapter

4 is devoted to the formulations and the reductions of Problem (P3). Firstly, we introduce the notion of plurisubharmonic functions. Using it, we prove Hartogs' separate analyticity theorem; in the course, we explain the Baire Category Theorem. Then, several kinds of pseudoconvexities of domains are defined. We discuss the equivalence and the relations of those pseudoconvexities, and formulate what is the pseudoconvexity problem. We then prove *Oka's Theorem of Boundary Distance Functions*. This serves the first important step toward the solution of Problem (P3). As an application we prove the *Tube Theorem* due to S. Bochner and K. Stein ($n = 2$).

Finally, in Chapter 5, we solve the Pseudoconvexity Problem (P3), the last of the Three Big Problems, which is formulated in the previous chapter. To begin with, we introduce the notion of a semi-normed space and a Fréchet space, and prove Banach's Open Mapping Theorem for mapppings from Fréchet spaces to Baire spaces. We then show the Oka Extension of the Joku-Iko Principle with estimate.

We give two proofs of the Steinness of strongly pseudoconvex domains (Levi's Problem); the first is K. Oka's and the second is the one due to H. Grauert; there is some similarity in the two proofs, which should be interesting for comparison. T. Nishino's book [37] presents the proof of Oka IX (1953) in more generalized form for complex spaces. In the proof we formulate a variant of the Fredholm integral equation of the second kind combined with the Oka Extension; we call it a *Fredholm integral equation of the second kind type*. This is the key of the proof. It is solved by a successive approximation and the convergence is obtained by the method of majorants. It is rather surprising to see such a difficult problem being solved by such an elementary method.

The second method due to Grauert is the well-known "bumping method" combined with *L. Schwartz's Fredholm Theorem*,[3] of which a short but complete proof is given; by making use of an idea of J.P. Demailly, we prove it in a slightly generalized form for Baire spaces as image spaces. For the purpose we introduce the first cohomology $H^1(\star, \mathscr{O})$.

With these preparations we finally prove *Oka's Pseudoconvexity Theorem* that pseudoconvex domains unramified over $\mathbf{C}^n$ are Stein.

At the end of each chapter some historical comments are made from the author's viewpoint and knowledge; it is expected to motivate the readers to confer other mentioned resources, but they are far from complete.

The present book is an outcome of the author [43], largely rewritten with a number of additions. It does not aim to give the full exposition of the fundamentals of analytic function theory of several variables or complex analysis in several variables. But, the related topics are mentioned in various places with references, which readers are encouraged to confer. And some of them are presented in Exercises, which readers are expected to solve by themselves. It is also recommended for readers to consult those books and monographs referred to throughout.

The author will be more than happy if readers get interested in the present subject through this book providing the elementary but self-contained proofs of the Three

---

[3] This term is due to A. Andreotti, according to a personal communication with A. Huckleberry (cf. Jahresber. Dtsch. Math.Ver. **115** (2013), 21-45).

Big Problems which form the basics of several complex variables, and if the original works of Kiyoshi Oka, full of creative ideas, are enjoyed and recognized more deeply.

While the author was writing the present book, he gave several talks at the weekly seminar on complex analysis and geometry, the University of Tokyo, Komaba, Tokyo (so-called Monday Morning Seminar), having a number of discussions with the members, which were very helpful and encouraging. In May 2017, he was kindly invited by Professor Sachiko Hamano of Osaka City University then to give an intensive one-week lecture course, based on the first draft of the book. In July of the same year, he gave a seminary talk on Oka's original method and related topics at Tor Vergata, Rome at the invitation of Professor Filippo Bracci. In March 2019, he gave a talk on the book at the Japan–Iceland Workshop, "Holomorphic Maps, Pluripotentials and Complex Geometry" at the invitation of Professor Masanori Adachi (Shizuoka University), and in May of that year he gave a series of talks at the invitation of Professor Steven Lu at Montreal; in July he gave a series of lectures based on this book at the Workshop, "Summer Program on Complex Geometry and Several Complex Variables" at Shanghai Center for Mathematical Sciences, Fudan University, Shanghai at the invitation of Professor Min Ru (University of Houston). The author learned a number of references on the pseudoconvexity problem from Professor Makoto Abe (Hiroshima University), and had many valuable suggestions and comments from Professors Hiroshi Yamaguchi (Shiga University, Emeritus; Nara Women University), Sachiko Hamano (Kyoto Sangyo University), Yohei Komori (Waseda Univerity, Tokyo), Shigeharu Takayama (University of Tokyo) and Jöel Merker (Université Paris-Saclay). Professor Viorel Vâjâitu (Université des Sciences et Technologies de Lille) kindly suggested useful information in Remarks at the end of Chap. 5. To all of them the author would like to express his sincere gratitude.

The author acknowledges with many thanks the support of JSPS KAKENHI Grant Number JP19K03511, which has been always helpful in carrying out the present project.

Kamakura
Spring, 2024
Junjiro Noguchi

Contents

1 Holomorphic Functions 1
1.1 Holomorphic Functions of Several Variables 1
1.1.1 Open Balls and Polydisks of $\mathbf{C}^n$ 1
1.1.2 Definition of Holomorphic Functions 3
1.1.3 Sequences and Series of Functions 6
1.1.4 Power Series of Several Variables 8
1.1.5 Elementary Properties of Holomorphic Functions of Several Variables 10
1.2 Analytic Continuation and Hartogs' Phenomenon 14
1.3 Runge Approximation on Convex Cylinder Domains 18
1.3.1 Cousin Integral 19
1.4 Implicit and Inverse Function Theorems 21
1.5 Analytic Subsets 24
Exercises 28

2 Coherent Sheaves and Oka's Joku-Iko Principle 31
2.1 Notion of Analytic Sheaves 31
2.1.1 Definitions of Rings and Modules 31
2.1.2 Analytic Sheaves 33
2.2 Coherent Sheaves 36
2.2.1 Locally Finite Sheaves 36
2.2.2 Coherent Sheaves 39
2.3 Oka's First Coherence Theorem 43
2.3.1 Weierstrass' Preparation Theorem 43
2.3.2 Oka's First Coherence Theorem 47
2.3.3 Coherence of Ideal Sheaves of Complex Submanifolds 52
2.4 Cartan's Merging Lemma 53
2.4.1 Matrices and Matrix-Valued Functions 53
2.4.2 Cartan's Matrix Decomposition 57
2.4.3 Cartan's Merging Lemma 61

2.5 Oka's Joku-Iko Principle 63
2.5.1 Oka Syzygy 63
2.5.2 Oka Extension of the Joku-Iko Principle 68
Exercises 71

3 Domains of Holomorphy 73
3.1 Definitions and Elementary Properties 73
3.1.1 Relatively Compact Hull 73
3.1.2 Domain of Holomorphy and Holomorphic Convexity 75
3.2 Cartan–Thullen Theorem 78
3.3 Analytic Polyhedron and Oka–Weil Approximation 83
3.3.1 Analytic Polyhedron 83
3.3.2 Oka–Weil Approximation Theorem 85
3.3.3 Runge Approximation Theorem (One Variable) 87
3.4 Cousin Problem 89
3.4.1 Cousin I Problem 91
3.4.2 Continuous Cousin Problem 92
3.4.3 Cousin I Problem — continued 97
3.4.4 Hartogs Extension over a Compact Subset 98
3.4.5 Mittag-Leffler Theorem (One Variable) 100
3.4.6 Cousin II Problem and Oka Principle 101
3.4.7 Weierstrass' Theorem (One Variable) 105
3.4.8 $\bar{\partial}$-Equation 108
3.5 Analytic Interpolation Problem 112
3.6 Unramified Domains over $\mathbf{C}^n$ 116
3.7 Stein Domains over $\mathbf{C}^n$ 122
3.8 Supplement: Ideal Boundary 124
Exercises 128

4 Pseudoconvex Domains I — Problem and Reduction 131
4.1 Plurisubharmonic Functions 131
4.1.1 Subharmonic Functions (One Variable) 131
4.1.2 Plurisubharmonic Functions 134
4.1.3 Smoothing 136
4.2 Hartogs' Separate Analyticity 139
4.2.1 Baire Category Theorem 139
4.2.2 Separate Analyticity 141
4.3 Pseudoconvexity 144
4.3.1 Pseudoconvexity Problem 144
4.3.2 Bochner's Tube Theorem 154
4.3.3 Pseudoconvex Boundary 157
4.3.4 Levi Pseudoconvexity 162
4.3.5 Strongly Pseudoconvex Boundary Points and Stein Domains 166
Exercises 173

5 Pseudoconvex Domains II — Solution ... 175
5.1 The Oka Extension with Estimate ... 175
5.1.1 Preparation from Topological Vector Spaces ... 175
5.1.2 The Oka Extension with Estimate ... 178
5.2 Strongly Pseudoconvex Domains ... 179
5.2.1 Oka's Method ... 179
5.2.2 Grauert's Method ... 189
5.3 Oka's Pseudoconvexity Theorem ... 202
Exercises ... 206

Afterword — Historical Comments ... 209

References ... 211

Index ... 215

Symbols ... 221

Conventions

(i) Theorems, equations etc. are numbered consecutively. Here an equation is numbered as (1.1.1) with parentheses; the first 1 stands for the chapter number and the second 1 for the section number.

(ii) The standard notation of set theory is assumed. For example, maps, images and inverse images are used without specific mention. Also, "$\forall x$" means "every or all x" and "$\exists x$" means "some x" or "existence of x".

(iii) The standard notion of the euclidean topology of $\mathbf{R}^n$ (also of $\mathbf{C}^n \cong \mathbf{R}^{2n}$) is assumed. For example, open sets, closed sets, the connectedness, and the boundary ∂U of a subset $U \subset \mathbf{R}^n$ are used without specific comments.

(iv) By $A := B$ we mean that A is defined to be B, or we put A to be B.

(v) A relation $x \sim y$ for all two elements $x, y \in S$ of a set S is called a *equivalence relation* if the following conditions are satisfied: (i) $x \sim x$ ($\forall x \in S$) (reflexive law); (ii) if two elements $x, y \in S$ satisfy $x \sim y$, then $y \sim x$ (symmetric law); (iii) if three elements $x, y, z \in S$ satisfy $x \sim y$ and $y \sim z$, then $x \sim z$ (transitive law).

(vi) For a set S, $|S|$ denotes its cardinality.

(vii) A map $f : X \to Y$ is said to be *injective* or an *injection* if $f(x_1) \neq f(x_2)$ for every distinct $x_1, x_2 \in X$, and to be *surjective* or a *surjection* if $f(X) = Y$. If f is injective and surjective, it is said to be *bijective* or called a *bijection*. The restriction of f to a subset $E \subset X$ is denoted by $f|_E$.

(viii) The set of natural numbers (positive integers) is denoted by $\mathbf{N}$, the set of integers by $\mathbf{Z}$, the set of rational numbers by $\mathbf{Q}$, the set of real numbers by $\mathbf{R}$, the set of complex numbers by $\mathbf{C}$, and the imaginary unit by i, as usual. We write $\mathbf{C}^* := \mathbf{C} \setminus \{0\}$. The set of non-negative integers (resp. numbers) is denoted by $\mathbf{Z}_+$ (resp. $\mathbf{R}_+$).

(ix) For a complex number $z = x + iy \in \mathbf{C}$ we set $\Re z = x$ and $\Im z = y$.

(x) *Monotone increasing* and *monotone decreasing* are used in the sense including the case of equality: e.g., a sequence of functions $\{\varphi_\nu(x)\}_{\nu=1}^{\infty}$ is said to be monotone increasing if for every point x of the defining domain $\varphi_\nu(x) \leq \varphi_{\nu+1}(x)$ for all $\nu = 1, 2, \ldots$.

(xi) For a (vector-valued) function f on a subset $U \subset \mathbf{R}^n$ (more generally, on a topological space U) the *support* denoted by $\operatorname{Supp} f$ is defined as the (topological) closure of the set $\{x \in U : f(x) \neq 0\}$ in U.

(xii) A function f defined on an open subset $U \subset \mathbf{R}^m$ is said to be of C^k-class ($1 \leq k \leq \infty$) if f is k-times continuously differentiable. $\mathscr{C}^k(U)$ denotes the set of all functions of C^k-class on U. $\mathscr{C}^k_0(U)$ stands for the set of all $f \in \mathscr{C}^k(U)$ with compact supports.

(xiii) For subsets $A \subset B \subset \mathbf{C}^n$ (or $\mathbf{R}^n$), "$A \Subset B$" means that the closure $\bar{A}$ is compact and $\bar{A} \subset B$; in this case, A is said to be *relatively compact* (in B).

(xiv) A continuous map $f : V \to W$ from a set $V \subset \mathbf{C}^m$ (or $\mathbf{R}^m$) to another $W \subset \mathbf{C}^n$ (or $\mathbf{R}^n$) (more generally, between locally compact topological spaces) is said to be *proper* if for every compact subset $K \subset W$, the inverse image $f^{-1}K$ is compact, too.

(xv) (Landau symbol) With a small variable $r > 0$ and a constant $\alpha \geq 0$, $o(r^\alpha)$ stands for a term or a quantity such that $\lim_{r \to +0} o(r^\alpha)/r^\alpha = 0$. In particular, if $\alpha = 0$, $\lim_{r \to +0} o(1) = 0$.

Chapter 1
Holomorphic Functions

We define holomorphic functions of n complex variables and present the elementary properties. A characteristic caused by increasing the number n of variables more than one is the so-called *Hartogs phenomenon* in the analytic continuation, such that all holomorphic functions in a domain of $\mathbf{C}^n$ $(n \geq 2)$ are simultaneously analytically continued over a strictly larger domain; such a phenomenon never occurs in one variable. We also prepare some basic notion necessary in the later chapters.

1.1 Holomorphic Functions of Several Variables

1.1.1 Open Balls and Polydisks of $\mathbf{C}^n$

Let $n \in \mathbf{N}$. We denote by $\mathbf{C}^n$ the complex vector space of the n-product of the complex plane $\mathbf{C}$ and by $z = (z_1, \ldots, z_n) \in \mathbf{C}^n$ the natural coordinate system. We define the *euclidean norm* by

(1.1.1) $$\|z\| = \sqrt{|z_1|^2 + \cdots + |z_n|^2} \ (\geq 0).$$

Put

(1.1.2) $$\mathrm{B}(a; r) = \{z \in \mathbf{C}^n : \|z - a\| < r\}, \quad \mathrm{B}(r) = \mathrm{B}(0; r),$$

which is called an *open ball* or simply a *ball* with center $a \in \mathbf{C}^n$ and radius r (> 0). In particular, $\mathrm{B}(1)$ is called the *unit ball.*

In the case of $n = 1$, we call it a *disk* and write

(1.1.3) $$\Delta(a; r) = \{z \in \mathbf{C} : |z - a| < r\}, \quad \Delta(r) = \Delta(0; r).$$

We call $\Delta(1)$ the *unit disk*. The closure of $\Delta(a; r)$ (resp. $\Delta(r)$) is denoted by $\bar{\Delta}(a; r)$ (resp. $\bar{\Delta}(r)$) and called a closed disk.

For a point $a = (a_1, \ldots, a_n) \in \mathbf{C}^n$ and positive numbers $r_j > 0$, $1 \leq j \leq n$, we put

J. Noguchi, *Basic Oka Theory in Several Complex Variables*, Universitext,
https://doi.org/10.1007/978-981-97-2056-9_1

(1.1.4) $$\mathrm{P}\Delta(a;(r_j)) = \{z = (z_j) \in \mathbf{C}^n : |z_j - a_j| < r_j,\ 1 \le j \le n\},$$

which is called a *polydisk* of *polyradius* (r_j) with center at a; for $a = 0$ we write

$$\mathrm{P}\Delta((r_j)) = \mathrm{P}\Delta(0;(r_j)).$$

With all $r_j = 1$ $(1 \le j \le n)$, $\mathrm{P}\Delta((1,\ldots,1))$ is called the *unit polydisk*.

We use the above notation all through the present book.

A connected open set $U \subset \mathbf{C}^n$ is called a *domain*. The closure $\bar{U}$ of U is called a *closed domain*. For example, a polydisk $\mathrm{P}\Delta(a;(r_j))$ is a domain, and its closure

$$\overline{\mathrm{P}\Delta}(a;(r_j)) = \{z = (z_j) \in \mathbf{C}^n : |z_j - a_j| \le r_j,\ 1 \le j \le n\}$$

is a closed domain, called a *closed polydisk*.

A subset $A \subset \mathbf{C}^n$ is called a *cylinder*, if there are subsets $A_j \subset \mathbf{C}$ $(1 \le j \le n)$ such that $A = \prod_{j=1}^n A_j$; in particular, if every A_j is a domain, A is called a *cylinder domain*,

A subset $B \subset \mathbf{R}^m$ of the real m-dimensional vector space $\mathbf{R}^m$ is said to be (affine) *convex* if for arbitrary two points $x, y \in B$ the line segment connecting them is contained in B, i.e.,

$$(1-t)x + ty \in B, \quad 0 \le t \le 1.$$

The smallest convex set containing B is called the *convex hull* of B (the existence is clear), and denoted by $\mathrm{ch}(B)$. A subset $A \subset \mathbf{C}^n$ is said to be convex if with the natural identification $\mathbf{C}^n \cong \mathbf{R}^{2n}$, A is convex in $\mathbf{R}^{2n}$.

Proposition 1.1.5. *Let $E \Subset \mathbf{C}$ be a compact convex subset with a neighborhood $U \Supset E$. Then there is an open convex polygon G such that $E \Subset G \Subset U$.*

Proof. Without loss of generality, U may be assumed to be bounded. Since E is convex, there is a unique line segment ℓ_0 of the shortest distance connecting a point $p \in \partial U$ and E (i.e., points of E). Let ℓ_1 be the line passing through the middle point of ℓ_0 and orthogonal to ℓ_0. Since $E \cap \ell_1 = \emptyset$, ℓ_1 divides the plane $\mathbf{C}$ to two half-planes, one of which contains E, and the other has no intersection with E (cf. Fig. 1.1). The latter is denoted by H, so that $H \cap \partial U \ni p$. Since ∂U is compact, there are finitely

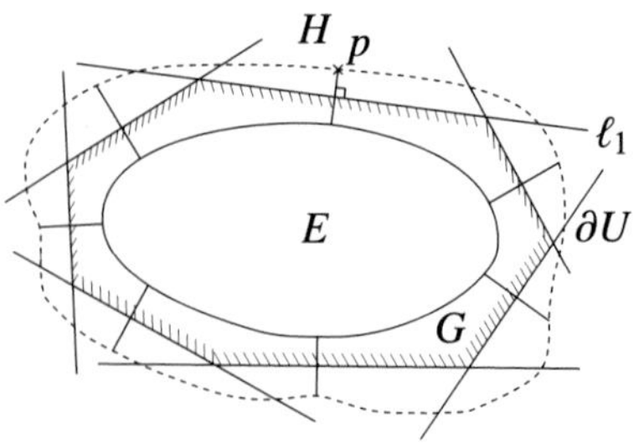

Fig. 1.1 Convex polygon neighborhood.

many such half-planes $H_1, \ldots, H_k$ satisfying

$$\partial U \Subset \bigcup_{h=1}^{k} H_h, \qquad E \Subset \bigcap_{h=1}^{k} (\mathbf{C} \setminus \bar{H}_h) \Subset U.$$

Then we get $G := \bigcap_{h=1}^{k} (\mathbf{C} \setminus \bar{H}_h)$ with $E \Subset G \Subset U$. □

1.1.2 Definition of Holomorphic Functions

We consider a function $\varphi : A \to \mathbf{C}$ defined on a subset $A \subset \mathbf{C}^n$. We say that φ is *continuous* if for every point $a \in A$ and every $\varepsilon > 0$ there is a polyradius (δ_j) satisfying

$$|\varphi(z) - \varphi(a)| < \varepsilon, \quad \forall z \in \mathrm{P}\Delta(a; (\delta_j)) \cap A.$$

Let $x_j = \Re z_j$ and $y_j = \Im z_j$ be the real and imaginary parts of the complex variables $z_j = x_j + i y_j, 1 \le j \le n$, respectively. Then the *holomorphic partial differential operators* and *anti-holomorphic partial differential operators* are defined respectively by

(1.1.6)
$$\frac{\partial}{\partial z_j} = \frac{1}{2}\left(\frac{\partial}{\partial x_j} + \frac{1}{i}\frac{\partial}{\partial y_j}\right), \quad 1 \le j \le n,$$
$$\frac{\partial}{\partial \bar{z}_j} = \frac{1}{2}\left(\frac{\partial}{\partial x_j} - \frac{1}{i}\frac{\partial}{\partial y_j}\right), \quad 1 \le j \le n.$$

A function $\varphi(z_1, \ldots, z_n)$ is said to be of C^l-class or a C^l function ($l \in \mathbf{N} \cup \{\infty\}$) when it is of C^l-class as a function of the real and imaginary parts, $(x_1, y_1, \ldots, x_n, y_n)$. Further, if $z_j(\xi) = z_j(\xi_1, \ldots, \xi_m)$ are C^1 functions of m complex variables $\xi = (\xi_1, \ldots, \xi_m)$, we have

(1.1.7)
$$\frac{\partial \varphi(z(\xi))}{\partial \xi_k} = \sum_{j=1}^{n} \left(\frac{\partial \varphi}{\partial z_j}(z(\xi)) \cdot \frac{\partial z_j}{\partial \xi_k}(\xi) + \frac{\partial \varphi}{\partial \bar{z}_j}(z(\xi)) \cdot \frac{\partial \bar{z}_j}{\partial \xi_k}(\xi) \right),$$
$$\frac{\partial \varphi(z(\xi))}{\partial \bar{\xi}_k} = \sum_{j=1}^{n} \left(\frac{\partial \varphi}{\partial z_j}(z(\xi)) \cdot \frac{\partial z_j}{\partial \bar{\xi}_k}(\xi) + \frac{\partial \varphi}{\partial \bar{z}_j}(z(\xi)) \cdot \frac{\partial \bar{z}_j}{\partial \bar{\xi}_k}(\xi) \right),$$

which follow from the formula of the differentiation of composed functions of real variables.

For a multi-index $\alpha = (\alpha_1, \ldots, \alpha_n) \in \mathbf{Z}_+^n$ we set

$$|\alpha| = \sum_{j=1}^{n} \alpha_j, \quad \alpha! = \prod_{j=1}^{n} \alpha_j!.$$

We define the holomorphic partial differential operator of (multi-) order α by

$$(1.1.8) \qquad \partial^\alpha = \partial_z^\alpha = \frac{\partial^{|\alpha|}}{\partial z_1^{\alpha_1} \cdots \partial z_n^{\alpha_n}}.$$

Let $\Omega \subset \mathbf{C}^n$ be an open subset.

Definition 1.1.9 (Holomorphic function). (i) A *holomorphic* function $f : \Omega \to \mathbf{C}$ on Ω is a function of C^1-class satisfying the so-called *Cauchy–Riemann equations*,

$$(1.1.10) \qquad \frac{\partial f}{\partial \bar{z}_j}(z) = 0, \quad 1 \leq j \leq n.$$

(ii) In general, a function $f : \Omega \to \mathbf{C}$ is *separately holomorphic* or *separately analytic* if in a neighborhood of every point $a = (a_1, \ldots, a_n) \in \Omega$, the function

$$z_j \longmapsto f(a_1, \ldots, a_{j-1}, z_j, a_{j+1}, \ldots, a_n)$$

of one variable z_j with fixed others is holomorphic in a neighborhood of $z_j = a_j$.

Proposition 1.1.11. *If $\varphi(z)$ and $z_j(\xi)$ in (1.1.7) are holomorphic, then the composed function $\varphi(z(\xi))$ is holomorphic in ξ, and satisfies*

$$\frac{\partial \varphi(z(\xi))}{\partial \xi_k} = \sum_{j=1}^{n} \frac{\partial \varphi}{\partial z_j}(z(\xi)) \cdot \frac{\partial z_j}{\partial \xi_k}(\xi), \quad 1 \leq k \leq m.$$

Proof. It is immediate by (1.1.7) and (1.1.10). □

Example 1.1.12. A polynomial $P(z)$ in variables $z = (z_1, z_2, \ldots, z_n)$ with coefficients in $\mathbf{C}$ defines naturally a holomorphic function in $\mathbf{C}^n$. We denote by $\mathbf{C}[(z_j)] := \mathbf{C}[z_1, \ldots, z_n]$ the set of of all polynomials in z with complex coefficients.

As shown in what follows, holomorphic functions are expanded to convergent power series in a neighborhood of each point where they are defined (Theorem 1.1.34), and hence analytic in all variables. Therefore, holomorphic functions are often called *analytic functions*; here we use mainly "holomorphic" functions.

We denote by $\mathscr{O}(\Omega)$ the set of all holomorphic functions on Ω. We then have the natural inclusions:

$$\mathbf{C}[z_1, \ldots, z_n] \subset \mathscr{O}(\mathbf{C}^n) \subset \mathscr{O}(\Omega).$$

When a point $a \in \mathbf{C}^n$ is contained in the a set A where a function f is defined, f being "holomorphic at a" means that there are a neighborhood $V\ (\subset \mathbf{C}^n)$ of a and $g \in \mathscr{O}(V)$ with $g|_{A \cap V} = f|_{A \cap V}$.

Theorem 1.1.13 (Cauchy's integral formula). *Let $f(z)$ be a continuous function in a neighborhood of a closed polydisk $\overline{\mathrm{P}\Delta}(a; (r_j))$. If $f(z)$ is separately holomorphic in $\mathrm{P}\Delta(a; (r_j))$, then the following integral formula holds:*

(1.1.14)
$$f(z_1,\ldots,z_n)=\left(\frac{1}{2\pi i}\right)^n\int\limits_{|\zeta_1-a_1|=r_1}\cdots\int\limits_{|\zeta_n-a_n|=r_n}\frac{f(\zeta_1,\ldots,\zeta_n)}{(\zeta_1-z_1)\cdots(\zeta_n-z_n)}d\zeta_1\cdots d\zeta_n,\quad z\in \mathrm{P}\Delta(a;(r_j)).$$

Proof. Repeat Cauchy's integral formula of one variable. □

The function
$$\frac{1}{(\zeta_1-z_1)\cdots(\zeta_n-z_n)}$$
in (1.1.14) is called the *Cauchy kernel* (of n variables).

Theorem 1.1.15. *Let $\Omega\subset\mathbf{C}^n$ be an open set.*

(i) *A continuous function $f:\Omega\to\mathbf{C}$ is holomorphic if and only if it is separately holomorphic.*

(ii) *Any holomorphic function $f\in\mathscr{O}(\Omega)$ is of C^∞-class.*

(iii) *Let $f\in\mathscr{O}(\Omega)$ and let $z\in\mathrm{P}\Delta(a;(r_j))\Subset\Omega$. For a multi-index $\alpha=(\alpha_j)$ we have*

(1.1.16)
$$\partial^\alpha f(z_1,\ldots,z_n)=\alpha!\left(\frac{1}{2\pi i}\right)^n\int\limits_{|\zeta_1-a_1|=r_1}\cdots\int\limits_{|\zeta_n-a_n|=r_n}\frac{f(\zeta_1,\ldots,\zeta_n)}{(\zeta_1-z_1)^{\alpha_1+1}\cdots(\zeta_n-z_n)^{\alpha_n+1}}d\zeta_1\cdots d\zeta_n.$$

Proof. (i), (ii): These are immediate from Theorem 1.1.13.

(iii): From integral formula (1.1.14) it follows, e.g., that

$$\frac{\partial}{\partial z_1}f(z_1,\ldots,z_n)=\left(\frac{1}{2\pi i}\right)^n\int\limits_{|\zeta_1-a_1|=r_1}\cdots\int\limits_{|\zeta_n-a_n|=r_n}\frac{f(\zeta_1,\ldots,\zeta_n)}{(\zeta_1-z_1)^2\cdots(\zeta_n-z_n)}d\zeta_1\cdots d\zeta_n.$$

Repeating this, we obtain (1.1.16). □

Remark 1.1.17. (i) In (i) of the above theorem the continuity condition for f is, in fact, unnecessary, but the proof is not as easy as above (see Hartogs' Theorem 4.2.9).

(ii) In the category of real analytic functions, the separate analyticity does not imply even the continuity. Consider the following function of $(x,y)\in\mathbf{R}^2$:

$$f(x,y)=\begin{cases}0, & (x,y)=(0,0),\\ \dfrac{xy}{x^2+y^2}, & (x,y)\neq(0,0).\end{cases}$$

Then, $f(x,y)$ is bounded and analytic in one variable with any fixed other, but it is not continuous at $(0,0)$. For $\lim_{x\to 0} f(x,kx) = k/(1+k^2)$ with $y = kx$ $(k \in \mathbf{R})$, while $f(0,0) = 0$ by definition.

(iii) Furthermore, in the category of real analytic functions, the continuity and the separate analyticity do not imply the analyticity in general; for example, we take

$$g(x,y) = \begin{cases} 0, & (x,y) = (0,0), \\ \dfrac{x^2y^2}{x^2+y^2}, & (x,y) \neq (0,0). \end{cases}$$

Then $g(x,y)$ is continuous in $\mathbf{R}^2$ and separately analytic. But it is not analytic at the origin. For, if $g(x,y)$ is analytic at $(0,0)$, then $g(0,y) = 0$. We would have $g(x,y) = xg_1(x,y)$, where $g_1(x,y)$ should be analytic (same in the sequel). Since $g_1(x,0) = 0$, we get similarly $g_1(x,y) = yg_2(x,y)$. Thus, $g(x,y) = xyg_2(x,y)$, and hence $g_2(x,y) = xy/(x^2+y^2)$. A similar argument would imply $g_2(x,y) = xyg_3(x,y)$, so that $g_3(x,y) = 1/(x^2+y^2)$ should be analytic; this is absurd.

1.1.3 Sequences and Series of Functions

For a *sequence of functions* $\{f_\nu\}_{\nu=0}^{\infty}$ defined on a subset $A \subset \mathbf{C}^n$ with

$$f_\nu : A \to \mathbf{C}, \quad \nu = 0, 1, \ldots,$$

we define the notions of point-convergence, Cauchy sequence, (locally) uniform convergence, (locally) uniform Cauchy sequence in the same way as in the case of one variable; when A is open, the locally uniform convergence is equivalent to the convergence uniform on every compact subset of A. In particular we state:

Theorem 1.1.18. *If a series $\{f_\nu\}_{\nu=0}^{\infty}$ of holomorphic functions defined on an open subset $A \subset \mathbf{C}^n$ converges locally uniformly, then the limit $f(z) = \lim_{\nu\to\infty} f_\nu(z)$ is a holomorphic function on A.*

Proof. Firstly, $f(z)$ is continuous. For an arbitrary point $a \in A$, we take a polydisk neighborhood $\mathrm{P}\Delta(a;(r_j)) \Subset A$ and apply (1.1.14) for $f_\nu(z)$. Since $\{f_\nu(z)\}_{\nu=0}^{\infty}$ converges to $f(z)$ uniformly on $\overline{\mathrm{P}\Delta}(a;(r_j))$, $f(z)$ also satisfies

$$f(z_1,\ldots,z_n) = \left(\frac{1}{2\pi i}\right)^n \int_{|\zeta_1-a_1|=r_1} \cdots \int_{|\zeta_n-a_n|=r_n} \frac{f(\zeta_1,\ldots,\zeta_n)}{(\zeta_1-z_1)\cdots(\zeta_n-z_n)} d\zeta_1\cdots d\zeta_n,$$
$$z \in \mathrm{P}\Delta(a;(r_j)).$$

The integrand of the right-hand side of the equation above is holomorphic in $z \in \mathrm{P}\Delta(a;(r_j))$, and so is $f(z)$. □

With a given sequence of functions $\{f_\nu(z)\}_{\nu=0}^{\infty}$ on A, the formal sum

$$\sum_{\nu=0}^{\infty} f_\nu(z)$$

is called a *series of functions* on A, and for $N \in \mathbf{Z}_+$

$$s_N(z) = \sum_{\nu=0}^{N} f_\nu(z), \quad z \in A$$

is called the *(Nth) partial sum*. When the sequence $\{s_N(z)\}_{N=0}^{\infty}$ of the partial sums converges, the series of functions $\sum_{\nu=0}^{\infty} f_\nu(z)$ is said to be convergent, and the limit is written as

$$\sum_{\nu=0}^{\infty} f_\nu(z) = \lim_{N\to\infty} s_N(z), \quad z \in A.$$

The (locally) uniform convergence is defined similarly. If $\{s_N(z)\}_{N=0}^{\infty}$ is a (uniform) Cauchy sequence, we say that the series of functions $\sum_{\nu=0}^{\infty} f_\nu(z)$ satisfies the (uniform) *Cauchy condition*.

We say that $\sum_{\nu=0}^{\infty} f_\nu(z)$ *converges absolutely*, if

$$\sum_{\nu=0}^{\infty} |f_\nu(z)|, \quad z \in A$$

converges; if so, the original series of functions converges and the limit is independent from the choice of orders of terms.

Let $\sum_{\nu=0}^{\infty} f_\nu(z)$ $(z \in A)$ be a series of functions. If non-negative constants $M_\nu \in \mathbf{R}_+, \nu = 0, 1, \ldots$ satisfy

$$|f_\nu(z)| \le M_\nu, \quad \forall z \in A, \ \nu = 0, 1, \ldots,$$

the series $\sum_{\nu=0}^{\infty} M_\nu$ is called a *majorant* of $\sum_{\nu=0}^{\infty} f_\nu(z)$. If $\sum_{\nu=0}^{\infty} M_\nu < \infty$ (convergent), we say that $\sum_{\nu=0}^{\infty} f_\nu(z)$ has a *convergent majorant*.

Theorem 1.1.19. *If a series of functions has a convergent majorant, it converges absolutely and uniformly.*

Proof. Use the uniform Cauchy condition. □

1.1.4 Power Series of Several Variables

With a point $a = (a_j) \in \mathbf{C}^n$ and a multi-index $\alpha = (\alpha_1, \ldots, \alpha_n) \in \mathbf{Z}_+^n$ we set

$$(1.1.20) \qquad (z-a)^\alpha = \prod_{j=1}^n (z_j - a_j)^{\alpha_j}.$$

For a polyradius $r = (r_1, \ldots, r_n)$ $(r_j > 0)$ we get

$$(1.1.21) \qquad r^\alpha = r_1^{\alpha_1} \cdots r_n^{\alpha_n}.$$

A series of functions of type

$$(1.1.22) \qquad f(z) = \sum_{\alpha \in \mathbf{Z}_+^n} c_\alpha (z-a)^\alpha, \qquad c_\alpha \in \mathbf{C}$$

is called a *power series* (of n variables) with center a. While the convergence of (1.1.22) does not make sense unless the order of terms is given, it is noticed that the absolute convergence is well-defined, independently from the choice of orders of terms.

For a moment, we set $a = 0$.

Lemma 1.1.23. (i) *If there is a point $z = (z_j) \in (\mathbf{C}^*)^n$ such that $\sum_{\alpha \in \mathbf{Z}_+^n} c_\alpha z^\alpha$ converges with respect to some order, then $\{c_\alpha z^\alpha : \alpha \in \mathbf{Z}_+^n\}$ is bounded.*

(ii) *If $\{c_\alpha w^\alpha : \alpha \in \mathbf{Z}_+^n\}$ is bounded for a point $w = (w_j) \in (\mathbf{C}^*)^n$, then $\sum_{\alpha \in \mathbf{Z}_+^n} c_\alpha z^\alpha$ converges absolutely and locally uniformly in the polydisk $\mathrm{P}\Delta((|w_j|)) = \{(z_j) : |z_j| < |w_j|, 1 \le j \le n\}$.*

Proof. (i) is clear.

(ii) By the assumption there is an $M > 0$ such that

$$|c_\alpha w^\alpha| < M, \quad \forall \alpha \in \mathbf{Z}_+^n.$$

With $0 < \theta < 1$ we set $|z_j| \le \theta |w_j|$ $(1 \le j \le n)$. Then,

$$\begin{aligned} \sum_{\alpha \in \mathbf{Z}_+^n} |c_\alpha z^\alpha| &\le \sum_{\alpha \in \mathbf{Z}_+^n} |c_\alpha w^\alpha| \cdot \theta^{|\alpha|} \\ &\le M \sum_{\alpha_1 \ge 0, \ldots, \alpha_n \ge 0} \theta^{\alpha_1} \cdots \theta^{\alpha_n} = M \left(\frac{1}{1-\theta} \right)^n < \infty. \end{aligned}$$

Hence by Theorem 1.1.19 $\sum_{\alpha \in \mathbf{Z}_+^n} c_\alpha z^\alpha$ converges absolutely and locally uniformly in $\mathrm{P}\Delta((|w_j|))$. □

If a power series (1.1.22) satisfies the condition of Lemma 1.1.23 (i), $f(z)$ is called a *convergent power series*: In this case we set

$$(1.1.24)\qquad \begin{aligned} \Omega^*(f) &= \{r = (r_j) \in (\mathbf{R}_+ \setminus \{0\})^n : |c_\alpha r^\alpha|, \alpha \in \mathbf{Z}_+^n, \text{are bounded}\}^\circ, \\ \Omega(f) &= \{(z_j) \in \mathbf{C}^n : \exists (r_j) \in \Omega^*(f), |z_j| < r_j, 1 \le j \le n\}, \\ \log \Omega^*(f) &= \{(\log r_j) \in \mathbf{R}^n : (r_j) \in \Omega^*(f)\}. \end{aligned}$$

Here $\{\cdot\}^\circ$ denotes the subset of the interior points. We call $\Omega(f)$ the *domain of convergence* of the power series $f(z)$.

Theorem 1.1.25. *Let $f(z) = \sum_{\alpha \in \mathbf{Z}_+^n} c_\alpha z^\alpha$ be a convergent power series. Then:*

(i) *$f(z)$ is holomorphic in $\Omega(f)$.*

(ii) (Fabry; logarithmic convexity) *$\log \Omega^*(f)$ is a convex set.*

(iii) *$f(z)$ is termwise partially differentiable in $\Omega(f)$, and satisfies*

$$(1.1.26)\qquad \frac{\partial f}{\partial z_j}(z) = \sum_{\alpha_1 \ge 0, \ldots, \alpha_j \ge 1, \ldots, \alpha_n \ge 0} \alpha_j c_{\alpha_1 \ldots \alpha_j \ldots \alpha_n} z_1^{\alpha_1} \cdots z_j^{\alpha_j - 1} \cdots z_n^{\alpha_n},$$

$$(1.1.27)\qquad \Omega(f) \subset \Omega\left(\frac{\partial f}{\partial z_j}\right).$$

Proof. (i) Note that $c_\alpha z^\alpha$ is holomorphic. It follows from Theorem 1.1.18 and Lemma 1.1.23 (ii) that $f(z)$ is holomorphic in $\Omega(f)$.

(ii) Take two points $(\log r_j), (\log s_j) \in \log \Omega^*(f)$. Since they are interior points, there exist some $r'_j > r_j, s'_j > s_j$ $(1 \le j \le n)$ and $M > 0$ such that with $r' = (r'_j)$ and $s' = (s'_j)$

$$|c_\alpha r'^\alpha| \le M, \quad |c_\alpha s'^\alpha| \le M, \quad \forall \alpha \in \mathbf{Z}_+^n.$$

It suffices to show that for every θ with $0 < \theta < 1$

$$(1.1.28)\qquad \theta(\log r_j) + (1-\theta)(\log s_j) \in \log \Omega^*(f).$$

For r' and s' we have

$$|c_\alpha r'^{\theta\alpha} s'^{(1-\theta)\alpha}| = (|c_\alpha| r'^\alpha)^\theta \cdot (|c_\alpha| s'^\alpha)^{1-\theta} \le M^\theta \cdot M^{1-\theta} = M.$$

Therefore, (1.1.28) follows.

(iii) The proof of termwise partial differentiability is the same as in the case of one variable (cf., e.g., [38] Theorem (3.1.10)).

We show (1.1.27). Take an arbitrary point $(z_k) \in \Omega(f)$. By definition there is some $(r_k) \in \Omega^*(f)$ and a constant $M > 0$ such that

$$|z_k| < r_k,\ 1 \le k \le n, \ \text{and}\ |c_\alpha r^\alpha| \le M, \quad \forall \alpha \in \mathbf{Z}_+^n.$$

Take (t_k) so that $|z_k| < t_k < r_k$ $(1 \le k \le n)$. For every $\alpha = (\alpha_k) \in \mathbf{Z}_+^n, \alpha_j \ge 1$, we have

$$|\alpha_j c_\alpha| t_1^{\alpha_1} \cdots t_j^{\alpha_j - 1} \cdots t_n^{\alpha_n} \le |c_\alpha| r^\alpha \frac{\alpha_j}{t_j} \left(\frac{t_j}{r_j}\right)^{\alpha_j}.$$

Because $0 < t_j/r_j < 1$, $\lim_{\alpha_j\to\infty} \alpha_j (t_j/r_j)^{\alpha_j} = 0$. Thus there is some $L > 0$ such that

$$\alpha_j \left(\frac{t_j}{r_j}\right)^{\alpha_j} \le L, \quad \forall \alpha_j \ge 1.$$

Hence

$$|\alpha_j c_\alpha| t_1^{\alpha_1} \cdots t_j^{\alpha_j - 1} \cdots t_n^{\alpha_n} \le \frac{LM}{t_j}.$$

It follows that $(t_k) \in \Omega^*\left(\frac{\partial f}{\partial z_j}\right)$, and $(z_k) \in \Omega\left(\frac{\partial f}{\partial z_j}\right)$. □

Because of the property (ii) above we say that $\Omega^*(f)$ is *logarithmically convex.*

1.1.5 Elementary Properties of Holomorphic Functions of Several Variables

Let Ω be a domain and let $f \in \mathcal{O}(\Omega)$. We take arbitrarily a closed polydisk $\overline{\mathrm{P}\Delta(a;(r_j))} \Subset \Omega$. For the sake of simplicity we assume $a = 0$ by a parallel translation of the coordinates. By the Cauchy integral formula (1.1.14),

$$f(z) = \left(\frac{1}{2\pi i}\right)^n \int\limits_{|\zeta_1|=r_1} \cdots \int\limits_{|\zeta_n|=r_n} \frac{f(\zeta_1, \dots, \zeta_n)}{(\zeta_1 - z_1)\cdots(\zeta_n - z_n)} d\zeta_1 \cdots d\zeta_n, \tag{1.1.29}$$

$$z = (z_1, \dots, z_n) \in \mathrm{P}\Delta((r_j)).$$

We expand the Cauchy kernel in the integrand above as follows:

$$\frac{1}{(\zeta_1 - z_1)\cdots(\zeta_n - z_n)} = \frac{1}{\zeta_1\left(1 - \frac{z_1}{\zeta_1}\right)\cdots\zeta_n\left(1 - \frac{z_n}{\zeta_n}\right)}$$

$$= \sum_{\alpha_1 \ge 0, \dots, \alpha_n \ge 0} \frac{1}{\zeta_1}\left(\frac{z_1}{\zeta_1}\right)^{\alpha_1} \cdots \frac{1}{\zeta_n}\left(\frac{z_n}{\zeta_n}\right)^{\alpha_n}. \tag{1.1.30}$$

Since $|z_j/\zeta_j| = |z_j|/r_j < 1$ $(1 \le j \le n)$, the power series (1.1.30) converges absolutely, and locally uniformly in $\mathrm{P}\Delta((r_j))$. Together with (1.1.29) we obtain

$$f(z) = \sum_{\alpha \in \mathbf{Z}_+^n} c_\alpha z^\alpha, \tag{1.1.31}$$

$$c_\alpha = \left(\frac{1}{2\pi i}\right)^n \int\limits_{|\zeta_1|=r_1} \cdots \int\limits_{|\zeta_n|=r_n} \frac{f(\zeta_1, \dots, \zeta_n)}{\zeta_1^{\alpha_1+1} \cdots \zeta_n^{\alpha_n+1}} d\zeta_1 \cdots d\zeta_n, \tag{1.1.32}$$

$$\alpha = (\alpha_j) \in \mathbf{Z}_+^n.$$

For a holomorphic partial differential operator ∂^α we have

$$\partial^\alpha f(0) = \alpha!\, c_\alpha, \quad \alpha \in \mathbf{Z}_+^n. \tag{1.1.33}$$

Therefore the coefficients c_α are uniquely determined by $f(z)$.

Theorem 1.1.34 (Power Series Expansion (or Development)). *Let $f \in \mathscr{O}(\Omega)$ and $\mathrm{P}\Delta(a;(r_j)) \subset \Omega$. Then $f(z)$ is uniquely expanded in $\mathrm{P}\Delta(a;(r_j))$ to an absolutely and locally uniformly convergent power series*

$$f(z) = \sum_{\alpha \in \mathbf{Z}_+^n} c_\alpha (z-a)^\alpha, \tag{1.1.35}$$
$$\partial^\alpha f(a) = \alpha!\, c_\alpha, \quad \alpha \in \mathbf{Z}_+^n.$$

Proof. Take arbitrarily $0 < r'_j < r_j$ $(1 \le j \le n)$. Then, $\overline{\mathrm{P}\Delta}(a;(r'_j)) \subset \Omega$, and so by (1.1.31) and (1.1.33) the theorem holds on $\mathrm{P}\Delta(a;(r'_j))$. Since the coefficients c_α of the power series expansion are unique and independent from the choices of r'_j, we deduce the theorem on $\mathrm{P}\Delta(a;(r_j))$ by letting $r'_j \nearrow r_j$. □

We present (1.1.35) as a series of homogeneous polynomials as follows:

$$P_\nu(z-a) = \sum_{\alpha \in \mathbf{Z}_+^n, |\alpha| = \nu} c_\alpha (z-a)^\alpha,$$
$$f(z) = \sum_{\nu=0}^{\infty} P_\nu(z-a), \tag{1.1.36}$$

which is called a *homogeneous polynomial expansion* of $f(z)$.

Remark 1.1.37. A homogeneous polynomial expansion is a series with a given order: It is noted that the domain of convergence of (1.1.36) can be different from that of (1.1.35) (cf. Exercise 4 at the end of this chapter).

A holomorphic function in $\mathbf{C}^n$ is called an *entire function*. By Theorem 1.1.34 we immediately see:

Corollary 1.1.38. *An entire function $f(z)$ is developed in $\mathbf{C}^n$ to*

$$f(z) = \sum_{\alpha \in \mathbf{Z}_+^n} c_\alpha z^\alpha, \tag{1.1.39}$$

where the convergence is absolute and locally uniform.

Theorem 1.1.40. *Let Ω be a domain. Let $K \Subset \Omega$ be a compact subset and let U be an open subset with $K \Subset U \Subset \Omega$. Then there is a positive constant C depending only on K, U and $\alpha \in \mathbf{Z}_+^n$ such that*

$$|\partial^\alpha f(a)| \le C \sup_{z \in U} |f(z)|, \quad \forall a \in K,\ \forall f \in \mathscr{O}(\Omega).$$

Proof. Take a sufficiently small polyradius $r = (r_j)$ so that

$$\mathrm{P}\Delta(a;(r_j)) \Subset U, \quad \forall a \in K.$$

With $a \in K$, (1.1.33) and (1.1.32) imply

$$(1.1.41) \qquad |\partial^\alpha f(a)| = \alpha! \left| \left(\frac{1}{2\pi i}\right)^n \int_{|\zeta_1 - a_1| = r_1} \cdots \int_{|\zeta_n - a_n| = r_n} \frac{f(\zeta_1, \ldots, \zeta_n)}{(\zeta_1 - a_1)^{\alpha_1+1} \cdots (\zeta_n - a_n)^{\alpha_n+1}} d\zeta_1 \cdots d\zeta_n \right|$$
$$\leq \frac{\alpha!}{r^\alpha} \sup_{z \in U} |f(z)|, \quad r^\alpha = r_1^{\alpha_1} \cdots r_n^{\alpha_n}.$$

Therefore it sufficient to set $C = \frac{\alpha!}{r^\alpha}$. □

Theorem 1.1.42 (Liouville)**.** *A bounded entire function is a constant.*

Proof. By Corollary 1.1.38 the entire function $f(z)$ is developed in $\mathbf{C}^n$ to an absolutely and locally uniformly convergent power series:

$$(1.1.43) \qquad f(z) = \sum_{\alpha \in \mathbf{Z}_+^n} c_\alpha z^\alpha.$$

If $|f(z)| \leq M$ $(z \in \mathbf{C}^n)$ with a constant M (> 0), (1.1.33) and (1.1.41) imply

$$(1.1.44) \qquad |c_\alpha| \leq \frac{M}{r_1^{\alpha_1} \cdots r_n^{\alpha_n}},$$

where $\mathrm{P}\Delta((r_j)) \subset \mathbf{C}^n$ is arbitrary. For any multi-index α $(|\alpha| > 0)$ there is an index j with $\alpha_j > 0$, and then by letting $r_j \nearrow \infty$ we see that $c_\alpha = 0$. Therefore, $f(z) = c_0$, and hence $f(z)$ is constant. □

Theorem 1.1.45 (Montel)**.** *If a sequence of holomorphic functions on Ω is uniformly bounded, then it has a subsequence which converges locally uniformly in Ω.*

Proof. Use the Ascoli–Arzelà Theorem and Theorem 1.1.40 with $|\alpha| = 1$ (similarly to the case of one variable: cf., e.g., [38] Theorem (6.4.2)). □

Theorem 1.1.46 (Identity Theorem)**.** *Let Ω be a domain and let $f \in \mathcal{O}(\Omega)$. Then the following three conditions are equivalent:*

(i) $f \equiv 0$.
(ii) *There is a non-empty open set $U \subset \Omega$ such that $f|_U \equiv 0$.*
(iii) *There is a point $a \in \Omega$ such that $\partial^\alpha f(a) = 0$ for all $\alpha \in \mathbf{Z}_+^n$.*

Proof. The implication relations (i)⇒(ii)⇒(iii) ($a \in U$) will be clear.

(iii)⇒(i) Take an arbitrary polydisk neighborhood $\mathrm{P}\Delta(a;(r_j)) \subset \Omega$ of a. Then Theorem 1.1.34 implies

(1.1.47) $$f(z) = \sum_{\alpha} \frac{\partial^\alpha f(a)}{\alpha!}(z-a)^\alpha \equiv 0, \quad z \in \mathrm{P}\Delta(a;(r_j)).$$

We set

$$V = \{z \in \Omega : \exists \mathrm{P}\Delta(z;(s_j)) \subset \Omega, f|_{\mathrm{P}\Delta(z;(s_j))} \equiv 0\}.$$

It follows from the definition and (1.1.47) that V is open and $V \neq \emptyset$. On the other hand, by (1.1.47) we write

$$V = \bigcap_{\alpha \in \mathbf{Z}_+^n} \{z \in \Omega : \partial^\alpha f(z) = 0\},$$

so that V is also closed. The connectedness of Ω implies $V = \Omega$. □

Theorem 1.1.48 (Maximum Principle). *Let Ω be a domain and let $f \in \mathcal{O}(\Omega)$. If $|f(z)|$ takes a maximal value at a point $a \in \Omega$ (in particular, the maximum value there), f is constant.*

Proof. For the sake of simplicity, we assume $a = 0$ by a parallel translation. With a sufficiently small polydisk $\overline{\mathrm{P}\Delta}((r_j)) \Subset \Omega$, $|f(0)|$ is the maximum value in $\overline{\mathrm{P}\Delta}((r_j))$. In a neighborhood of $\overline{\mathrm{P}\Delta}((r_j))$, $f(z)$ is expanded to a power series:

$$f(z) = \sum_{(\alpha_j) \in \mathbf{Z}_+^n} c_\alpha z_1^{\alpha_1} \cdots z_n^{\alpha_n}.$$

With $z_j = r_j e^{i\theta_j}$ we have

(1.1.49) $$\left(\frac{1}{2\pi}\right)^n \int_0^{2\pi} d\theta_1 \cdots \int_0^{2\pi} d\theta_n \left| \sum_{(\alpha_j) \in \mathbf{Z}_+^n} c_\alpha r_1^{\alpha_1} e^{i\alpha_1\theta_1} \cdots r_n^{\alpha_n} e^{i\alpha_n\theta_n} \right|^2$$
$$= \sum_{(\alpha_j) \in \mathbf{Z}_+^n} |c_\alpha|^2 r_1^{2\alpha_1} \cdots r_n^{2\alpha_n} \geq |c_{(0,\ldots,0)}|^2 = |f(0)|^2.$$

On the other hand, since $|f(0)|^2$ is the maximum of $|f(z)|^2$ in $z \in \overline{\mathrm{P}\Delta}((r_j))$,

$$\sum_{(\alpha_j) \in \mathbf{Z}_+^n} |c_\alpha|^2 r_1^{2\alpha_1} \cdots r_n^{2\alpha_n} \leq |f(0)|^2.$$

Therefore it follows that

$$c_\alpha = 0, \quad \forall \alpha \in \mathbf{Z}_+^n, \ |\alpha| > 0.$$

By the Identity Theorem 1.1.46, $f(z) \equiv f(0)$ on Ω. □

1.2 Analytic Continuation and Hartogs' Phenomenon

Let Ω be a domain of $\mathbf{C}^n$.

Definition 1.2.1 (Analytic Continuation). (i) Let $f \in \mathcal{O}(\Omega)$. Let $V \not\subset \Omega$ be a domain such that $\Omega \cap V \neq \emptyset$. Assume that there are an element $g \in \mathcal{O}(V)$ and a connected component W of $\Omega \cap V$ such that $f|_W = g|_W$ on W. Then we say that f is *analytically continued* over V (to g) (through W). Also g is called an *analytic continuation* of f (through W).

(ii) The notion of analytic continuation along curves is defined in the same way as in the case of one variable (cf., e.g., [38] §5.2).

Remark 1.2.2. The above analytic continuation g of f, if it exists, is unique (Theorem 1.1.46). Also, the above V may have in general a non-empty intersection with Ω other than W, and $g(z)$ may have different values to those of the original $f(z)$. Allowing multivalues, we may define a multivalued function $\tilde{f}$ to be f on Ω, and g on V; $\tilde{f}$ is called an *analytic continuation* of f, too.

Remark 1.2.3. Let $n = 1$. Given V, W as above, we take a boundary point $b \in \partial\Omega \cap W$. Then $f(z) = 1/(z-b) \in \mathcal{O}(\Omega)$, so that there is no analytic continuation $g \in \mathcal{O}(V)$ of f.

In the case of $n \geq 2$, however, the issue of analytic continuation appears to be totally different. In fact, according to the shape of $\partial\Omega$ the following may happen:

Definition 1.2.4 (Hartogs' phenomenon). We say that *Hartogs' phenomenon* occurs for Ω if there exist a domain $V \not\subset \Omega$ with $V \cap \Omega \neq \emptyset$ and a connected component W of $V \cap \Omega$ such that all holomorphic functions in Ω are analytically continued over V through W.

We observe such Hartogs' phenomena in the sequel, assuming $n \geq 2$.

(a) Let $t_j > 0$ $(1 \leq j \leq n)$ and $0 < s_j < t_j$ $(j = 1,2)$ be given. We consider the polydisk $\mathrm{P}\Delta(a;(t_j))$ of polyradius (t_j) with center at $a = (a_j)$, and the closed subset

$$(1.2.5) \qquad F = \bigl\{(z_1, z_2, z'') \in \mathrm{P}\Delta(a;(t_j)) : |z_j - a_j| \leq s_j, j = 1,2, \\ z'' = (z_3, \ldots, z_n) \in \mathrm{P}\Delta(a''; r'')\bigr\},$$

where $a'' = (a_3, \ldots, a_n)$ and $r'' = (r_3, \ldots, r_n)$. Then F contains a as an interior point.

Lemma 1.2.6 (Hartogs). *Every holomorphic function in* $\mathrm{P}\Delta(a;(t_j)) \setminus F$ *is necessarily analytically continued on* $\mathrm{P}\Delta(a;(t_j))$.

Proof. We may assume $a = 0$ by a parallel translation. Take any element $f \in \mathcal{O}(\mathrm{P}\Delta((t_j)) \setminus F)$. For a point $z = (z_j) \in \mathrm{P}\Delta((t_j))$ we choose ρ_1 with $\max\{|z_1|, s_1\} < \rho_1 < t_1$ and set

$$\tilde{f}(z_1, z') = \frac{1}{2\pi i} \int_{|\zeta_1| = \rho_1} \frac{f(\zeta_1, z')}{\zeta_1 - z_1} d\zeta_1, \qquad z' = (z_2, \ldots, z_n).$$

It is easy to see that this is independent of the choice of ρ_1, and so $\tilde{f} \in \mathcal{O}(\mathrm{P}\Delta((t_j)))$. For z_2 with $s_2 < |z_2| < t_2$, $\tilde{f}(z_1, z_2, z'') = f(z_1, z_2, z'')$. The uniqueness of analytic continuation implies that $\tilde{f} = f$ on $\mathrm{P}\Delta((t_j)) \setminus F$. □

Even from this simple theorem the following facts which feature differences of n (≥ 2) variables and one variable are deduced:

Theorem 1.2.7. *Set $S = \{(z_j) \in \mathbf{C}^n : z_1 = z_2 = 0\}$. Then every holomorphic function in $\mathrm{P}\Delta((r_j)) \setminus S$ is analytically continued on $\mathrm{P}\Delta((r_j))$.*

Proof. Use Lemma 1.2.6. □

Theorem 1.2.8. *Let Ω be a cylinder domain $\mathrm{P}\Delta(a;(r_j)) = \prod_{j=1}^n \Delta(a_j; r_j)$ $(\subset \mathbf{C}^n)$ with allowing some $r_j = \infty$, for which $\Delta(a_j;\infty) = \mathbf{C}$. Let $K \Subset \Omega$ be a compact subset such that $\Omega \setminus K$ is connected. Then every holomorphic function in $\Omega \setminus K$ is analytically continued to a unique holomorphic function on Ω.*

Proof. It is easily reduced to the case of a (bounded) polydisk $\Omega = \mathrm{P}\Delta(a;(r_j)) \Supset K$. We choose a polydisk $\mathrm{P}\Delta(a;(s_j))$ such that

$$K \Subset \mathrm{P}\Delta(a;(s_j)) \Subset \mathrm{P}\Delta(a;(r_j)).$$

Take any function $f \in \mathcal{O}(\mathrm{P}\Delta(a;(r_j)) \setminus K)$. Then f is holomorphic in $\mathrm{P}\Delta(a;(r_j)) \setminus \overline{\mathrm{P}\Delta}(a;(s_j))$. It follows from Lemma 1.2.6 that f is analytically continued to a unique holomorphic function $\tilde{f}$ in $\mathrm{P}\Delta(a;(r_j))$. Since $\mathrm{P}\Delta(a;(r_j)) \setminus K$ is connected, $\tilde{f}|_{\mathrm{P}\Delta(a;(r_j))\setminus K} = f$ by the uniqueness of analytic continuation. □

This will be extended to general domains of $\mathbf{C}^n$ (see §3.4.4).

Corollary 1.2.9. *The zero set $\{z \in \Omega : f(z) = 0\}$ of a holomorphic function f in a domain Ω contains no isolated point.*

Proof. Set $\Sigma = \{z \in \Omega : f(z) = 0\}$. Suppose that $a \in \Sigma$ is isolated. We take a small polydisk $\mathrm{P}\Delta(a;(r_j)) \subset \Omega$ such that $\mathrm{P}\Delta(a;(r_j)) \cap \Sigma = \{a\}$. Then $g := 1/f \in \mathcal{O}(\mathrm{P}\Delta(a;(r_j)) \setminus \{a\})$. It follows from Theorem 1.2.8 with $K = \{a\}$ that $g \in \mathcal{O}(\mathrm{P}\Delta(a;(r_j)))$ and $f(z) \cdot g(z) = 1$ on $\mathrm{P}\Delta(a;(r_j))$. At $z = a$, $f(a) \cdot g(a) = 1$, which contradicts $f(a) = 0$. □

Remark 1.2.10. In function theory of one variable it is a well-known fact that the zero set of a non-constant holomorphic function consists only of isolated points. And in the case of real n (≥ 1) variables the function

$$f(x_1, \ldots, x_n) = x_1^\nu + \cdots + x_n^\nu$$

with even natural number n has the zero set $\{x = (x_j) \in \mathbf{R}^n : f(x) = 0\} = \{0\}$, and so the zero point of f is isolated. The corollary above shows that the figure of the zero set or a constant surface $\{f = c\}$ $(c \in \mathbf{C})$ of an analytic function f of complex variables more than one is considerably different to that of the one-variable case or the real analytic case.

(b) In the above (a), the case of $n = 2$ is essential, and in that case the set over which functions are analytically continued is contained in the outside larger domain as a relatively compact subset. In what follows we give an example such that the analytic continuation takes place on a relatively non-compact subset.

We keep $n \geq 2$. Let $0 < s_j < t_j, 1 \leq j \leq n$, be given. The center can be an arbitrary point of $\mathbf{C}^n$, to say, 0. We put (cf. Fig. 1.2)

$$\begin{aligned}(1.2.11)\qquad &\Omega_1 = \{(z_j) : |z_1| < t_1, |z_j| < s_j,\ 2 \leq j \leq n\},\\ &\Omega_2 = \{(z_j) : s_1 < |z_1| < t_1, |z_j| < t_j,\ 2 \leq j \leq n\},\\ &\Omega_{\mathrm{H}} = \Omega_1 \cup \Omega_2.\end{aligned}$$

The domain Ω_{H} is classically known as a *Hartogs domain*.

Write $z' = (z_2, \ldots, z_n)$ and let $f(z_1, z') \in \mathcal{O}(\Omega_{\mathrm{H}})$. By Cauchy's integral formula we have

$$f(z_1, z') = \frac{1}{2\pi i} \int_{|\zeta| = r_1} \frac{f(\zeta, z')}{\zeta - z_1} d\zeta$$

for $(z_1, z') \in \Omega_1$, where $|z_1| < r_1 < t_1$; furthermore, since r_1 can be chosen arbitrarily close to t_1, we see by the integral formula that $f(z_1, z')$ is analytically continued on $\mathrm{P}\Delta((t_j))$. Hence we obtain:

Theorem 1.2.12. *Every holomorphic function in Ω_{H} is analytically continued on the whole polydisk $\mathrm{P}\Delta((t_j))$.*

(c) Let Ω be a domain. For a point $a = (a_j) \in \Omega$ we put

$$a + \mathbf{R}^n = \{z = (z_j) \in \mathbf{C}^n : \Im(z_j - a_j) = 0,\ 1 \leq j \leq n\},$$

which is called a *totally real subspace* of Ω through a.

Theorem 1.2.13 (Removability of totally real subspaces)**.** *With the notation above, every $f \in \mathcal{O}(\Omega \setminus (a + \mathbf{R}^n))$ is analytically continued on Ω.*

Proof. It is sufficient to prove the analytic continuation in a neighborhood of an arbitrarily given point $c \in \Omega \cap (a + \mathbf{R}^n)$. Since $a + \mathbf{R}^n = c + \mathbf{R}^n$, we may put $c = a$. By

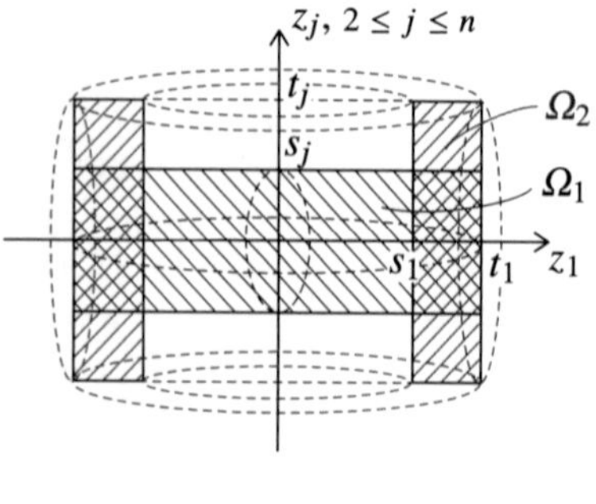

Fig. 1.2 Hartogs domain Ω_{H}.

translations and multiplications (by non-zero numbers) of the coordinates we may assume that

$$\begin{aligned} \mathrm{P}\Delta &= \mathrm{P}\Delta((2,\dots,2)) \subset \Omega, \\ \mathrm{P}\Delta_j &= \{(z_1,\dots,z_n) \in \mathrm{P}\Delta : |z_j| < 1\}, \\ \omega &= \bigcup_{j=1}^{n} \mathrm{P}\Delta_j, \\ a &= (\rho i,\dots,\rho i) \in \mathrm{P}\Delta, \quad 1 < \rho < \sqrt{2}. \end{aligned}$$

If $(w_j) \in a+\mathbf{R}^n$, then $|w_j| \geq \rho > 1$ and so $\omega \cap (a+\mathbf{R}^n) = \emptyset$. Take a function $f \in \mathscr{O}(\Omega \setminus (a+\mathbf{R}^n))$. Since $\mathrm{P}\Delta((1,\dots,1)) \subset \omega$, $f(z)$ is developed in the polydisk $\mathrm{P}\Delta((1,\dots,1))$ to a power series

$$f(z) = \sum_{\alpha \in \mathbf{Z}_+^n} c_\alpha z^\alpha. \tag{1.2.14}$$

The domain of convergence satisfies

$$\Omega(f) \supset \omega.$$

Set

$$\begin{aligned} \omega^* &= \{(z_j) \in \omega : z_j \neq 0,\ 1 \leq j \leq n\}, \\ \log \omega^* &= \{(\log |z_j|) \in \mathbf{R}^n : (z_j) \in \omega^*\}, \\ U &= \{(z_j) \in \mathrm{P}\Delta : 0 < |z_1| \cdots |z_n| < 2^{n-1}\}, \\ \log U &= \{(\log |z_j|) : (z_j) \in U\}, \\ \log a &= (\log \rho, \dots, \log \rho) \in \mathbf{R}^n. \end{aligned}$$

Note that $\log U$ is the convex hull of $\log \omega^*$ (cf. Fig. 1.3). Since the domain $\Omega^*(f)$ is logarithmically convex by Theorem 1.1.25 (ii), we see that

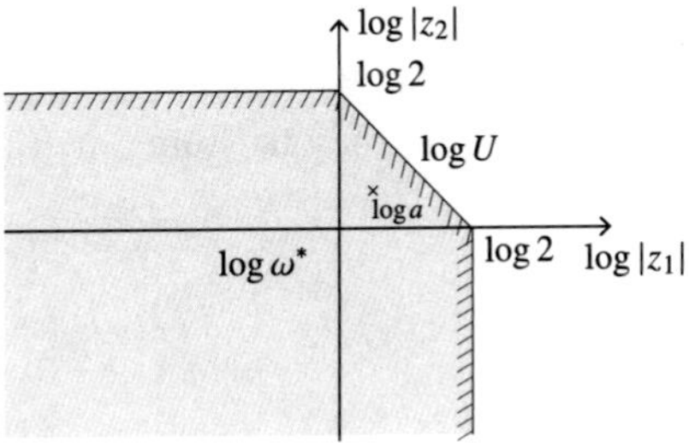

Fig. 1.3 $\log \omega^* \subset \log U$ (the case of $n = 2$)

$$U = \{(z_j) \in \mathrm{P}\Delta : 0 < |z_1| \cdots |z_n| < 2^{n-1}\} \subset \Omega^*(f).$$

For a we have

$$|a_1| \cdots |a_n| = \rho^n < 2^{n/2} \le 2^{n-1} \quad (n \ge 2).$$

Therefore, $a \in \Omega^*(f)$, so that (1.2.14) converges absolutely and uniformly in a neighborhood of a; i.e., $f(z)$ is analytically continued on a neighborhood of a. □

1.3 Runge Approximation on Convex Cylinder Domains

In general, we define the *sup-norm* (supremum norm) of a function $f : A \to \mathbf{C}$ on $A \subset \mathbf{C}^n$ by

$$\|f\|_A = \sup\{|f(z)| : z \in A\}. \tag{1.3.1}$$

Theorem 1.3.2. *Let $\Omega \subset \mathbf{C}^n$ be a bounded convex cylinder domain. Then a holomorphic function $f(z)$ in a neighborhood of the closure $\bar{\Omega}$ is approximated uniformly on $\bar{\Omega}$ by polynomials, i.e., for every $\varepsilon > 0$ there is a polynomial $P(z)$ such that*

$$\|f - P\|_{\bar{\Omega}} < \varepsilon.$$

Proof. Let $\Omega = \prod_{j=1}^n \Omega_j$ with convex open sets $\Omega_j \Subset \mathbf{C}$, $1 \le j \le n$. We take convex open polygons $E_j \Supset \Omega_j$ so that $f(z)$ is holomorphic in a neighborhood of the convex closed cylinder $\prod_{j=1}^n \bar{E}_j$ (Proposition 1.1.5). The boundary $\partial E_j = C_j = \bigcup_{k=1}^{l_j} C_{jk}$ of each E_j consists of a finite number of line segments C_{jk}. By Cauchy's integral formula we have

$$\begin{aligned} f(z) &= \left(\frac{1}{2\pi i}\right)^n \int_{C_1} d\zeta_1 \cdots \int_{C_n} d\zeta_n \frac{f(\zeta_1, \ldots, \zeta_n)}{(\zeta_1 - z_1) \cdots (\zeta_n - z_n)} \\ &= \sum_{k_1, \ldots, k_n} \left(\frac{1}{2\pi i}\right)^n \int_{C_{1k_1}} d\zeta_1 \cdots \int_{C_{nk_n}} d\zeta_n \frac{f(\zeta_1, \ldots, \zeta_n)}{(\zeta_1 - z_1) \cdots (\zeta_n - z_n)} \end{aligned} \tag{1.3.3}$$

for $z \in \bar{\Omega}$, where the summation runs over $1 \le k_1 \le l_1, \ldots, 1 \le k_n \le l_n$. Take a point ξ_{jk_j} on the line passing through the middle point of each line segment C_{jk_j}, vertical to it in the same side as Ω_j and sufficiently far from C_{jk_j} (cf. Fig. 1.4). Then there is a constant $\theta > 0$ such that

$$\left|\frac{z_j - \xi_{jk_j}}{\zeta_j - \xi_{jk_j}}\right| < \theta < 1, \quad \zeta_j \in C_{jk_j},\ z_j \in \Omega_j,\ 1 \le k_j \le l_j, 1 \le j \le n. \tag{1.3.4}$$

For these z_j and ζ_j we have

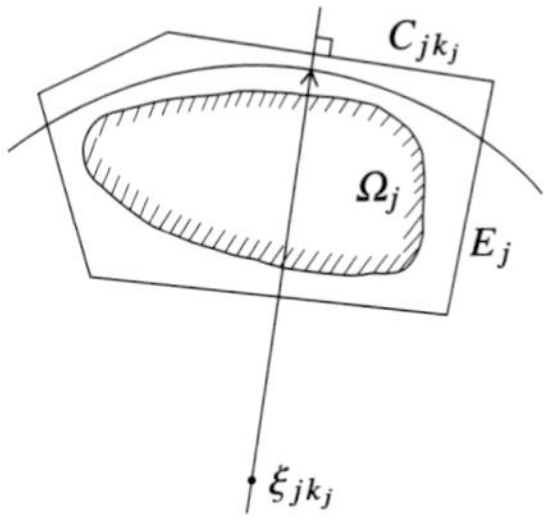

Fig. 1.4 Convex domain Ω_j.

(1.3.5)
$$\frac{1}{\zeta_j - z_j} = \frac{1}{\zeta_j - \xi_{jk_j} - (z_j - \xi_{jk_j})} = \sum_{\alpha_j=0}^{\infty} \frac{(z_j - \xi_{jk_j})^{\alpha_j}}{(\zeta_j - \xi_{jk_j})^{\alpha_j+1}}.$$

By (1.3.4) the series of the right-hand side above is of majorant convergence. It follows from (1.3.3) and (1.3.5) that

(1.3.6)
$$f(z) = \sum_{(\alpha_j)\in \mathbf{Z}_+^n} \sum_{k_1,\ldots,k_n} \left(\frac{1}{2\pi i}\right)^n \int_{C_{1k_1}} d\zeta_1 \cdots \cdots \int_{C_{nk_n}} d\zeta_n\, f(\zeta) \prod_{j=1}^n \frac{(z_j - \xi_{jk_j})^{\alpha_j}}{(\zeta_j - \xi_{jk_j})^{\alpha_j+1}}, \quad z \in \bar{\Omega}.$$

Since $f(\zeta)$ ($\zeta \in \prod_j \bar{E}_j$) is bounded, we see by (1.3.4) that the right-hand side of (1.3.6) is of majorant convergence. Hence for every $\varepsilon > 0$ there is a sufficiently large number $N \in \mathbf{N}$ such that the polynomial

$$P(z) = \sum_{|(\alpha_j)|\le N} \sum_{k_1,\ldots,k_n} \left(\frac{1}{2\pi i}\right)^n \int_{C_{1k_1}} d\zeta_1 \cdots \cdots \int_{C_{nk_n}} d\zeta_n\, f(\zeta) \prod_{j=1}^n \frac{(z_j - \xi_{jk_j})^{\alpha_j}}{(\zeta_j - \xi_{jk_j})^{\alpha_j+1}}$$

of degree at most N satisfies

$$\|f(z) - P(z)\|_{\bar{\Omega}} < \varepsilon.$$

□

1.3.1 Cousin Integral

In $\mathbf{C}^n$ ($\ni z = (z_1, \ldots, z_n)$) we consider a cuboid F (always bounded) containing a part of the boundary:

$$F = F' \times \{z_n \in \mathbf{C} : a < \Re z_n < b, |\Im z_n| \le c\},$$
$$F^\circ = F' \times \{z_n \in \mathbf{C} : a < \Re z_n < b, |\Im z_n| < c\},$$

where $c > 0$, F' is an open cuboid in $\mathbf{C}^{n-1}$ $(\ni z' = (z_1, \ldots, z_{n-1}))$ and F° is the interior of F.

Let $\varphi(z)$ be a continuous function on F, holomorphic in F°. With $t = (a+b)/2$ we denote by ℓ the oriented line segment in z_n-plane from $z_n = t - ic$ to $z_n = t + ic$. The following path integral of $\varphi(z)$ is called the *Cousin integral*:

$$\Phi(z', z_n) = \frac{1}{2\pi i} \int_\ell \frac{\varphi(z', \zeta)}{\zeta - z_n} d\zeta. \tag{1.3.7}$$

We first consider it as $\Phi(z', z_n) \in \mathscr{O}(F' \times \{z_n : \Re z_n < t, |\Im z_n| < c\})$. With $t < b' < b$ and $0 < c' < c$ the part of the oriented boundary of the domain $\{z_n : t < \Re z_n < b', |\Im z_n| < c'\}$ which is a part of ℓ is denoted by $-\ell'$, and the other part of the boundary is denoted by ℓ'' (cf. Fig. 1.5). By Cauchy's integral formula,

$$\frac{1}{2\pi i} \int_{\ell'} \frac{\varphi(z', \zeta)}{\zeta - z_n} d\zeta = \frac{1}{2\pi i} \int_{\ell''} \frac{\varphi(z', \zeta)}{\zeta - z_n} d\zeta$$

for $\Re z_n < t, |\Im z_n| < c$ and $z' \in F'$. Therefore

$$\Phi(z', z_n) = \frac{1}{2\pi i} \int_{\ell \setminus \ell'} \frac{\varphi(z', \zeta)}{\zeta - z_n} d\zeta + \frac{1}{2\pi i} \int_{\ell''} \frac{\varphi(z', \zeta)}{\zeta - z_n} d\zeta. \tag{1.3.8}$$

The right-hand side above is holomorphic in $F' \times \{z_n : \Re z_n < b', |\Im z_n| < c'\}$, and so $\Phi(z', z_n)$ is analytically continued on $F' \times \{z_n : \Re z_n < b', |\Im z_n| < c'\}$. Letting $b' \to b$ and $c' \to c$, we see that $\Phi(z', z_n)$ is analytically continued on

$$U_1 := F' \times \{z_n : \Re z_n < b, |\Im z_n| < c\},$$

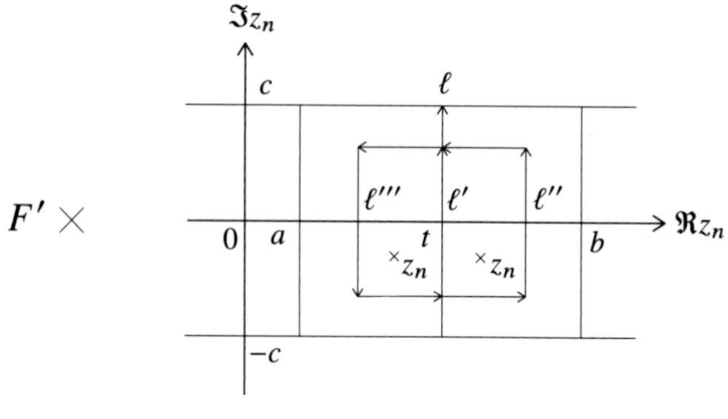

Fig. 1.5 Cousin integral.

and write $\Phi_1(z', z_n) \in \mathscr{O}(U_1)$ for it.

Next, we consider (1.3.7) in $F' \times \{z_n : \Re z_n > t, |\Im z_n| < c\}$. With $a < a' < t, 0 < c' < c$, the part of the oriented boundary of the domain $\{z_n : a' < \Re z_n < t, |\Im z_n| < c'\}$ contained in ℓ is denoted by ℓ', and the other part of the boundary is denoted by ℓ''' (cf. Fig. 1.5). In the same way as in the case of Φ_1, we have

$$\frac{1}{2\pi i}\int_{\ell'} \frac{\varphi(z',\zeta)}{\zeta - z_n} d\zeta = -\frac{1}{2\pi i}\int_{\ell'''} \frac{\varphi(z',\zeta)}{\zeta - z_n} d\zeta,$$

$$\Phi(z', z_n) = \frac{1}{2\pi i}\int_{\ell\setminus\ell'} \frac{\varphi(z',\zeta)}{\zeta - z_n} d\zeta - \frac{1}{2\pi i}\int_{\ell'''} \frac{\varphi(z',\zeta)}{\zeta - z_n} d\zeta \tag{1.3.9}$$

for $\Re z_n > t, |\Im z_n| < c$ and $z' \in F'$. The right-hand side above is holomorphic in $F' \times \{z_n : \Re z_n > a', |\Im z_n| < c'\}$, and so $\Phi(z', z_n)$ is analytically continued on $F' \times \{z_n : \Re z_n > a', |\Im z_n| < c'\}$. As $a' \to a$ and $c' \to c$, $\Phi(z', z_n)$ is analytically continued on

$$U_2 := F' \times \{z_n : \Re z_n > a, |\Im z_n| < c\}.$$

We denote it by $\Phi_2(z', z_n) \in \mathscr{O}(U_2)$.

Lemma 1.3.10 (Cousin decomposition). $\Phi_j(z', z_n) \in \mathscr{O}(U_j)$ $(j = 1, 2)$ *satisfy*

$$\varphi(z', z_n) = \Phi_1(z', z_n) - \Phi_2(z', z_n), \quad (z', z_n) \in U_1 \cap U_2 \ (= F^\circ). \tag{1.3.11}$$

Proof. For $(z', z_n) \in U_1 \cap U_2$ we take a', b', c' so that

$$a < a' < \min\{t, \Re z_n\}, \ \max\{t, \Re z_n\} < b' < b, \ |\Im z_n| < c' < c.$$

Then by Cauchy's integral formula

$$\varphi(z', z_n) = \frac{1}{2\pi i}\int_{\ell''} \frac{\varphi(z',\zeta)}{\zeta - z_n} d\zeta + \frac{1}{2\pi i}\int_{\ell'''} \frac{\varphi(z',\zeta)}{\zeta - z_n} d\zeta.$$

From this together with (1.3.8) and (1.3.9) we obtain (1.3.11). □

1.4 Implicit and Inverse Function Theorems

Here, readers may consult any textbook ready to hand on the implicit function theorem of real functions of real variables. We consider simultaneous equations with holomorphic functions

$$f_j(z_1, \ldots, z_n, w_1, \ldots, w_m) = 0, \quad 1 \le j \le m. \tag{1.4.1}$$

The *complex Jacobi matrix* and *complex Jacobian* for $f = (f_1, \ldots, f_m)$ are defined respectively by

(1.4.2)
$$\left(\frac{\partial f_j}{\partial w_k}\right)_{1\le j,k\le m}, \qquad \frac{\partial f}{\partial w} := \det\left(\frac{\partial f_j}{\partial w_k}\right)_{1\le j,k\le m}.$$

For the real and imaginary parts of f_j and w_k we write respectively

$$f_j = f_{j1} + i f_{j2}, \quad w_k = w_{k1} + i w_{k2}.$$

Then (1.4.1) is equivalent to the following:

(1.4.3)
$$\begin{aligned} f_{j1}(z_1,\ldots,z_n,w_{11},w_{12},\ldots,w_{m1},w_{m2}) &= 0, \quad 1\le j\le m,\\ f_{j2}(z_1,\ldots,z_n,w_{11},w_{12},\ldots,w_{m1},w_{m2}) &= 0, \quad 1\le j\le m. \end{aligned}$$

The real Jacobian of (1.4.3) is

(1.4.4)
$$\frac{\partial(f_{j1},f_{j2})}{\partial(w_{k1},w_{k2})} = \begin{vmatrix} \frac{\partial f_{11}}{\partial w_{11}} & \frac{\partial f_{11}}{\partial w_{12}} & \frac{\partial f_{11}}{\partial w_{21}} & \frac{\partial f_{11}}{\partial w_{22}} & \cdots \\ \frac{\partial f_{12}}{\partial w_{11}} & \frac{\partial f_{12}}{\partial w_{12}} & \frac{\partial f_{12}}{\partial w_{21}} & \frac{\partial f_{12}}{\partial w_{22}} & \cdots \\ \vdots & \vdots & \vdots & \vdots & \end{vmatrix}.$$

Lemma 1.4.5. *The real Jacobian and the complex Jacobian associated with holomorphic functions $f_j(z,w), 1\le j\le m$, with $z=(z_1,\ldots,z_n)$ and $w=(w_1,\ldots,w_m)$ are related by*

$$\frac{\partial(f_{j1},f_{j2})}{\partial(w_{k1},w_{k2})} = \left|\det\left(\frac{\partial f_j}{\partial w_k}\right)\right|^2.$$

Proof. Cf. [39] Lemma 1.2.39. □

Theorem 1.4.6 (Implicit function). *In a neighborhood of a pint $(a,b)\in \mathbf{C}^n\times\mathbf{C}^m$ we consider simultaneous equations* (1.4.1) *defined by holomorphic functions $f_j(a,b)=0, 1\le j\le m$. Assume that*

(1.4.7)
$$\det\left(\frac{\partial f_j}{\partial w_k}(a,b)\right)_{1\le j,k\le m} \neq 0.$$

Then there are uniquely holomorphic solutions (1.4.1) *in a neighborhood of* (a,b)

$$(w_j) = (g_j(z_1,\ldots,z_n)), \quad b=(g_j(a)) \quad (1\le j\le m).$$

Proof. Cf. [39] Theorem 1.2.41. □

A map from an open set $U\subset \mathbf{C}^n$ to $\mathbf{C}^m$

$$f: z\in U \to (f_1(z),\ldots,f_m(z))\in \mathbf{C}^m$$

is called a *holomorphic map (or mapping)* if all elements $f_j(z)$ are holomorphic functions.

Theorem 1.4.8 (Inverse function). *If a holomorphic map between neighborhoods U, V of the origin of $\mathbf{C}^n$*

$$f : z = (z_k) \in U \to (f_j(z)) \in V, \quad f(0) = 0$$

has a non-zero complex Jacobian $\frac{\partial f}{\partial z}(0) \neq 0$ at 0, then f with U and V shrunk smaller if necessary, has the holomorphic inverse $f^{-1} : V \to U$.

Proof. Cf. [39] Theorem 1.2.43. □

If a holomorphic map $f : U \to V$ between two open sets U and V has the holomorphic inverse $f^{-1} : V \to U$, f is called a *biholomorphic* map; in this case, we say that U and V are *holomorphically isomorphic* and write $U \cong V$.

In the case of $n = 1$ the following theorem which standardizes simply connected domains is well known.

Theorem 1.4.9 (Riemann's Mapping Theorem). *Let $\Omega \subset \mathbf{C}$ be a simply connected domain. If $\partial\Omega \neq \emptyset$, then there exists a biholomorphic mapping $f : \Omega \to \Delta(1)$ (unit disk). Furthermore, for an arbitrarily fixed $a \in \Omega$, f is unique with $f(a) = 0$ and $f'(a) > 0$.*

Proof. Cf. [38] Theorem (6.4.4). □

There is no such simple standardization of topologically simple domains in several variables; this was proved by H. Poincaré in 1907. A polydisk and an open ball of $\mathbf{C}^n$ $(n \geq 2)$ are the natural generalizations of a disk of $\mathbf{C}$, which are topologically cells and homeomorphic to each other. But, they are not biholomorphic:

Theorem 1.4.10 (Poincaré). *Let $n \geq 2$. Then there is no biholomorphic map $f : \Delta(1)^n \to \mathrm{B}(1)$ from the unit polydisk $\Delta(1)^n$ onto the unit ball $\mathrm{B}(1)$ in $\mathbf{C}^n$.*

Proof. We write $(z, w, v) \in \Delta(1) \times \Delta(1) \times \Delta(1)^{n-2}$ for the variables. Fix $v = v_0 \in \Delta(1)^{n-2}$ arbitrarily. We consider a proper map $g(z, w) := f(z, w, v_0)$ from $\Delta(1)^2$ into $\mathrm{B}(1)$. Let $w_\lambda \in \Delta(1)$ $(\lambda = 1, 2, \ldots)$ be any sequence with $\lim_{\lambda\to\infty} w_\lambda \in \partial\Delta(1)$. By Montel's Theorem 1.1.45 the sequence $\{g(z, w_\lambda)\}_\lambda$ $(z \in \Delta(1))$ of uniformly bounded (vector-valued) functions has a subsequence $\{g(z, w_{\lambda_\mu})\}_\mu$ which converges locally uniformly to $h : \Delta(1) \to \bar{\mathrm{B}}(1)$, where $\bar{\mathrm{B}}(1)$ denotes the closure of $\mathrm{B}(1)$. But in fact, $h(\Delta(1)) \subset \partial\mathrm{B}(1)$. Therefore, $\|h(z)\|^2 \equiv 1$. Since $\frac{\partial^2}{\partial z \partial \bar{z}} \|h(z)\|^2 = \|h'(z)\|^2$, $h'(z) \equiv 0$. We consider a holomorphic function $\frac{\partial g}{\partial z}(z, w)$ and see that the sequence $\{\frac{\partial g}{\partial z}(z, w_{\lambda_\mu})\}_\mu$ of partial derivatives converges locally uniformly to $h'(z) = 0$.

It follows from the above arguments that $\frac{\partial g}{\partial z}(z, w)$ is holomorphic in $(z, w) \in \Delta(1) \times \Delta(1)$ and extends continuously over $\Delta(1) \times \bar{\Delta}(1)$ so that $\frac{\partial g}{\partial z}(z, w) = 0$ for $|z| < 1$ and $|w| = 1$. By the Maximum Principle (Theorem 1.1.48), $\frac{\partial g}{\partial z}(z, w) \equiv 0$. Since $\frac{\partial g}{\partial z}(z, w) = \frac{\partial f}{\partial z}(z, w, v_0)$ and $v_0 \in \Delta(1)^{n-2}$ is arbitrary, we see that $\frac{\partial f}{\partial z}(z, w, v) \equiv 0$. Then the complex Jacobian of f must be identically 0: Contradiction. □

1.5 Analytic Subsets

We describe the definition of analytic subsets and the preliminary properties. Let $U \subset \mathbf{C}^n$ be an open set.

Definition 1.5.1. A subset $A \subset U$ is called an *analytic (sub)set* if for every point $a \in U$ there are a neighborhood $V \subset U$ of a and finitely many holomorphic functions $f_j \in \mathscr{O}(V)$, $1 \le j \le l$, satisfying

$$A \cap V = \{z \in V : f_1(z) = \cdots = f_l(z) = 0\}.$$

In particular, if at every $a \in A$, $l = 1$ and $A \cap V \ne V$ for a connected neighborhood V of a, A is called a *complex hypersurface*.

By definition an analytic subset of U is closed in U (cf. Exercise 9 at the end of the present chapter).

Theorem 1.5.2. *If an analytic subset A of a domain Ω contains an interior point, then $A = \Omega$; hence, if $A \ne \Omega$, A is a nowhere dense closed subset of Ω.*

Proof. Let A° be the set of interior points of A. By the assumption, $A^\circ \ne \emptyset$. Take a point $a \in \overline{A^\circ} \cap \Omega$. Then there are a connected neighborhood V of a in Ω, and finitely many $f_j \in \mathscr{O}(V), 1 \le j \le l$, such that

$$A \cap V = \{f_1 = \cdots = f_l = 0\}.$$

For a point $b \in V \cap A^\circ$, there is a neighborhood $W \subset A \cap V$ of b with $W \cap A = W$; i.e., $f_j|_W(z) \equiv 0, 1 \le j \le l$. By the Identity Theorem 1.1.46, $f_j(z) \equiv 0, 1 \le j \le l$. Therefore $V \cap A = V$ and $a \in A^\circ$. We see that $A^\circ (\subset \Omega)$ is open and closed. Since Ω is connected, $A^\circ = \Omega$, so that $A = \Omega$. □

Remark 1.5.3. In the case $n = 1$, an analytic subset of U without interior point is the same as a closed discrete subset of U (i.e., a discrete subset of U with no accumulation point in U).

We consider a holomorphic function $f \in \mathscr{O}(\mathrm{P}\Delta(a;(r_j)))$ in a polydisk $\mathrm{P}\Delta(a;(r_j))$. Assume that $f \ne 0$ ($f(z) \not\equiv 0$). Then $f(z)$ is expanded to a series of homogeneous polynomials as follows:

$$(1.5.4) \qquad f(z) = \sum_{\lambda} c_\lambda (z-a)^\lambda = \sum_{\nu=\nu_0}^{\infty} P_\nu(z-a),$$
$$P_\nu(z-a) = \sum_{|\lambda|=\nu} c_\lambda (z-a)^\lambda, \quad P_{\nu_0}(z-a) \not\equiv 0.$$

The *order of zero* of f at a is defined by

$$(1.5.5) \qquad \mathrm{ord}_a f := \nu_0.$$

For $f = 0$ we set $\mathrm{ord}_a f = \infty$.

Similarly, for a holomorphic function $g \in \mathcal{O}(\mathrm{P}\Delta(a;(r_j)))$ with $g \neq 0$ we have the homogeneous polynomial expansion $g(z) = \sum_{\nu=\nu_1}^{\infty} Q_\nu(z-a)$ with $Q_{\nu_1}(z-a) \not\equiv 0$. Therfore, it follows that

$$
\begin{aligned}
(1.5.6) \qquad f(z)g(z) &= P_{\nu_0}(z-a)Q_{\nu_1}(z-a) + \big(\text{term of order } \geq (\mu_1+\mu_2+1)\big),\\
& P_{\nu_0}(z-a)Q_{\nu_1}(z-a) \neq 0,
\end{aligned}
$$

and so

$$
(1.5.7) \qquad \mathrm{ord}_a(fg) = \mathrm{ord}_a f + \mathrm{ord}_a g.
$$

For the sake of simplicity we set $a = 0$ by a parallel translation, and assume $f(0) = 0$ ($\nu_0 \geq 1$). We take a vector $v \in \mathbf{C}^n \setminus \{0\}$ with $P_{\nu_0}(v) \neq 0$. Then for small $\zeta \in \mathbf{C}$ we have

$$
f(\zeta v) = \sum_{\nu=\nu_0}^{\infty} \zeta^\nu P_\nu(v) = \zeta^{\nu_0}(P_{\nu_0}(v) + \zeta P_{\nu_0+1}(v) + \cdots).
$$

By a linear transform of the coordinates we choose a new coordinate system $z = (z_1, \ldots, z_n)$ with $v = (0, \ldots, 0, 1)$. With a polyradius $r = (r_1, \ldots, r_n)$ we write

$$
\begin{aligned}
(1.5.8) \qquad \mathrm{P}\Delta((r_j)) &= \mathrm{P}\Delta_{n-1} \times \Delta(0; r_n) \subset \mathbf{C}^{n-1} \times \mathbf{C},\\
\mathrm{P}\Delta_{n-1} &= \{z' = (z_1, \ldots, z_{n-1}) \in \mathbf{C}^{n-1} : |z_j| < r_j, 1 \leq j \leq n-1\}
\end{aligned}
$$

for the polydisks and for the coordinates

$$
z = (z', z_n) \in \mathrm{P}\Delta_{n-1} \times \Delta(r_n), \quad 0 = (0,0).
$$

Since $f(0,0) = 0$ and $f(0, z_n) \not\equiv 0$, for a small $r_n > 0$ there is a $\delta > 0$ such that

$$
\begin{aligned}
&\{z_n : |z_n| \leq r_n, f(0, z_n) = 0\} = \{0\},\\
&|f(0, z_n)| > \delta, \qquad |z_n| = r_n.
\end{aligned}
$$

Therefore with sufficiently small $r_j > 0, 1 \leq j \leq n-1$,

$$
|f(z', z_n)| > \delta, \qquad z' \in \mathrm{P}\Delta_{n-1},\ |z_n| = r_n.
$$

Summarizing the above, we have:

Lemma 1.5.9. *For a holomorphic function $f(z)$ ($\not\equiv 0$) in a connected neighborhood of $0 \in \mathbf{C}^n$, there are a coordinate system $z = (z_1, z_2, \ldots, z_n) = (z', z_n)$ and a polydisk $\mathrm{P}\Delta((r_j)) = \mathrm{P}\Delta_{n-1} \times \Delta(r_n)$ satisfying the following:*

(i) *$f(z)$ is holomorphic on the closed polydisk $\overline{\mathrm{P}\Delta((r_j))}$, and the homogeneous polynomial expansion $f(z) = \sum_{\nu=\nu_0}^{\infty} P_\nu(z)$ satisfies that $P_{\nu_0}(0,1) \neq 0$ and*

$$f(0,z_n) = z_n^{\nu_0}(P_{\nu_0}(0,1) + z_n P_{\nu_0+1}(0,1) + \cdots).$$

(ii) *For a sufficiently small* $r_n > 0$, $\{|z_n| \le r_n : f(0,z_n) = 0\} = \{0\}$.

(iii) *For sufficiently small* $r_1, \ldots, r_{n-1} > 0$ *dependent on* $r_n > 0$, *the roots* z_n *of* $f(z', z_n) = 0$ *with every fixed* $z' \in \overline{\mathrm{P}\Delta_{n-1}}$, *are contained in* $\Delta(r_n)$. *In particular,* $|f(z', z_n)| > 0$ *for all* $(z', z_n) \in \overline{\mathrm{P}\Delta_{n-1}} \times \{|z_n| = r_n\}$.

Definition 1.5.10. We call the above $\mathrm{P}\Delta((r_j)) = \mathrm{P}\Delta_{n-1} \times \Delta(0; r_n)$ the *standard polydisk* of f, and $z = (z', z_n)$ the *standard coordinate system* of f.

Theorem 1.5.11 (Riemann's Extension). *Let Ω be a domain and let $A \subsetneqq \Omega$ be an analytic set. If a holomorphic function $f \in \mathcal{O}(\Omega \setminus A)$ is locally bounded about every point $a \in A$, that is, there is a neighborhood V of a such that the restriction $f|_{V \setminus A}$ is bounded, then f is analytically continued uniquely on Ω.*

Proof. In the case of $n = 1$, the theorem follows from Remark 1.5.3 and Riemann's Extension Theorem of one variable ([38] Theorem (5.1.1)).

Let $n \ge 2$, and let $a \in A$. By a parallel translation we may set $a = 0$. By the assumption there are a neighborhood V of 0 and $\phi \in \mathcal{O}(V) \setminus \{0\}$ such that $A \cap V \subset \{\phi = 0\}$. For V we take the standard polydisk $\mathrm{P}\Delta = \mathrm{P}\Delta_{n-1} \times \Delta(r_n)$ of ϕ. Then, $\phi(z', z_n) \ne 0$ for all $z' \in \mathrm{P}\Delta_{n-1}$ and $|z_n| = r_n$, i.e., $(\mathrm{P}\Delta_{n-1} \times \{|z_n| = r_n\}) \cap A = \emptyset$. With a fixed $z' \in \mathrm{P}\Delta_{n-1}$, $\phi(z', z_n) = 0$ has at most finitely many zeros in $\Delta(r_n)$, and $f(z', z_n)$ is locally bounded about those zeros. By the case of $n = 1$ above, $f(z', z_n)$ is holomorphic in $z_n \in \Delta(r_n)$. Therefore we have

$$f(z', z_n) = \frac{1}{2\pi i} \int_{|\zeta_n| = r_n} \frac{f(z', \zeta_n)}{\zeta_n - z_n} d\zeta_n, \quad |z_n| < r_n.$$

Since $f(z', \zeta_n)$ with $|\zeta_n| = r_n$ is holomorphic in z', the integral expression above implies $f(z', z_n)$ being holomorphic in $\mathrm{P}\Delta$. □

Theorem 1.5.12. *Let Ω be a domain and let $A \subsetneqq \Omega$ be an analytic set. Then $\Omega \setminus A$ is a domain, too.*

Proof. Suppose that $\Omega \setminus A$ is not connected. There are non-empty open subsets V_1, V_2 of $\Omega \setminus A$ such that

$$\Omega \setminus A = V_1 \cup V_2, \qquad V_1 \cap V_2 = \emptyset.$$

Put $f \in \mathcal{O}(\Omega \setminus A)$ as follows:

$$f(z) = \begin{cases} 0, & z \in V_1, \\ 1, & z \in V_2. \end{cases}$$

It follows from Theorem 1.5.11 that f is analytically continued to a unique $\tilde{f} \in \mathcal{O}(\Omega)$. Since $\tilde{f}|_{V_1} \equiv 0$, the Identity Theorem 1.1.46 implies $\tilde{f}(z) \equiv 0$; this is absurd. □

Let $U \subset \mathbf{C}^n$ be an open set and let $A \subset U$ be an analytic set.

Definition 1.5.13 ((non-)singular point)**.** A point $a \in A$ is called a *non-singular point* or a *smooth point* of A if the following property holds: There are a neighborhood $V(\subset U)$ of a and finitely many holomorphic functions $f_j \in \mathscr{O}(V)$ $(1 \le j \le l)$ such that

$$A \cap V = \{z \in V : f_j(z) = 0, 1 \le j \le l\}, \tag{1.5.14}$$

and

$$\operatorname{rank}\left(\frac{\partial f_j}{\partial z_k}(a)\right)_{1 \le j \le l, 1 \le k \le n} = l. \tag{1.5.15}$$

Necessarily, $l \le n$, and (1.5.15) holds for all $z \in V$ after shrinking V if necessary. A point of A which is *not* non-singular is called a *singular point.*

The set $\Sigma(A)$ of all singular points of A is closed in U by definition, but moreover it is known that $\Sigma(A)$ is an analytic subset (this is non-trivial; cf., e.g., [39] Theorem 6.5.10).

Definition 1.5.16 (complex submanifold)**.** An analytic set without singular point is called a *complex submanifold.*

N.B. The connectedness is not assumed in the definition.

Theorem 1.5.17. *Let $S \subset U$ be a complex submanifold. For every point $a \in S$ there are a neighborhood V (resp. W) of a (resp. $0 \in \mathbf{C}^n$) and a biholomorphic map $\varphi : z \in V \to w = f(z) \in W$ such that*

$$\varphi(S \cap V) = \{w = (w_1, \ldots, w_l, w_{l+1}, \ldots w_n) \in W : w_1 = \cdots = w_l = 0\}. \tag{1.5.18}$$

Proof. We may set $a = 0$ by a parallel transformation. After changing the order of coordinates we have by (1.5.15)

$$\operatorname{rank}\left(\frac{\partial f_j}{\partial z_k}(z)\right)_{1 \le j,k \le l} = l, \quad z \in V. \tag{1.5.19}$$

We then consider the following holomorphic map:

$$\varphi : z \in V \to (f_1(z), \ldots, f_l(z), z_{l+1}, \ldots, z_n) = w \in \mathbf{C}^n. \tag{1.5.20}$$

It follows from (1.5.19) that the Jacobian of (1.5.20) satisfies $\frac{\partial \varphi}{\partial z}(0) \ne 0$. By the inverse function Theorem 1.4.8 we see with suitably chosen neighborhoods V of a and W of 0 that $\varphi : V \to W$ is biholomorphic. By the definition, (1.5.18) holds. □

Through (1.5.18) we may identify $S \cap V$ with $\varphi(S \cap V)$. Then a point of $S \cap V$ is parameterized by $(w_{l+1}, \ldots, w_n)$, where $(0, \ldots, 0, w_{l+1}, \ldots, w_n) \in W$; in this sense we call $(w_{l+1}, \ldots, w_n)$ a *holomorphic local coordinate system* (of S) in $S \cap V$ or simply about a.

Definition 1.5.21. If a function $g : S \to \mathbf{C}$ on a complex submanifold $S (\subset U)$ is holomorphic in a holomorphic local coordinate system $(w_{l+1}, \ldots, w_n)$ about every $a \in S$, g is called a *holomorphic function* on S. This property is independent of the choice of the holomorphic local coordinate system, and we write $\mathcal{O}(S)$ for the set of all holomorphic functions on S.

Remark 1.5.22. Let $g : S \to \mathbf{C}$ be a holomorphic function on a complex submanifold $S (\subset U)$. For every point $a \in S$ there are a neighborhood $V(\subset U)$ of a and $\tilde{g} \in \mathcal{O}(V)$ with $\tilde{g}|_{S\cap V} = g|_{S\cap V}$; $\tilde{g}$ is called a *local extension* of g. The local extension is, of course, not unique. With the notation of (1.5.14), if $\tilde{g}$ is a local extension of g, then

$$\tilde{g}(z) + \sum_{j=1}^{l} c_j(z) f_j(z) \quad (c_j(z) \in \mathcal{O}(V))$$

is also a local extension of g.

Note. In the beginning of the 19th century, Theorem 1.1.25 (ii) was proved by E. Fabry [15] (1902); it was an important finding in the early time of the theory of analytic functions of several variables that the shape of singularities of analytic functions of variables more than one is *not* arbitrary and its nature is different to the case of one variable.

The method of Lemma 1.3.10 to express a holomorphic function $\varphi(z)$ defined in the overlapped part $U_1 \cap U_2 = F^\circ$ of two adjacent open sets U_1 and U_2 by the difference of elements of $\mathcal{O}(U_1)$ and $\mathcal{O}(U_2)$ was used by P. Cousin [12] (1895), when he resolved the problem that carries his name for cylinder domains, and also used in Oka Theory repeatedly. Therefore it is convenient to put a name to it; here, we call equation (1.3.11) the *Cousin decomposition* of $\varphi(z)$.

H. Poincaré [52] (1907) proved Theorem 1.4.10 in the case of $n = 2$ by studying the holomorphic automorphism groups of the domains: In the proof, a boundary regularity was assumed, and it was removed later by K. Reinhardt [54] (1921) ($n = 2$). The present proof is found in R. Range [53], Chap. I §2; the technique seems to be rather old (cf. ibid., p. 41).

Exercises

1. Show (1.1.7).
2. (i) Expand $e^{z_1} e^{z_2}$ to a homogeneous polynomial series.
   (ii) Expand $\sin z_1 \cos z_2$ to a homogeneous polynomial series.
3. Draw the figure of Fig. 1.3 in the case of $n = 3$.
4. a. What is the domain $\Omega(f)$ of convergence of the series

$$f(z,w) = \sum_{\nu,\mu=0}^{\infty} \binom{\nu+\mu}{\nu} z^\nu w^\mu$$

in two variables z, w?

b. Let $f(z,w) = \sum_{\nu=0}^{\infty}(z+w)^\nu$ be the expansion of $f(z,w)$ to a series of homogeneous polynomials. What is the domain of convergence of the series of homogeneous polynomials?

5. If a domain $\Omega \subset \mathbf{C}^n$ is invariant by poly-rotations $(z_j) \mapsto (e^{i\theta_j} z_j)$ $(\theta_j \in \mathbf{R}, 1 \le j \le n)$ (i.e., $(e^{i\theta_j} z_j) \in \Omega$ for all $(z_j) \in \Omega$ and $(\theta_j) \in \mathbf{R}^n$), Ω is called a *Reinhardt domain*. Let Ω be a Reinhardt domain containing the origin 0.
   Show that every $f \in \mathcal{O}(\Omega)$ is expanded to a series of polynomials, $f(z) = \sum_{\alpha \in \mathbf{Z}_+^n} c_\alpha z^\alpha$ $(z \in \Omega)$. (Cf., e.g., [39] §5.2.)
6. Let $n, m, l \in \mathbf{N}$ such that $n = m + l \ge 2$. Let $\mathrm{B}_n(1)$, $\mathrm{B}_m(1)$ and $\mathrm{B}_l(1)$ be the unit balls of $\mathbf{C}^n$, $\mathbf{C}^m$ and $\mathbf{C}^l$, respectively.
   Show that $\mathrm{B}_m(1) \times \mathrm{B}_l(1)$ is not biholomorphic to $\mathrm{B}_n(1)$.
7. Let $\Omega \subset \mathbf{C}^n$ be a domain, let $a = (a_j) \in \Omega$ and let $f \in \mathcal{O}(\Omega)$. Assume that there is a neighborhood U of a such that

$$f(z) = 0, \quad \forall z \in U \cap \{(z_j) \in \mathbf{C}^n : \Im z_j = \Im a_j, 1 \le j \le n\}.$$

   Then, show that $f(z) \equiv 0$ on Ω.
8. (Schwarz's Lemma in several variables)
   a. With the notation of (1.1.2) we consider a bounded holomorphic function $f(z)$ in $\mathrm{B}(1)$ such that $|f(z)| \le M$ $(z \in \mathrm{B}(1))$ and $f(0) = 0$.
      Show that $|f(z)| \le M\|z\|$ $(z \in \mathrm{B}(1))$.
   b. With $|z|_{\max} = \max_{1 \le j \le n} |z_j|$ for $z = (z_1, \ldots, z_n) \in \mathbf{C}^n$ we write $\mathrm{P}\Delta_1 = \{|z|_{\max} < 1\}$ for the unit polydisk. Let $f(z)$ be a bounded holomorphic function in $\mathrm{P}\Delta_1$ such that $|f(z)| \le M$ $(z \in \mathrm{P}\Delta_1)$ and $f(0) = 0$.
      Show that $|f(z)| \le M|z|_{\max}$ $(z \in \mathrm{P}\Delta_1)$.
9. Prove that an analytic subset of an open set U of $\mathbf{C}^n$ is closed in U.
10. Let $\Omega \subset \mathbf{C}^n$ be a domain, and let $f_j \in \mathcal{O}(\Omega)$ $(1 \le j \le m)$ be finitely many holomorphic functions. Show that the graph

$$\Sigma = \{(z, (w_j)) \in \Omega \times \mathbf{C}^m : w_j = f_j(z), 1 \le j \le m\}$$

   is a complex submanifold of $\Omega \times \mathbf{C}^m$.

Chapter 2
Coherent Sheaves and Oka's Joku-Iko Principle

Beginning with the definition of analytic sheaves, we introduce the notion of coherence. We prove the coherence of $\mathscr{O}_{\mathbf{C}^n}$ (Oka's First Coherence Theorem), and then prove the coherence of ideal sheaves of complex submanifolds (a special case of Oka's Second Coherence Theorem). We then prove Cartan's Merging Lemma, the Oka Syzygy for coherent sheaves, and the Oka Extension of the Joku-Iko Principle, which is the key to Oka Theory.

2.1 Notion of Analytic Sheaves

2.1.1 Definitions of Rings and Modules

We begin with the definitions of algebraic terminologies of rings and modules. Those who know the definitions of these terminologies may skip this section.

Let R be a set. Assume that for every two elements $a, b \in R$ there is associated the third element $a + b \in R$ satisfying condition (1) below; we call this an algebraic operation. If an algebraic operation satisfies condition (2) below, it is called an *addition*. Furthermore, if it satisfies (3) and (4), R is called an *additive group* (or commutative group, also abelian group):

(1) For every $c \in R$, $(a+b)+c = a+(b+c)$ (associative law).
(2) $a+b = b+a$ (commutative).
(3) There is a special element $0 \in R$, called a *zero (element)* satisfying $a+0 = a$ for all $a \in R$.
(4) For every element $a \in R$ there exists a unique element $-a \in R$ called the *inverse* of a such that $a+(-a) = 0$.

For example, $\mathbf{Z}$ is an additive group by the natural addition; $\mathbf{N}$ carries the natural addition, but is *not* an additive group.

Assume that for every two elements $a, b \in R$, there is the associated third element $a \cdot b \in R$, and the following condition is satisfied:

J. Noguchi, *Basic Oka Theory in Several Complex Variables*, Universitext,
https://doi.org/10.1007/978-981-97-2056-9_2

(5) For every $c \in R$, $(a \cdot b) \cdot c = a \cdot (b \cdot c)$ (associative law).

The operation to associate $a \cdot b$ with a and b is called a *multiplication*.

If there exists an element $1 \in R$ with $a \cdot 1 = 1 \cdot a = a$ ($\forall a \in R$), 1 is called a *unit element* for multiplication, which is unique if it exists.

Definition 2.1.1 (ring). If a set R carries an addition "+" and a multiplication "·" and the following conditions are satisfied, then R is called a *ring*:

(i) R has a unit element 1 for multiplication.
(ii) For arbitrary three elements $a, b, c \in R$, the so-called distributive laws hold:

$$a \cdot (b+c) = (a \cdot b) + (a \cdot c) \ (= a \cdot b + a \cdot c, \quad \text{so written}),$$
$$(b+c) \cdot a = (b \cdot a) + (c \cdot a) \ (= b \cdot a + c \cdot a).$$

In particular, if the multiplication is commutative ($a \cdot b = b \cdot a$), R is called a *commutative ring* The multiplication $a \cdot b$ is often simplified to ab without "·".

Remark 2.1.2. Here we deal with only commutative rings, and so a ring is always assumed to be commutative all through the present textbook.

Let R be a ring, and let $a \in R$. If there is an element $b \in R$ with $ab = 1$, a is called a *unit* and b is denoted by a^{-1}.

A subset I of R is called an *ideal* if

$$a \cdot (b+c) \in I, \quad \forall a \in R, \ \forall b, c \in I.$$

In particular, $0 \in I$, and $-a \in I$ for all $a \in I$.

Definition 2.1.3 (module). Let R be a ring, and let M be an additive group. If for arbitrary elements $a \in R$ and $x \in M$, there is associated an element $a \cdot x \in M$, and the following conditions are satisfied, then M is called a *module* over R or a *R-module* : that is, for $a, b \in R$, $x, y \in M$:

(i) $1 \cdot x = x$;
(ii) $(ab) \cdot x = a \cdot (b \cdot x)$;
(iii) $(a+b) \cdot x = a \cdot x + b \cdot x$;
(iv) $a \cdot (x+y) = a \cdot x + a \cdot y$.

A subset $M' \subset M$ is called a *submodule*, if M' forms a module over R by itself with respect to the algebraic operations induced from those of M.

Example 2.1.4. (i) Any additive group may be regarded as a module over $\mathbf{Z}$.

(ii) The qth product $\mathbf{Z}^q$ with $q \in \mathbf{N}$ forms an additive group by the element-wise addition, and by the element-wise multiplication:

$$a \cdot x = (ax_1, \dots, ax_q) \in \mathbf{Z}^q, \quad a \in \mathbf{Z}, \ x = (x_1, \dots, x_q) \in \mathbf{Z}^q.$$

$\mathbf{Z}^q$ is a module over $\mathbf{Z}$.

(iii) Let $p \in \mathbf{N}$ and let $p\mathbf{Z}$ be the set of all multiples of p. Then, $p\mathbf{Z}$ is an ideal of $\mathbf{Z}$.

(iv) The set $\mathscr{O}(U)$ of all holomorphic functions in an open subset $U \subset \mathbf{C}^n$ is a ring by the natural addition and multiplication. As in (ii), $\mathscr{O}(U)^q$ is an $\mathscr{O}(U)$-module.

(v) Let $X \subset U$ be a subset of an open subset $U \subset \mathbf{C}^n$. Let $I(X)$ denote the set of all $f \in \mathscr{O}(U)$ such that $f|_X \equiv 0$. Then, $I(X)$ is an ideal of $\mathscr{O}(U)$.

(vi) In general, with a ring R given and with finitely many indeterminates $X_1, \ldots, X_N$, we denote by $R[X_1, \ldots, X_N]$ the set of all polynomials in X_j $(1 \le j \le N)$ with coefficients in R. By the natural algebraic operations, $R[X_1, \ldots, X_N]$ forms a ring, called an *R-polynomial ring* (of N variables), which is also an R-module.

2.1.2 Analytic Sheaves

Here we leave the general treatment of the notion of sheaves to other books (e.g., [39] §1.3; more generally, cf. Godeman [20]), we restrict ourselves to dealing with necessary holomorphic (analytic) functions.

We consider a holomorphic function $f(z)$ in a connected neighborhood of $a \in \mathbf{C}^n$. With a fixed coordinate system $z = (z_j)_{1 \le j \le n}$, $f(z)$ is uniquely expressed by a convergent power series

$$f(z) = \sum_{\nu \in \mathbf{Z}_+^n} c_\nu (z-a)^\nu. \tag{2.1.5}$$

We identify holomorphic functions about a whose power series expansions (2.1.5) are the same, and then we denote by $\mathscr{O}_a$ all of such equivalence classes. We write $\underline{f}_a \in \mathscr{O}_a$ for an element of $\mathscr{O}_a$. An element with all $c_\nu = 0$ $(\nu \in \mathbf{Z}_+^n)$ in (2.1.5) is denoted by $\underline{0}_a$ (often abbreviated to 0). An element with $c_{(0,\ldots,0)} = 1$ and $c_\nu = 0$ $(|\nu| > 0)$ is denoted by $\underline{1}_a$ (often abbreviated to 1). $\mathscr{O}_a$ forms a ring with the natural addition and multiplication, which is called an *analytic local ring*.

Proposition 2.1.6 (Integral domain). *If $\underline{f}_a \cdot \underline{g}_a = 0$ for two elements $\underline{f}_a, \underline{g}_a$ of $\mathscr{O}_a$, then either $\underline{f}_a = 0$ or $\underline{g}_a = 0$.*

Proof. It is immediate by (1.5.6). □

For a domain $\Omega \subset \mathbf{C}^n$ we put

$$\mathscr{O}_\Omega = \bigsqcup_{a \in \Omega} \mathscr{O}_a, \tag{2.1.7}$$

where "$\sqcup$" stands for a disjoint union. We call $\mathscr{O}_\Omega$ the *sheaf of holomorphic (or analytic) functions over Ω*. A holomorphic function $f \in \mathscr{O}(U)$ on an open subset $U \subset \Omega$ induces a map

$$\underline{f} : a \in U \longrightarrow \underline{f}_a \in \mathscr{O}_U, \tag{2.1.8}$$

which is called a *section* of $\mathscr{O}_\Omega$ (or $\mathscr{O}_{\mathbf{C}^n}$) over U. We denote by $\Gamma(U,\mathscr{O}_\Omega)$ (or $\Gamma(U,\mathscr{O}_{\mathbf{C}^n})$) the set of all sections of $\mathscr{O}_\Omega$ over U. Then $\Gamma(U,\mathscr{O}_\Omega)$ is a ring with the natural algebraic structure of addition and multiplication.

Remark 2.1.9. By definition we can identify $\mathscr{O}(U)$ and $\Gamma(U,\mathscr{O}_U)$ through

$$f \in \mathscr{O}(U) \longrightarrow \underline{f} \in \Gamma(U,\mathscr{O}_U).$$

It is noticed, however, that for $a \in U$, $f(a)$ is a *value* (complex number), and $\underline{f}_a$ is a convergent power series about a.

Let $q \in \mathbf{N}$. Taking the qth direct product $\mathscr{O}_a^q = \overbrace{\mathscr{O}_a \times \cdots \times \mathscr{O}_a}^{q}$ $(a \in \Omega)$ of $\mathscr{O}_a$, we define the qth *direct product sheaf* of $\mathscr{O}_\Omega$ by

$$\mathscr{O}_\Omega^q = \bigsqcup_{a\in\Omega} \mathscr{O}_a^q. \tag{2.1.10}$$

Note that it is *not* the qth direct product $\overbrace{\mathscr{O}_\Omega \times \cdots \times \mathscr{O}_\Omega}^{q}$ of $\mathscr{O}_\Omega$ as sets. For $\underline{f}_a \in \mathscr{O}_a$ $(a \in \Omega)$ and $(\underline{g_j}_a), (\underline{h_j}_a) \in \mathscr{O}_a^q$ $(1 \le j \le q)$ the following algebraic structure is naturally introduced:

$$\begin{aligned} \underline{f}_a \cdot (\underline{g_j}_a) &= (\underline{f}_a \, \underline{g_j}_a) \in \mathscr{O}_a^q, \\ \underline{f}_a \cdot \left((\underline{g_j}_a) \pm (\underline{h_j}_a)\right) &= \underline{f}_a \cdot (\underline{g_j}_a) \pm \underline{f}_a \cdot (\underline{h_j}_a). \end{aligned} \tag{2.1.11}$$

By the algebraic structure above, $\mathscr{O}_a^q$ is an $\mathscr{O}_a$-module.

Definition 2.1.12 (Analytic sheaf)**.** A subset $\mathscr{F} \subset \mathscr{O}_\Omega^q$ is called an *analytic sheaf* over Ω if it has the following structure:

(i) For every $a \in \Omega$ an associated $\mathscr{O}_a$-submodule $\mathscr{F}_a \subset \mathscr{O}_a^q$ exists.
(ii) $\mathscr{F} = \bigsqcup_{a\in\Omega} \mathscr{F}_a$.

In particular, an analytic sheaf $\mathscr{F}$ with $q = 1$ is called a *sheaf of ideals* of $\mathscr{O}_\Omega$.

Naturally, $\mathscr{O}_\Omega^q$ itself is an analytic sheaf. Although it is a trivial example, $\mathscr{F}$ with $\mathscr{F}_a = \{0\}$ $(\forall a \in \Omega)$ is an analytic sheaf called the *zero sheaf* and written as 0. The restriction of an analytic sheaf $\mathscr{F}$ to a subset $U \subset \Omega$ is defined by

$$\mathscr{F}|_U := \bigsqcup_{a\in U} \mathscr{F}_a.$$

Remark 2.1.13. It is common to define analytic sheaves by making use of a more abstract notion of sheaves (e.g., cf. [29], [26], [30], and [39], etc.). Here, for the purpose to solve the "Three Big Problems" it is sufficient to deal with this simple case.

A q-vector-valued holomorphic function $g = (g_1, \ldots, g_q) : U \to \mathbf{C}^q$ on an open subset $U \subset \Omega$ induces a map

$$(2.1.14) \qquad \underline{g} = \left(\underline{g_j}\right) : a \in U \longrightarrow \underline{g}_a := \left(\underline{g_1}_a, \ldots, \underline{g_q}_a\right) \in \mathcal{O}^q_\Omega.$$

We call this a *section* of $\mathcal{O}^q_\Omega$ (or $\mathcal{O}^q_{\mathbf{C}^n}$) over U, and the set of all of them is denoted by $\Gamma(U, \mathcal{O}^q_\Omega)$. For $f \in \mathcal{O}(U)$ ($\underline{f} \in \Gamma(U, \mathcal{O}_\Omega)$) and $\underline{g} \in \Gamma(U, \mathcal{O}^q_\Omega)$ we put

$$(2.1.15) \qquad \underline{f} \cdot \underline{g} : a \in U \longrightarrow \underline{f}_a \cdot \underline{g}_a \in \mathcal{O}^q_\Omega,$$

so that $\Gamma(U, \mathcal{O}^q_\Omega)$ forms a module over the ring $\mathcal{O}(U)$ $(= \Gamma(U, \mathcal{O}_\Omega))$.

Let $\mathscr{F} \subset \mathcal{O}^q_\Omega$ be an analytic sheaf over Ω. If $\underline{g}$ in (2.1.14) satisfies $\underline{g}_a \in \mathscr{F}_a$ for all $a \in U$, $\underline{g}$ is called a *section* of $\mathscr{F}$ over U. We denote the *section space* of all of them by

$$(2.1.16) \qquad \Gamma(U, \mathscr{F}) = \left\{\underline{g} \in \Gamma(U, \mathcal{O}^q_\Omega) : \underline{g}_a \in \mathscr{F}_a, \forall a \in U\right\}.$$

By (2.1.15), $\Gamma(U, \mathscr{F})$ is a module over the ring $\mathcal{O}(U)$ (or, $\Gamma(U, \mathcal{O}_\Omega)$), and also a vector space over $\mathbf{C}$.

A section $s \in \Gamma(U, \mathscr{F})$ may be regarded as a map from U to $\mathscr{F}$, for which we write

$$s : a \in U \longrightarrow s(a) \in \mathscr{F}.$$

For a closed set E we denote by $\Gamma(E, \mathscr{F})$ the set of all sections of $\mathscr{F}$ over a neighborhood of E, where the neighborhood may vary.

Remark 2.1.17. Throughout the present text we often consider holomorphic functions and sections of analytic sheaves over a neighborhood of a closed set, and the neighborhood will depend on each case. Therefore, by the term, holomorphic functions or sections of an analytic sheaf $\mathscr{F}$ over E, we understand that they are defined in some neighborhoods of E, and denote the set of all of them by $\mathcal{O}(E)$ (or $\Gamma(E, \mathscr{F})$); in cases where the neighborhood must be clarified, we will state it precisely.

Remark 2.1.18. If a knowledge of "general topology" is assumed, it is standard to define a sheaf with the topology such that the sections defined above are continuous. Then, the definition of analytic sheaves may take a more general abstract form. Here, we do not need a general theory of sheaves, so we prefer the style above: Cf., e.g., [39] §1.3.

2.2 Coherent Sheaves

2.2.1 Locally Finite Sheaves

We begin with the definition of the local finiteness of an analytic sheaf. Let $n, p \in \mathbf{N}$. Let $\Omega \subset \mathbf{C}^n$ be an open set and let $\mathscr{F} \subset \mathscr{O}_\Omega^p$ be an analytic sheaf.

Definition 2.2.1. (i) Let $U \subset \Omega$ be an open subset. A finite family $\{\underline{\sigma_j}\}_{j=1}^l \subset \Gamma(U, \mathscr{F})$ is said to generate $\mathscr{F}$ over U, if

$$\mathscr{F}|_U = \sum_{j=1}^{l} \mathscr{O}_U \cdot \underline{\sigma_j}.$$

That is, we have

$$\mathscr{F}_z = \sum_{j=1}^{l} \mathscr{O}_z \cdot \underline{\sigma_j}_z, \qquad \forall z \in U.$$

In this case, $\{\underline{\sigma_j}\}_{j=1}^l$ is called a *finite generator system* of $\mathscr{F}$ over U.

(ii) $\mathscr{F}$ is said to be *locally finite* about a point $a \in \Omega$, if there is a neighborhood $U(\subset \Omega)$ of a with a finite generator system $\{\underline{\sigma_j}\}_{j=1}^l$ of $\mathscr{F}$ over U. In this case, $\{\underline{\sigma_j}\}_{j=1}^l$ is called a *locally finite generator system* of $\mathscr{F}$ about a.

(iii) $\mathscr{F}$ is said to be *locally finite* (in Ω), if $\mathscr{F}$ is locally finite about every $a \in \Omega$.

Example 2.2.2. We give a simple example of an analytic sheaf which is not locally finite (hence not coherent as defined in the next subsection). Let $n = 1$, and let $\Delta(1) \Subset \mathbf{C}$ be the unit disk. We define an analytic sheaf $\mathscr{F} \subset \mathscr{O}_\mathbf{C}$ over $\mathbf{C}$ by

$$\mathscr{F}_a = \begin{cases} 0, & a \in \mathbf{C} \setminus \Delta(1), \\ \mathscr{O}_a, & a \in \Delta(1). \end{cases}$$

Here, 0 means a submodule consisting only of the zero element of $\mathscr{O}_a$, and equals $\{\underline{0}_a\}$ as set.[1] Over $\Delta(1)$, $\underline{1} \in \Gamma(\Delta(1), \mathscr{O}_\mathbf{C})$ is a finite generator system of $\mathscr{F}$, and over $\mathbf{C} \setminus \bar{\Delta}(1)$, $\underline{0} \in \Gamma(\mathbf{C} \setminus \bar{\Delta}(1), \mathscr{O}_\mathbf{C})$ is a generator system of $\mathscr{F}$. But, $\mathscr{F}$ is not locally finite about any point of $\partial\Delta(1)$. In fact, if $\mathscr{F}$ is locally finite about a point $a \in \partial\Delta(1)$, then there would be a connected neighborhood $U \ni a$ and finitely many $f_j \in \mathscr{O}(U)$ $(1 \leq j \leq N)$ such that

$$\mathscr{F}_b = \sum_{j=1}^{N} \mathscr{O}_b \cdot \underline{f_j}_b, \qquad \forall b \in U.$$

[1] The zero module or a zero element is often simplified in this way; similarly, for instance, a vector space E consisting only of the zero vector is written as $E = 0$.

Since $\mathscr{F}_b = 0$ for $b \in U \setminus \Delta(1)$, $\underline{f_j}_b = 0$ for all j. Thus, $f_j \equiv 0$ (identically) in a sufficiently small neighborhood of b. The Identity Theorem 1.1.46 would imply $f_j \equiv 0$ on U, and hence $\mathscr{F}_b = 0$ ($\forall\, b \in U$). On the other hand, $\mathscr{F}_b = \mathscr{O}_b$ for $b \in U \cap \Delta(1)\ (\neq \emptyset)$; this is absurd.

The above example is easily extended to the case of $n \geq 2$.

Example 2.2.3. We introduce an example of an analytic sheaf due to Oka VII [48] which is not locally finite.[2] We consider a complex hyperplane $X = \{z = w\}$ in $\mathbf{C}^2$ with variables z, w. Taking open balls $\mathrm{B}_i = \{|z|^2 + |w|^2 < r_i^2\}$ with $0 < r_1 < r_2$, we set $X_0 = X \cap \mathrm{B}_2 \setminus \mathrm{B}_1$. Let $\Gamma_1 = \partial \mathrm{B}_1$ be the boundary sphere.

Let $\mathscr{B}_a$ denote the set of all those power series $\underline{f}_a$ about $a \in \mathrm{B}_2$ of a holomorphic function $f(z,w)$ in a neighborhood U of a such that $f(z,w)/(z-w)$ is holomorphic at every point of $U \cap X_0$, and set $\mathscr{B} = \bigsqcup_{a \in \mathrm{B}_2} \mathscr{B}_a$. Then, $\mathscr{B}$ is an analytic subsheaf of $\mathscr{O}_{\mathrm{B}_2}$. By the construction we have

$$\mathscr{B}_a = \begin{cases} \mathscr{O}_a \cdot \underline{(z-w)}_a, & a \in X_0, \\ \mathscr{O}_a, & a \in \mathrm{B}_2 \setminus X_0. \end{cases} \tag{2.2.4}$$

Now, $\mathscr{B}$ is not locally finite at any point $a \in X_0 \cap \Gamma_1$. If otherwise, there are a polydisk neighborhood U of a and finitely many holomorphic functions $f_j \in \mathscr{O}(U)$, $\underline{f_j} \in \Gamma(U, \mathscr{B})$, $1 \leq j \leq N$, such that

$$\mathscr{B}_b = \sum_{j=1}^{N} \mathscr{O}_b \cdot \underline{f_j}_b, \qquad \forall\, b \in U.$$

However, since $f_j(z,z) \equiv 0$, $\mathscr{B}_b \neq \mathscr{O}_b$ at $b \in U \cap X \setminus X_0$; this contradicts (2.2.4).

Finally, although trivial, the zero sheaf 0 is coherent.

Let $\Omega \subset \mathbf{C}^n$ be an open set, and let $A \subset \Omega$ be a subset. For $\underline{f}_a \in \mathscr{O}_a$ ($a \in \Omega$) we write

$$f|_A = 0,$$

if f is holomorphic in a polydisk neighborhood $\mathrm{P}\Delta$ of a, and $f|_{\mathrm{P}\Delta \cap A} = 0$, i.e.,

$$f(z) = 0, \qquad \forall z \in \mathrm{P}\Delta \cap A.$$

Put

$$\mathscr{I}\langle A \rangle_a = \left\{ \underline{f}_a \in \mathscr{O}_a : f|_A = 0 \right\}, \quad \mathscr{I}\langle A \rangle = \bigsqcup_{a \in \Omega} \mathscr{I}\langle A \rangle_a. \tag{2.2.5}$$

$\mathscr{I}\langle A \rangle$ is a sheaf of ideals of $\mathscr{O}_\Omega$, and is called the *ideal sheaf of the subset* A.

---

[2] H. Cartan put an emphasis on this counter-example in his comment on Oka VII [48].

It is an important case of $\mathscr{I}\langle A\rangle$ when A is an analytic subset of Ω; in that case $\mathscr{I}\langle A\rangle$ is called the *ideal sheaf of the analytic subset* A.[3] If A is non-singular, i.e., a complex submanifold, we call $\mathscr{I}\langle A\rangle$ the *ideal sheaf of the complex submanifold* A.

In general, $\mathscr{I}\langle A\rangle = \mathscr{I}\langle \bar{A}\rangle$ for any subset A $(\subset \Omega)$ and its topological closure $\bar{A}$ in Ω.

Lemma 2.2.6. *Let $A \subset \Omega$ be a closed subset. Assume that there is a finite generator system $\left\{\underline{\alpha_j} \in \Gamma(\Omega, \mathscr{I}\langle A\rangle) : 1 \le j \le L\right\}$ of $\mathscr{I}\langle A\rangle$ over Ω. Then*

$$A = \{z \in \Omega : \alpha_j(z) = 0,\ 1 \le j \le L\}. \tag{2.2.7}$$

In particular, A is an analytic subset.

Proof. We denote the right-hand side of (2.2.7) by B, which is an analytic subset by definition. By the choice of α_j, $B \supset A$. Let $a \in \Omega \setminus A$ be an arbitrary point. Since A is closed, there is a neighborhood $U \ni a$ with $A \cap U = \emptyset$. Therefore $\mathscr{I}\langle A\rangle_a = \mathscr{O}_a \ni \underline{1}_a$ and then there are $\underline{f_j}_a \in \mathscr{O}_a$ $(1 \le j \le L)$ such that

$$\underline{1}_a = \underline{f_1}_a \cdot \underline{\alpha_1}_a + \cdots + \underline{f_L}_a \cdot \underline{\alpha_L}_a.$$

That is, there is a neighborhood V of a such that f_j, α_j are holomorphic in V, and

$$1 = f_1(z) \cdot \alpha_1(z) + \cdots + f_L(z) \cdot \alpha_L(z), \quad z \in V.$$

Setting $z = a$, we have some $\alpha_j(a) \ne 0$. Therefore $a \in \Omega \setminus B$, and $A \supset B$, so that $A = B$. □

The next theorem relates the local finiteness of $\mathscr{I}\langle A\rangle$ with the analyticity of A, which is basic and a special case of a more general support theorem ([39] Theorem 6.9.10).

Theorem 2.2.8. *Let $A \subset \Omega$ be a closed set. If $\mathscr{I}\langle A\rangle$ is locally finite, then A is an analytic subset.*

Proof. Let $a \in \Omega$ be an arbitrary point. Suppose that there are a neighborhood $U \subset \Omega$ of a and a generator system $\underline{\sigma_j} \in \Gamma(U, \mathscr{I}\langle A\rangle)$, $1 \le j \le l$, of $\mathscr{I}\langle A\rangle$ over U. It follows from Lemma 2.2.6 that

$$A \cap U = \{z \in U : \sigma_j(z) = 0, 1 \le j \le l\}.$$

Therefore A is an analytic subset. □

Remark 2.2.9. In fact, the converse of the above theorem holds by Oka's Second Coherence Theorem mentioned in the next subsection.

---

[3] Oka (VIII) called $\mathscr{I}\langle A\rangle$ "l'idéal géométrique de domaines indéterminés attaché à A"; so we may call it the *geometric ideal sheaf of* A.

2.2.2 Coherent Sheaves

Let $\mathscr{F}$ be an analytic sheaf over an open set $\Omega \subset \mathbf{C}^n$ as above.

Definition 2.2.10. A *relation sheaf* of $\mathscr{F}$ is an analytic sheaf $\mathscr{R}\left((\underline{\tau_j})_{1\le j\le q}\right)$ defined as follows:

(i) Let $U \subset \Omega$ be an open set.
(ii) Let $\underline{\tau_j} \in \Gamma(U, \mathscr{F})$, $1 \le j \le q$, be finitely many sections.
(iii) We consider an element $\left(\underline{a_1}_z, \dots, \underline{a_q}_z\right) \in \mathscr{O}_z^q$ $(z \in U)$ satisfying the linear relation

$$\underline{a_1}_z \cdot \underline{\tau_1}_z + \cdots + \underline{a_q}_z \cdot \underline{\tau_q}_z = 0, \tag{2.2.11}$$

and denote all of them by $\mathscr{R}\left(\underline{\tau_1}, \dots, \underline{\tau_q}\right)_z = \mathscr{R}\left((\underline{\tau_j})_{1\le j\le q}\right)_z$. We put

$$\mathscr{R}\left(\underline{\tau_1}, \dots, \underline{\tau_q}\right) = \mathscr{R}\left(\left(\underline{\tau_j}\right)_{1\le j\le q}\right) = \bigsqcup_{z\in U} \mathscr{R}\left(\left(\underline{\tau_j}\right)_{1\le j\le q}\right)_z. \tag{2.2.12}$$

Definition 2.2.13 (Coherent sheaf)**.** An analytic sheaf $\mathscr{F}$ over Ω is called a *coherent analytic sheaf* or simply a *coherent sheaf* if the following two conditions are satisfied:

(i) $\mathscr{F}$ is locally finite.
(ii) Every relation sheaf of $\mathscr{F}$ is locally finite.

K. Oka proved three fundamental coherence theorems (Oka's Three Coherence Theorems [48] Oka VII, VIII; cf. Noguchi [39]). The implication of the series of Oka's Coherence Theorems is broad and deep, and it is impossible to explain the importance in a few lines. The first is stated as follows:

Oka's First Coherence Theorem (Oka VII, 1948) *$\mathscr{O}_{\mathbf{C}^n}$ is coherent.*

We will prove this in the next section. When $n = 1$, the proof is easy and left to the readers (Exercise 3 at the end of the chapter).

Oka's Second Coherence Theorem (Oka VIII, 1951, Cartan [10], 1950)[4] *The ideal sheaf of any analytic subset is coherent.*

This is the converse of Theorem 2.2.8, so that:

Theorem 2.2.14. *A closed subset $A \subset \Omega$ is analytic if and only if $\mathscr{I}\langle A\rangle$ is locally finite.*

The case of non-singular A is easy and will be proved in §2.3.3 (cf. [39] §6.5 in general).

---

[4] The main part of the proof was done in the paper of Oka's First Coherence Theorem (Oka VII, 1948), and this result was announced there. In 1950, H. Cartan gave another proof of it by making use of Oka's result (1948). Cf. [39] Chap. 9, On Coherence.

Since it requires a rather long preparation to state the third one, we have to allow ourselves to use some terminologies without precise definitions for a moment. Let A be an analytic subset of Ω. Then by Oka's Second Coherence Theorem $\mathscr{I}\langle A\rangle$ is coherent, and the quotient sheaf $\mathscr{O}_A := \mathscr{O}_\Omega/\mathscr{I}\langle A\rangle$ is defined and gives rise to a coherent sheaf, which is called the structure sheaf of A as a complex space. Then the so-called "normalization sheaf $\hat{\mathscr{O}}_A$" is defined:

Oka's Third Coherence Theorem (Oka VIII, 1951) *The normalization sheaf $\hat{\mathscr{O}}_A$ is coherent.*

Cf., e.g, [39] §6.10 for details.

For general properties of coherent sheaves we have:

Proposition 2.2.15. *Let $\mathscr{F}$ be a coherent sheaf over Ω.*

(i) *An analytic subsheaf of $\mathscr{F}$ is coherent if and only if it is locally finite.*

(ii) *$\mathscr{F}^N$ $(N \in \mathbf{N})$ are coherent. That is, every simultaneous relation sheaf of $\mathscr{F}$ is coherent.*

(iii) *Let $\{\underline{\gamma_j}\}_{j=1}^L$ be a finite family of sections of $\mathscr{F}$ over Ω. If $\{\underline{\gamma_j}_a\}_{j=1}^L$ generates $\mathscr{F}_a$ at a point $a \in \Omega$, then there is a neighborhood V $(\subset \Omega)$ of a such that $\{\underline{\gamma_j}_z\}_{j=1}^L$ generates $\mathscr{F}_z$ at every point $z \in V$.*

Proof. (i) It remains to show the local finiteness of relation sheaves, but they are so since $\mathscr{F}$ is coherent.

(ii) We shall proceed by induction on N. The case of $N = 1$ is the assumption.

Let $N \geq 2$, and suppose that it holds for $N - 1$. The local finiteness of $\mathscr{F}^N$ follows from that of $\mathscr{F}$. Let $U \subset \Omega$ be an open subset, and let $F_i \in \Gamma(U, \mathscr{F}^N)$, $1 \leq i \leq q$, be a finite number of sections. It suffices to show the local finiteness of the relation sheaf

$$\mathscr{R} = \left\{(a_i) \in \mathscr{O}_z^q : \sum_{i=1}^q a_i F_i(z) = 0,\ z \in U\right\} \subset \mathscr{O}_U^q.$$

With the expressions $F_i = (F_{i1}, \ldots, F_{iN})$, $\mathscr{R}$ $(\subset (\mathscr{O}|_U)^q)$ is determined by

$$(a_i) \in \mathscr{O}_z^q, \quad \sum_{i=1}^q a_i \underline{F_{ij}}_z = 0, \quad 1 \leq j \leq N,\ z \in U.$$

We first consider the case for $j = 1$. Denote by $\mathscr{R}_1 \subset (\mathscr{O}|_U)^q$ the relation sheaf defined by

$$(a_i) \in \mathscr{O}_z^q, \quad \sum_{i=1}^q a_i \underline{F_{i1}}_z = 0, \quad z \in U.$$

Then $\mathscr{R} \subset \mathscr{R}_1$, and since $\mathscr{F}$ is coherent, $\mathscr{R}_1$ is locally finite. For every point $a \in U$ there are a neighborhood $V \subset U$ of a and a locally finite generator system $\{\phi^{(\lambda)}\}_{\lambda=1}^L$ of $\mathscr{R}_1|_V$ with $\phi^{(\lambda)} \in \Gamma(V, \mathscr{R}_1)$. Set $\phi^{(\lambda)} = (\phi_i^{(\lambda)})_{1 \leq i \leq q}$. At every point $z \in V$ an element of $\mathscr{R}_{1z}$,

$$(a_i) = \left(\sum_\lambda \underline{c_\lambda}_z \cdot \phi_i^{(\lambda)}(z)\right), \qquad \underline{c_\lambda}_z \in \mathscr{O}_z$$

belongs to $\mathscr{R}_z$ if and only if

(2.2.16) $$\sum_i \sum_\lambda \underline{c_\lambda}_z \cdot \phi_i^{(\lambda)}(z) \cdot \underline{F_{ij}}_z = 0, \quad 1 \le j \le N.$$

We consider this as a linear relation on $\left(\underline{c_\lambda}_z\right)$. For $j = 1$ it is already satisfied because of the choice of $\phi_i^{(\lambda)}$. Therefore, simultaneous relation (2.2.16), in fact, consists of $N - 1$ relations. The induction hypothesis implies that in a neighborhood $W(\subset V)$ of a such $\left(\underline{c_\lambda}_z\right)$ is written by a linear sum of finitely many sections $\gamma^{(\nu)} = \left(\gamma_\lambda^{(\nu)}\right)$ with $\gamma_\lambda^{(\nu)} \in \Gamma(W, \mathscr{O})$. Therefore, the sections

$$\left(a_i^{(\nu)}\right) = \sum_\lambda \gamma_\lambda^{(\nu)} \cdot \phi_i^{(\lambda)}$$

generate $\mathscr{R}_z$ at every $z \in W$.

(iii) Because of the coherence of $\mathscr{F}$ there is a finite generator system $\{\underline{\sigma_h}\}_{h=1}^N$ of $\mathscr{F}$ over a neighborhood U $(\subset \Omega)$ of a. By the assumption, there are elements $\underline{f_{hj}}_a \in \mathscr{O}_a$ with

$$\underline{\sigma_h}_a = \sum_{j=1}^L \underline{f_{hj}}_a \underline{\gamma_j}_a, \quad 1 \le h \le N.$$

It follows that there is a neighborhood $V \subset U$ of a, where

$$\sigma_h(z) = \sum_{j=1}^L f_{hj}(z)\gamma_j(z), \quad \forall z \in V, \quad 1 \le h \le N.$$

Hence,

$$\underline{\sigma_h}_z = \sum_{j=1}^L \underline{f_{hj}}_z \underline{\gamma_j}_z, \quad \forall z \in V, \quad 1 \le h \le N.$$

Therefore, $\{\underline{\gamma_j}\}_{j=1}^L$ generates $\mathscr{F}$ over V. □

Proposition 2.2.17. *Let $\mathscr{F}$ be a coherent sheaf over Ω. If $\mathscr{F}_i \subset \mathscr{F}, i = 1, 2$, are coherent analytic subsheaves of $\mathscr{F}$ over Ω, so is the intersection $\mathscr{F}_1 \cap \mathscr{F}_2$.*

Proof. For every point $a \in \Omega$ there are a neighborhood $U(\subset \Omega)$ of a and locally finite generator systems of $\mathscr{F}_i, i = 1, 2$,

$$\alpha_j \in \Gamma(U, \mathscr{F}_1), \quad 1 \le j \le l,$$
$$\beta_k \in \Gamma(U, \mathscr{F}_2), \quad 1 \le k \le m.$$

At any point $z \in U$, $\gamma \in \mathscr{F}_z$ belongs to $\mathscr{F}_{1z} \cap \mathscr{F}_{2z}$ if and only if it is written as

$$\gamma = \sum_j a_j \alpha_j(z) = \sum_k b_k \beta_k(z), \quad a_j,\ b_k \in \mathscr{O}_z.$$

This is equivalent to

$$(2.2.18) \qquad \sum_j a_j \alpha_j(z) + \sum_k b_k(-\beta_k(z)) = 0,$$
$$\gamma = \sum_j a_j \alpha_j(z).$$

The above expression defines a relation sheaf of $\mathscr{F}$ with (a_j, b_k) being unknowns, which is denoted by $\mathscr{R} := \mathscr{R}(\ldots,\alpha_j,\ldots,-\beta_k,\ldots)$. Since $\mathscr{F}$ is coherent, $\mathscr{R}$ is locally finite. Taking a smaller $U \ni a$ if necessary, we may assume that $\mathscr{R}|_U$ is generated by a finite number of $\eta^{(h)} \in \Gamma(U,\mathscr{F})$, $1 \le h \le L$. With $\eta^{(h)} := \left(a_j^{(h)}, b_k^{(h)}\right)$, $(\mathscr{F}_1 \cap \mathscr{F}_2)|_U$ is generated by

$$\xi^{(h)} := \sum_j a_j^{(h)} \alpha_j = \sum_k b_k^{(h)} \beta_k \in \Gamma(U, \mathscr{F}_1 \cap \mathscr{F}_2), \quad 1 \le h \le L. \qquad \square$$

Example 2.2.19. The analytic sheaf $\mathscr{B}$ of Example 2.2.3 is, of course, not coherent. We give an example of a "relation sheaf" by non-analytic sections which is not locally finite.

We take an open ball B of $\mathbf{C}^n$ with center at the origin. Denote by Γ its boundary. Let $\chi(a)$ denote the characteristic function of B; that is, on B, $\chi = 1$, and $\chi = 0$ on $\mathbf{C}^n \setminus B$. Set

$$\mathscr{R} = \left\{ \underline{f}_a \in \mathscr{O}_{n,a} : \underline{f}_a \cdot \underline{\chi}_a = 0, \quad a \in \mathbf{C}^n \right\}.$$

Then,

$$\mathscr{R}_a = \begin{cases} 0, & a \in B \cup \Gamma, \\ \mathscr{O}_{n,a}, & a \notin B \cup \Gamma. \end{cases}$$

Therefore $\mathscr{R}$ is not coherent about $a \in \Gamma$.

In the above example, the length of the relation is one. To make the length two, we set $\phi(a) = 1 - \chi(a)$ and

$$\mathscr{S} = \left\{ \underline{f}_a \oplus \underline{g}_a \in \mathscr{O}_{n,a} \oplus \mathscr{O}_{n,a};\ \underline{f}_a \cdot \underline{\chi}_a + \underline{g}_a \cdot \underline{\phi}_a = 0, \quad a \in \mathbf{C}^n \right\} \subset \mathscr{O}_n^2.$$

Then,

$$\mathscr{S}_a = \begin{cases} 0 \oplus \mathscr{O}_{n,a}, & a \in B, \\ 0 \oplus 0, & a \in \Gamma, \\ \mathscr{O}_{n,a} \oplus 0, & a \notin B \cup \Gamma. \end{cases}$$

Therefore, $\mathscr{S}$ is not locally finite about any point of Γ.

If one requires the differentiability for χ and ϕ, it suffices to take a C^∞ function such that

$$\chi(a) > 0, \quad a \in B; \ \chi(a) = 0, \quad a \notin B.$$

Then, the same conclusion is obtained.

2.3 Oka's First Coherence Theorem

2.3.1 Weierstrass' Preparation Theorem

We first prove the titled theorem by the residue theorem in one variable and the relations of roots and coefficients of algebraic equations.

Let $\underline{f}_a \in \mathscr{O}_{\mathbf{C}^n,a}$. For a moment we assume that $a = 0$ and $\underline{f}_0 \neq 0$. Then f is a holomorphic function in a neighborhood of 0. By Lemma 1.5.9 we may take the standard polydisk $\mathrm{P}\Delta(0;r) = \mathrm{P}\Delta_{n-1} \times \Delta(0;r_n)$ $(\subset \mathbf{C}^{n-1} \times \mathbf{C})$ with $f \in \mathscr{O}(\mathrm{P}\Delta(0;r))$ and with the standard coordinate system $z = (z', z_n)$. We call $\mathrm{P}\Delta(0;r) = \mathrm{P}\Delta_{n-1} \times \Delta(0;r_n)$ the *standard polydisk* of $\underline{f}_0$, and $z = (z', z_n)$ the *standard coordinate system* of $\underline{f}_0$ (see Definition 1.5.10).

We note:

2.3.1. (i) The standard polydisks of $\underline{f}_0$ form a basis of neighborhoods about 0, because $r_n > 0$ can be chosen arbitrarily small and then, depending on it, $r_j, 1 \leq j \leq n-1$, are chosen arbitrarily small.

(ii) Let P_{ν_0} $(\not\equiv 0)$ be the first term in the homogeneous polynomial expansion of f about 0 (see (1.5.4)). Since $\{v \in \mathbf{C}^n : P_{\nu_0}(v) = 0\}$ contains no interior point, the standard coordinate system and the standard polydisk can be chosen to be the same for finitely many $\underline{f_k}_0 \in \mathscr{O}_{\mathbf{C}^n,0} \setminus \{0\}$, $1 \leq k \leq l$.

Theorem 2.3.2 (Weierstrass' Preparation Theorem). *Let $\underline{f}_0 \in \mathscr{O}_{\mathbf{C}^n,0} \setminus \{0\}$ with $p = \mathrm{ord}_0 f > 0$. Let $\mathrm{P}\Delta = \mathrm{P}\Delta_{n-1} \times \Delta(0;r_n)$ $(\ni z = (z', z_n))$ be the standard polydisk of $\underline{f}_0$.*

(i) *There exist unique holomorphic functions, $a_\nu \in \mathscr{O}(\overline{\mathrm{P}\Delta_{n-1}})$ with $a_\nu(0) = 0, 1 \leq \nu \leq p$, and zero-free $u \in \mathscr{O}(\overline{\mathrm{P}\Delta})$ such that*

$$f(z) = f(z', z_n) = u(z)\left(z_n^p + \sum_{\nu=1}^{p} a_\nu(z') z_n^{p-\nu}\right), \tag{2.3.3}$$

$$z = (z', z_n) \in \overline{\mathrm{P}\Delta_{n-1}} \times \bar{\Delta}(0;r_n).$$

(ii) *For every $\varphi \in \mathscr{O}(\mathrm{P}\Delta)$ there are unique holomorphic functions, $a \in \mathscr{O}(\mathrm{P}\Delta)$ and $b_\nu \in \mathscr{O}(\mathrm{P}\Delta_{n-1}), 1 \leq \nu \leq p$, satisfying*

$$\varphi(z) = af + \sum_{\nu=1}^{p} b_\nu(z') z_n^{p-\nu}, \quad z = (z', z_n) \in \mathrm{P}\Delta_{n-1} \times \Delta(0;r_n). \tag{2.3.4}$$

Proof. (i) For $k \in \mathbf{Z}_+$ we set

(2.3.5) $$\sigma_k(z') = \frac{1}{2\pi i}\int_{|z_n|=r_n} z_n^k \frac{\frac{\partial f}{\partial z_n}(z', z_n)}{f(z', z_n)} dz_n, \quad z' \in \overline{\mathrm{P}\Delta_{n-1}}.$$

It follows that $\sigma_k \in \mathscr{O}(\overline{\mathrm{P}\Delta_{n-1}})$. By the argument principle $\sigma_0(z') \equiv p \in \mathbf{N}$. Therefore, for every $z' \in \mathrm{P}\Delta_{n-1}$, the number of roots of $f(z', z_n) = 0$ with counting multiplicities is identically p. We write $\zeta_1(z'), \ldots, \zeta_p(z')$ for them with counting multiplicities. From the residue theorem we get

$$\sigma_k(z') = \sum_{j=1}^{p} (\zeta_j(z'))^k, \quad k = 0, 1, \ldots.$$

We set the elementary symmetric polynomial of degree ν in $\zeta_1(z'), \ldots, \zeta_p(z')$, multiplied by $(-1)^\nu$,

$$a_\nu(z') = (-1)^\nu \sum_{1 \le j_1 < \cdots < j_\nu \le p} \zeta_{j_1}(z') \cdots \zeta_{j_\nu}(z').$$

Then, $a_\nu(z') \in \mathbf{Q}[\sigma_1(z'), \ldots, \sigma_\nu(z')]$ (cf. Ex. 1 at the end of this chapter). For instance, when $p = 2$,

$$a_2(z') = \zeta_1(z')\zeta_2(z') = \frac{1}{2}\sigma_1(z')^2 - \frac{1}{2}\sigma_2(z').$$

Since all $\zeta_j(0) = 0$, we have $a_\nu \in \mathscr{O}(\overline{\mathrm{P}\Delta_{n-1}})$ and $a_\nu(0) = 0, \nu \ge 1$. We put

(2.3.6) $$W(z', z_n) = \prod_{j=1}^{p} (z_n - \zeta_j(z')) = z_n^p + \sum_{\nu=1}^{p} a_\nu(z') z_n^{p-\nu}.$$

Then, $W(z', z_n) \in \mathscr{O}(\overline{\mathrm{P}\Delta_{n-1}})[z_n] \subset \mathscr{O}(\overline{\mathrm{P}\Delta_{n-1}} \times \mathbf{C})$. With $s_n > r_n$ sufficiently close to r_n, and with every $z' \in \overline{\mathrm{P}\Delta_{n-1}}$ fixed, the roots of $W(z', z_n) = 0$ and $f(z', z_n) = 0$ in $|z_n| < s_n$ are the same with counting multiplicities, so that

$$u(z', z_n) := \frac{f(z', z_n)}{W(z', z_n)}$$

is zero-free on $\overline{\mathrm{P}\Delta}$. We have the following integral formula:

$$u(z', z_n) = \int_{|\zeta_n|=s_n} \frac{f(z', \zeta_n)}{W(z', \zeta_n)} \cdot \frac{d\zeta_n}{\zeta_n - z_n}.$$

It follows that $u \in \mathscr{O}(\overline{\mathrm{P}\Delta_{n-1}} \times \{|z_n| \le r_n\})$ (zero-free), and that

$$f(z', z_n) = u(z)\left(z_n^p + \sum_{\nu=1}^{p} a_\nu(z') z_n^{p-\nu}\right) = u(z) W(z', z_n),$$

where $a_\nu \in \mathscr{O}(\overline{\mathrm{P}\Delta_{n-1}})$ and $a_\nu(0) = 0$.

We confirm that $u(z)$ and $W(z', z_n)$ are uniquely determined as elements of $\mathscr{O}_0$. Suppose that

$$\underline{f}_0 = \underline{u}_0 \cdot \underline{\left(z_n^p + \textstyle\sum_{\nu=1}^p a_\nu(z')z_n^{p-\nu}\right)}_0 = \underline{\tilde{u}}_0 \cdot \underline{\left(z_n^p + \textstyle\sum_{\nu=1}^p \tilde{a}_\nu(z')z_n^{p-\nu}\right)}_0 .$$

Let $\widetilde{\mathrm{P}\Delta}_{n-1} \times \Delta(0; \tilde{r}_n)$ be a standard polydisk for which the above expressions make sense. For each fixed $z' \in \widetilde{\mathrm{P}\Delta}_{n-1}$, the roots of two equations

$$z_n^p + \sum_{\nu=1}^p a_\nu(z')z_n^{p-\nu} = 0, \quad z_n^p + \sum_{\nu=1}^p \tilde{a}_\nu(z')z_n^{p-\nu} = 0$$

are identical with counting multiplicities, and then

$$a_\nu(z') = \tilde{a}_\nu(z'), \quad 1 \le \nu \le p.$$

Hence, $u(z) = \tilde{u}(z)$ follows.

(ii) We may assume that

$$\begin{aligned} f(z', z_n) = W(z', z_n) &= z_n^p + \sum_{\nu=1}^p a_\nu(z')z_n^{p-\nu} \\ &= \sum_{\nu=0}^p a_\nu(z')z_n^{p-\nu} \in \mathscr{O}(\overline{\mathrm{P}\Delta_{n-1}})[z_n]. \end{aligned}$$

Here, we put $a_0(z') = 1$. For $\varphi \in \mathscr{O}(\mathrm{P}\Delta_{n-1} \times \Delta(0; r_n))$ we set

$$a(z', z_n) = \frac{1}{2\pi i} \int_{|\zeta_n| = t_n} \frac{\varphi(z', \zeta_n)}{W(z', \zeta_n)} \frac{d\zeta_n}{\zeta_n - z_n}, \quad (z', z_n) \in \mathrm{P}\Delta_{n-1} \times \Delta(0; r_n), \tag{2.3.7}$$

where $|z_n| < t_n < r_n$. Since $a(z', z_n)$ is independent of the choice of t_n close to r_n, $a(z', z_n) \in \mathscr{O}(\mathrm{P}\Delta_{n-1} \times \Delta(0; r_n))$ is determined. For $z' \in \mathrm{P}\Delta_{n-1}$ and $|z_n| < t_n$ we write

$$\begin{aligned} &\varphi(z', z_n) - a(z', z_n)W(z', z_n) \\ &= \frac{1}{2\pi i} \int_{|\zeta_n| = t_n} \varphi(z', \zeta_n) \frac{d\zeta_n}{\zeta_n - z_n} - \frac{W(z', z_n)}{2\pi i} \int_{|\zeta_n| = t_n} \frac{\varphi(z', \zeta_n)}{W(z', \zeta_n)} \frac{d\zeta_n}{\zeta_n - z_n} \\ &= \frac{1}{2\pi i} \int_{|\zeta_n| = t_n} \varphi(z', \zeta_n) \left\{ 1 - \frac{W(z', z_n)}{W(z', \zeta_n)} \right\} \frac{d\zeta_n}{\zeta_n - z_n} \\ &= \frac{1}{2\pi i} \int_{|\zeta_n| = t_n} \varphi(z', \zeta_n) \frac{\sum_{\nu=0}^{p-1} a_\nu(z')(\zeta_n^{p-\nu} - z_n^{p-\nu})}{W(z', \zeta_n)(\zeta_n - z_n)} d\zeta_n \end{aligned} \tag{2.3.8}$$

(continued)

$$= \frac{1}{2\pi i}\int_{|\zeta_n|=t_n} \frac{\varphi(z',\zeta_n)}{W(z',\zeta_n)} \left\{ \sum_{\nu=0}^{p-1} a_\nu(z')(\zeta_n^{p-\nu-1} + \zeta_n^{p-\nu-2} z_n + \cdots + z_n^{p-\nu-1}) \right\} d\zeta_n$$

$$= b_1(z')z_n^{p-1} + b_2(z')z_n^{p-2} + \cdots + b_p(z'),$$

where $b_\nu(z')$ are given by

$$b_\nu(z') = \frac{1}{2\pi i}\int_{|\zeta_n|=t_n} \frac{\varphi(z',\zeta_n)}{W(z',\zeta_n)} \left(\sum_{h=0}^{\nu-1} a_h(z')\zeta_n^{\nu-1-h} \right) d\zeta_n. \tag{2.3.9}$$

It follows from this expression that $b_\nu(z') \in \mathscr{O}(\mathrm{P}\Delta_{n-1}), 1 \le \nu \le p$ (independent of t_n). Therefore we obtain

$$\varphi(z',z_n) = a(z',z_n)W(z',z_n) + \sum_{\nu=1}^{p} b_\nu(z')z_n^{p-\nu}. \tag{2.3.10}$$

Next, we show the uniqueness. Suppose that

$$\varphi(z',z_n) = \tilde{a}(z',z_n)W(z',z_n) + \sum_{\nu=1}^{p} \tilde{b}_\nu(z')z_n^{p-\nu}. \tag{2.3.11}$$

Subtracting both sides of (2.3.10) and (2.3.11) and shifting terms, we assume that

$$(a(z',z_n) - \tilde{a}(z',z_n))\, W(z',z_n) = \sum_{\nu=1}^{p} \left(\tilde{b}_\nu(z') - b_\nu(z')\right) z_n^{p-\nu} \not\equiv 0.$$

Then for a fixed $z' \in \widetilde{\mathrm{P}\Delta_{n-1}}$ the left-hand side has at least p roots with counting multiplicities. The right-hand side has at most $p-1$ roots with counting multiplicities; this is absurd. Hence,

$$\tilde{b}_\nu(z') = b_\nu(z'), \qquad \tilde{a}(z',z_n) = a(z',z_n). \qquad \square$$

Definition 2.3.12. (i) Letting $\mathrm{P}\Delta_{n-1} \subset \mathbf{C}^{n-1}$, we call the z_n-polynomial with coefficients in $\mathscr{O}(\overline{\mathrm{P}\Delta_{n-1}})$

$$W(z',z_n) = z_n^p + \sum_{\nu=1}^{p} a_\nu(z') \cdot z_n^{p-\nu}, \quad a_\nu \in \mathscr{O}(\overline{\mathrm{P}\Delta_{n-1}}),\ a_\nu(0) = 0,$$

a *Weierstrass polynomial* (in z_n). Considering the induced germ

$$W = z_n^p + \sum_{\nu=1}^{p} \underline{a_\nu}_0 \cdot z_n^{p-\nu} \in \mathscr{O}_{\mathrm{P}\Delta_{n-1},0}[z_n],$$

we also call this a Weierstrass polynomial (in z_n).

(ii) Write (2.3.3) as $f(z',z_n) = uW(z',z_n)$ with unit u (i.e., $\exists u^{-1}$) and Weierstrass polynomial $W(z',z_n)$. We call $f(z',z_n) = uW(z',z_n)$ the *Weierstrass decomposition* of f at 0, which is unique.

Lemma 2.3.13. *Let $Q(z',z_n) \in \mathscr{O}_{n-1,0}[z_n]$ be a Weierstrass polynomial, and let $R \in \mathscr{O}_{n-1,0}[z_n]$. If $R = Q \cdot \underline{g}_0$ with $\underline{g}_0 \in \mathscr{O}_{n,0}$, then $\underline{g}_0 \in \mathscr{O}_{n-1,0}[z_n]$.*

Proof. Assume that all the functions above are holomorphic in a neighborhood of a closed polydisk $\overline{\mathrm{P}\Delta}$. Since the leading coefficient of $Q(z',z_n)$ as a polynomial in z_n is 1, the Euclid division algorithm implies that

$$
\begin{aligned}
&R = \varphi Q + \psi, \qquad \varphi, \psi \in \mathscr{O}_{n-1,0}[z_n], \\
&\deg_{z_n} \psi < p = \deg_{z_n} Q.
\end{aligned} \tag{2.3.14}
$$

We may assume that $\mathrm{P}\Delta = \mathrm{P}\Delta_{n-1} \times \Delta_{(n)}$ ($\Delta_{(n)} = \{|z_n| < r_n\}$) is a standard polydisk for Q. With every $z' \in \mathrm{P}\Delta_{n-1}$ fixed, $Q(z',z_n) = 0$ has p zeros with counting multiplicities. Therefore, R has at least p zeros with counting multiplicities. By (2.3.14) ψ has at least p zeros with counting multiplicities, too. Since $\deg \psi < p$, $\psi \equiv 0$. Therefore we obtain

$$
\begin{aligned}
R &= \varphi Q = gQ, \\
(\varphi - g)Q &= 0, \qquad Q \neq 0.
\end{aligned}
$$

Since $\mathscr{O}_{n,0}$ is a ring of integral domain, $\varphi - g = 0$. Thus we see that $g = \varphi \in \mathscr{O}_{n-1,0}[z_n]$. □

Remark 2.3.15. The following properties of the integral domain $\mathscr{O}_a$ follow from Weierstrass' Preparation Theorem 2.3.2:

(i) $\mathscr{O}_a$ is a unique factorization domain (cf., e.g., [39] Theorem 2.2.12).

(ii) $\mathscr{O}_a$ is a noetherian ring (cf., e.g., [39] Theorem 2.2.20).

2.3.2 Oka's First Coherence Theorem

Theorem 2.3.16 (Oka's First Coherence). *The sheaf $\mathscr{O}_{\mathbf{C}^n}$ is coherent; hence, all $\mathscr{O}_{\mathbf{C}^n}^N$ ($N \geq 1$) are coherent.*

Proof. We proceed by induction on $n \geq 0$ with general $N \geq 1$. We write $\mathscr{O}_{\mathbf{C}^n} = \mathscr{O}_n$.

(a) $n = 0$: In this case, it is a matter of a finite-dimensional vector space over $\mathbf{C}$.

(b) $n \geq 1$: Suppose that $\mathscr{O}_{n-1}^N$ is coherent for every $N \geq 1$.

By Proposition 2.2.15 (ii) it suffices to prove the coherence of $\mathscr{O}_n$.

The problem is local and it is sufficient to prove Definition 2.2.13 (ii). Taking an open subset $\Omega \subset \mathbf{C}^n$ and $\tau_j \in \mathscr{O}(\Omega) \cong \Gamma(\Omega, \mathscr{O}_n)$, $1 \leq j \leq q$, we consider the relation sheaf $\mathscr{R}(\tau_1, \ldots, \tau_q)$ defined by

(2.3.17) $$\underline{f_1}_z \underline{\tau_1}_z + \cdots + \underline{f_q}_z \underline{\tau_q}_z = 0, \quad \underline{f_j}_z \in \mathscr{O}_{n,z}, \; z \in \Omega.$$

What we want to show is:

Claim 2.3.18. *For every point $a \in \Omega$ there are a neighborhood $V \subset \Omega$ of a and finitely many sections $s_k \in \Gamma(V, \mathscr{R}(\tau_1, \ldots, \tau_q))$, $1 \le k \le l$, such that*

$$\mathscr{R}(\tau_1, \ldots, \tau_q)_b = \sum_{k=1}^{l} \mathscr{O}_{n,b} \cdot s_k(b), \quad \forall b \in V.$$

For the proof we may assume that $a = 0$. The case of $q = 1$ is trivial, and so we assume $q \ge 2$. If an element $\underline{\tau_j}_0 = 0$, then the jth component of $\mathscr{R}(\tau_1, \ldots, \tau_q)|_V \subset (\mathcal{O}_V)^q$ is just $\mathcal{O}_V$ in a neighborhood V of 0; so we may assume $\underline{\tau_j}_0 \neq 0$, $1 \le j \le q$.

We set

(2.3.19) $$T_{i,j} = \bigl(0, \ldots, 0, \overset{i\text{th}}{-\underline{\tau_j}}, 0, \ldots, 0, \overset{j\text{th}}{\underline{\tau_i}}, 0, \ldots, 0\bigr), \quad 1 \le i < j \le q,$$

which trivially satisfy (2.3.17), and call $T_{i,j}$ the *trivial solutions*.

If there is an element τ_j, to say, τ_1 with $\tau(0) \neq 0$, then (2.3.17) is solved as

(2.3.20) $$\underline{f_1}_b = -\sum_{j=2}^{q} \frac{\underline{\tau_j}_b}{\underline{\tau_1}_b} \underline{f_j}_b;$$

that is, $\mathscr{R}$ is generated by $\{T_{1,j}\}_{j=2}^{q}$ about 0. So, we assume all $\tau_j(0) = 0$ $(1 \le j \le q)$.

Let p_j be the order of zero of τ_j at 0, and set

$$p = \max_{1 \le j \le q} p_j, \quad p' = \min_{1 \le j \le q} p_j \ge 0.$$

After reordering the indices, we may assume that $p' = p_1$.

To avoid the notational complications we use f, z_n etc. for their sections $\underline{f}$, $\underline{z_n}$ etc., unless confusion occurs.

Take a common standard polydisk $\mathrm{P}\Delta = \mathrm{P}\Delta_{n-1} \times \Delta_{(n)}$ with $\Delta_{(n)} = \{|z_n| < r_n\}$ for all τ_j. By Weierstrass's Preparation Theorem 2.3.2 at 0 one can transfer a unit factor of τ_j to f_j in (2.3.17) so that all τ_j may be assumed to be Weierstrass polynomials:

(2.3.21) $$\tau_j(z) = P_j(z', z_n) = \sum_{\nu=0}^{p_j} a_{j\nu}(z') z_n^{\nu} = \sum_{\nu=0}^{p} a_{j\nu}(z') z_n^{\nu} \in \mathscr{O}(\mathrm{P}\Delta_{n-1})[z_n],$$
$$a_{j\nu}(0) = 0 \; (\nu < p_j), \quad a_{jp_j} = 1, \quad a_{j\nu} = 0 \; (p_j < \nu \le p).$$

Now we have $\mathscr{R} = \mathscr{R}(P_1, \ldots, P_q)$ defined by the equation

(2.3.22) $$f_1 P_1 + \cdots + f_q P_q = 0.$$

The trivial solutions are

(2.3.23) $$T_{i,j} = \bigl(0,\dots,0,\overset{i\text{th}}{-P_j},0,\dots,0,\overset{j\text{th}}{P_i},0,\dots,0\bigr), \quad 1 \le i < j \le q,$$

We perform a kind of division algorithm for an unknown q-tuple $\alpha = (\alpha_j) \in \mathscr{R}$ with respect to the trivial solutions $T_{i,j}$ (cf. (2.3.29), (2.3.30)).

Take an arbitrary point $b = (b', b_n) \in \mathrm{P}\Delta_{n-1} \times \Delta_{(n)}$. We call an element of $\mathscr{O}_{n-1,b'}[z_n]$ a *z_n-polynomial-like germ*. In the same way, we call $\alpha = (\alpha_1, \dots, \alpha_q) \in \mathscr{O}^q_{n,b}$ consisting of z_n-polynomial-like germs α_j a *polynomial-like element*, and $f = (f_j)$ with $(f_j)_{1 \le j \le q} \in (\mathscr{O}(\mathrm{P}\Delta_{n-1})[z_n])^q$ a *z_n-polynomial-like section*. We set

$$\deg \alpha = \deg_{z_n} \alpha = \max_j \deg_{z_n} \alpha_j,$$
$$\deg f = \deg_{z_n} f = \max_j \deg_{z_n} f_j.$$

Then we have:

2.3.24. *The trivial solutions $T_{i,j}$ are z_n-polynomial-like sections of* $\deg T_{i,j} \le p$.

We now show the following:

Lemma 2.3.25 (Degree structure). *Let the notation be as above. Then an element of $\mathscr{R}_b$ is written as a finite linear sum of the trivial solutions, $T_{1,j}$ $(2 \le j \le q)$, and a z_n-polynomial-like element $\alpha = (\alpha_1, \alpha_2, \dots, \alpha_q)$ of $\mathscr{R}_b$ with coefficients in $\mathscr{O}_{n,b}$ such that*

$$\deg \alpha_1 < p, \quad \deg \alpha_j < p' \quad (2 \le j \le q).$$

$\because$) By making use of Weierstrass' Preparation Theorem 2.3.2 at $b = (b', b_n)$ we decompose P_1 into a unit u and a Weierstrass polynomial Q at b:

(2.3.26) $$P_1(z', z_n) = u \cdot Q(z', z_n - b_n), \qquad \deg Q = d \le p_1,$$

where $Q \in \mathscr{O}_{n-1,b'}[z_n - b_n] = \mathscr{O}_{n-1,b'}[z_n]$. Lemma 2.3.13 implies that

$$u \in \mathscr{O}_{n-1,b'}[z_n - b_n] = \mathscr{O}_{n-1,b'}[z_n].$$

It follows that

(2.3.27) $$\deg_{z_n} u = p_1 - d.$$

Take an arbitrary $f = \bigl(f_1, \dots, f_q\bigr) \in \mathscr{R}_b$. By Weierstrass' Preparation Theorem 2.3.2 (ii) we have

(2.3.28) $$f_i = c_i Q + \beta_i, \quad c_i \in \mathscr{O}_{n,b}, \quad \beta_i \in \mathscr{O}_{n-1,b'}[z_n - b_n] = \mathscr{O}_{n-1,b'}[z_n], \\ \deg_{z_n} \beta_i \le d - 1 \quad (1 \le i \le q).$$

Since $u \in \mathscr{O}_{n,b}$ is a unit, with $\tilde{c}_i := c_i u^{-1}$ we get

(2.3.29) $$f_i = \tilde{c}_i P_1 + \beta_i, \qquad 1 \le i \le q.$$

By making use of this we perform the following calculation:

$$
\begin{aligned}
(2.3.30)\qquad & (f_1,\dots,f_q)-\tilde{c}_2T_2-\cdots-\tilde{c}_qT_q\\
&=(\tilde{c}_1P_1+\beta_1,\tilde{c}_2P_1+\beta_2,\dots,\tilde{c}_qP_1+\beta_q)\\
&\quad+(\tilde{c}_2P_2,-\tilde{c}_2P_1,0,\dots,0)+\cdots\\
&\quad+(\tilde{c}_qP_q,0,\dots,0,-\tilde{c}_qP_1)\\
&=\left(\sum_{i=1}^{q}\tilde{c}_iP_i+\beta_1,\beta_2,\dots,\beta_q\right)\\
&=(g_1,\beta_2,\dots,\beta_q),
\end{aligned}
$$

where $g_1 := \sum_{i=1}^{q}\tilde{c}_iP_i+\beta_1 \in \mathscr{O}_{n,b}$; note that it is *unknown if* $g_1 \in \mathscr{O}_{n-1,b'}[z_n]$. It is noticed that $\beta_i \in \mathscr{O}_{n-1,b'}[z_n]$, $2 \le i \le q$. Since $(g_1,\beta_2,\dots,\beta_q) \in \mathscr{R}_b$,

$$
(2.3.31)\qquad g_1P_1=-\beta_2P_2-\cdots-\beta_qP_q \in \mathscr{O}_{n-1,b'}[z_n].
$$

It follows from the expression of the right-hand side of (2.3.31) that

$$
\deg_{z_n} g_1P_1 \le \max_{2\le i\le q}\deg_{z_n}\beta_i+\max_{2\le i\le q}\deg_{z_n}P_i \le d+p-1.
$$

On the other hand, $g_1P_1=(g_1u)Q$ and Q is a Weierstrass polynomial at b. Again by Lemma 2.3.13 we see that

$$
\begin{aligned}
& \alpha_1 := g_1u \in \mathscr{O}_{n-1,b'}[z_n],\\
(2.3.32)\qquad & \deg_{z_n}\alpha_1=\deg_{z_n} g_1P_1-\deg_{z_n}Q\\
& \qquad\le d+p-1-d=p-1.
\end{aligned}
$$

Setting $\alpha_i=u\beta_i(\in \mathscr{O}_{n-1,b'}[z_n])$ for $2\le i\le q$, we have by (2.3.27) and (2.3.28) that

$$
(2.3.33)\qquad \deg_{z_n}\alpha_i \le p_1-d+d-1=p_1-1=p'-1,\quad 2\le i\le q,
$$

and then by (2.3.30) that

$$
(2.3.34)\qquad f=\sum_{i=2}^{q}\tilde{c}_iT_i+u^{-1}(\alpha_1,\alpha_2,\dots,\alpha_q). \qquad \triangle
$$

Until now we have not used the induction hypothesis. Now we are going to use it to prove the existence of a locally finite generator system of those $(\alpha_1,\dots,\alpha_q)$ appearing in (2.3.34). We write

$$
(2.3.35)\qquad \alpha_1=\sum_{\nu=0}^{p-1}\underline{c_{1\nu}(z')}_{b'}z_n^{\nu},\quad \underline{c_{1\nu}(z')}_{b'} \in \mathscr{O}_{n-1,b'},
$$

$$\alpha_i = \sum_{\nu=0}^{p'-1} \underline{c_{i\nu}(z')}_{b'} z_n^\nu, \quad \underline{c_{i\nu}(z')}_{b'} \in \mathscr{O}_{n-1,b'},\ 2 \le i \le q.$$

By $\mathscr{S}$ we denote the sheaf of all $(\alpha_1, \dots, \alpha_q)$ over $\mathrm{P}\Delta = \mathrm{P}\Delta_{n-1} \times \Delta_{(n)}$ written as in (2.3.35), which satisfy

$$\alpha_1 P_1 + \alpha_2 P_2 + \cdots + \alpha_q P_q = 0. \tag{2.3.36}$$

The left-hand side above is a z_n-polynomial-like element of degree at most $p + p' - 1$, and relation (2.3.36) is equivalent to the nullity of all $p + p'$ coefficients. With the expression in (2.3.21) we have

$$\sum_{i=1}^{q} \sum_{k+h=\nu}{}' \underline{c_{ih}(z')}_{b'} \cdot \underline{a_{ik}(z')}_{b'} = 0 \in \mathscr{O}_{n-1,b'}, \quad 0 \le \nu \le p + p' - 1. \tag{2.3.37}$$

Here, $\sum'$ stands for the sum over those indices h, k to which some elements $\underline{a_{ik}(z')}_{b'}$, $\underline{c_{ih}(z')}_{b'}$ correspond. Then (2.3.37) defines a $(p + p')$-simultaneous relation sheaf $\tilde{\mathscr{S}}$ in $\mathscr{O}_{\mathrm{P}\Delta_{n-1}}^{p+p'(q-1)}$ with $p + p'(q-1)$ unknowns, c_{ih}'s. The induction hypothesis on n implies that $\tilde{\mathscr{S}}$ is coherent, and hence there is a locally finite generator system of $\tilde{\mathscr{S}}$ over a polydisk neighborhood $\widetilde{\mathrm{P}\Delta}_{n-1} \subset \mathrm{P}\Delta_{n-1}$ of 0. We set

$$\widetilde{\mathrm{P}\Delta} = \widetilde{\mathrm{P}\Delta}_{n-1} \times \Delta_{(n)} \subset \mathrm{P}\Delta_{n-1} \times \Delta_{(n)} = \mathrm{P}\Delta.$$

Together with (2.3.35) we then infer that $\mathscr{S}$ has a locally finite generator system $\{\pi_\mu\}_{\mu=1}^M$ $(\subset \mathscr{O}(\widetilde{\mathrm{P}\Delta}_{n-1})[z_n])$ over $\widetilde{\mathrm{P}\Delta}$.

Thus, the finite system $\{T_{1,j}\}_{j=2}^q \cup \{\pi_\mu\}_{\mu=1}^M$ generates $\mathscr{R}$ over $\widetilde{\mathrm{P}\Delta}$. □

Once the coherence of $\mathscr{O}_n$ is proved, it is immediate but important to see:

Corollary 2.3.38. *Every relation sheaf of a coherent sheaf is coherent.*

Remark 2.3.39. (i) In the proof of Theorem 2.3.16 it is essential to convert the expression of f_i in (2.3.28) with Weierstrass' polynomial Q at b to that in (2.3.29) with P_1 after multiplying the unit u to Q. If we keep the expression with Q, we could not advance the arguments, nor have an expression effective about 0.

To see it, we consider a simple example, $P_1(z_1, z_2) = z_2^2 - z_1$ in $(z_1, z_2) \in \mathbf{C}^2$. Let $b = (b_1, b_2) \in \mathbf{C}^2$ such that $P_1(b_1, b_2) = 0$ and $b \ne 0$. Let $\sqrt{z_1}$ be a branch about b_1 with $\sqrt{b_1} = b_2$. Then $\underline{P_1}_b = \underline{(z_2 - \sqrt{z_1})}_b \cdot \underline{(z_2 + \sqrt{z_1})}_b$ and $u = \underline{(z_2 + \sqrt{z_1})}_b$ is a unit of $\mathscr{O}_{2,b}$. We have the decomposition $\underline{P_1}_b = uQ$ (cf. (2.3.26)) with Weierstrass' polynomial at b,

$$Q(z_1, z_2 - b_2) = (z_2 - b_2) - \sqrt{z_1} + \sqrt{b_1} \in \mathscr{O}_{1,b_1}[z_2 - b_2] = \mathscr{O}_{1,b_1}[z_2].$$

This expression of Q does not make sense about 0. After multiplying u to Q again, we obtain $P_1 = uQ$, which provides an expression effective about 0.

(ii) In Lemma 2.3.25 the role of the minimum degree p' was observed in [40]. In the former arguments, the minimum degree p' was not used and only the maximum one p was used (cf. Oka [48] VII, Cartan [10], Hörmander [30] §6.4). The proof of Lemma 2.3.25 works even in the case of $p' = 0$, when the argument is reduced to the simplest case of (2.3.20); this reduction is not implied by the arguments only with the maximum degree p.

2.3.3 Coherence of Ideal Sheaves of Complex Submanifolds

Theorem 2.3.40. *The ideal sheaf of a complex submanifold is coherent.*

Proof. Let $S \subset \Omega$ be a submanifold of an open set $\Omega \subset \mathbf{C}^n$. By Proposition 2.2.15 (i) it suffices to show the local finiteness of $\mathscr{I}\langle S\rangle$.

Take a point $a \in \Omega$.

Case of $a \notin S$: Since S is a closed set, there is a neighborhood $U \subset \Omega$ of a such that $U \cap S = \emptyset$. Since

$$\mathscr{I}\langle S\rangle_z = \mathscr{O}_z = 1 \cdot \mathscr{O}_z, \quad \forall z \in U,$$

$\{1\}$ is a finite generator system of $\mathscr{I}\langle S\rangle$ over U.

Case of $a \in S$: There is a neighborhood U of a with a local coordinate system $z = (z_1, \ldots, z_n)$ such that

$$\begin{aligned} &a = (0,\ldots,0) \in U = \mathrm{P}\Delta((r_j)), \\ &S \cap U = \{z = (z_j) \in U : z_1 = \cdots = z_q = 0\} \quad (1 \le \exists q \le n). \end{aligned} \tag{2.3.41}$$

We take arbitrarily an element $\underline{f}_b \in \mathscr{I}\langle S\rangle_b$ $(b \in U \cap S)$. Set $b = (b_j)$ in coordinates (z_j). Then, $b = (0,\ldots,0,b_{q+1},\ldots,b_n)$, and f is uniquely expanded to a power series

$$f(z) = \sum_{\nu=(\nu_1,\nu')\in \mathbf{Z}_+^n, |\nu|>0} c_\nu (z-b)^\nu.$$

We decompose this as follows:

$$\begin{aligned} f(z) &= \sum_{\nu=(\nu_1,\nu')\in \mathbf{Z}_+^n, \nu_1>0} c_\nu (z-b)^\nu + \sum_{\nu=(\nu_1,\nu')\in \mathbf{Z}_+^n, \nu_1=0} c_\nu (z-b)^\nu \\ &= \left(\sum_{\nu=(\nu_1,\nu')\in \mathbf{Z}_+^n, \nu_1>0} c_\nu z_1^{\nu_1-1} (z'-b')^{\nu'} \right) z_1 + \sum_{\nu'\in \mathbf{Z}_+^{n-1}} c_{0\nu'} (z'-b')^{\nu'}, \end{aligned}$$

where $\nu' = (\nu_2,\ldots,\nu_n)$, $z' = (z_2,\ldots,z_n)$, and $b' = (b_2,\ldots,b_n)$. Setting

$$h_1(z_1,z') = \left(\sum_{\nu=(\nu_1,\nu')\in \mathbf{Z}_+^n,\nu_1>0} c_\nu z_1^{\nu_1-1}(z'-b')^{\nu'}\right),$$
$$g_1(z') = \sum_{\nu'\in \mathbf{Z}_+^{n-1}} c_{0\nu'}(z'-b')^{\nu'},$$

we write

$$f(z_1,z') = h_1(z_1,z')\cdot z_1 + g_1(z'). \tag{2.3.42}$$

Considering a similar decomposition of $g_1(z')$ with respect to variable z_2, we get

$$g_1(z') = h_2\cdot z_2 + g_2(z''), \quad z'' = (z_3,\dots,z_n).$$

Repeating this procedure, we obtain

$$f(z) = \sum_{j=1}^{q} h_j(z)\cdot z_j + g_q(z_{q+1},\dots,z_n).$$

Since $f(z)=0$ for $z_1=\cdots=z_q=0$, we deduce $g_q(z_{q+1},\dots,z_n)=0$. It follows that

$$f(z) = \sum_{j=1}^{q} h_j(z)\cdot z_j.$$

Therefore we have a finite generation

$$\mathscr{I}\langle S\rangle|_U = \sum_{j=1}^{q} \mathscr{O}_U\cdot \underline{z_j}. \tag{2.3.43}$$

□

2.4 Cartan's Merging Lemma

Let $\mathscr{F}\to\Omega$ be a locally finite analytic sheaf over a domain $\Omega\subset\mathbf{C}^n$. For given finite generator systems of $\mathscr{F}$ over adjacent closed subdomains E', $E''(\Subset\Omega)$, we would like to construct a finite generator system of $\mathscr{F}$ over $E'\cup E''$. We begin with the elementary facts of matrices.

2.4.1 Matrices and Matrix-Valued Functions

We describe some basic properties of sequences, series, infinite products of matrices, and matrix-valued functions, which will be necessary in the forthcoming discussions.

In general, we may consider two norms for a $p\,(\in \mathbf{N})$th (order) square (complex) matrix $A = (a_{ij})$:

$$\|A\|_\infty = \max_{i,j}\{|a_{ij}|\},$$
$$\|A\| = \max\{\|A\xi\| : \xi \in \mathbf{C}^p, \|\xi\| = 1\}.$$

We call $\|A\|$ the *operator norm* of A. By taking

$$\xi = {}^t(0,\ldots,0,1,0,\ldots,0)$$

we easily see that

$$\|A\|_\infty \le \|A\| \le p\,\|A\|_\infty.$$

Hence the convergence is the same with respect to either of them, so that the norms are complete. As for multiplication, $\|A\|$ has a better property than $\|A\|_\infty$, and henceforth we use $\|A\|$.

If $A = A(z)$ is a square matrix valued function on a subset $E \subset \mathbf{C}^n$, we write

$$\|A\|_E = \sup\{\|A(z)\| : z \in E\}.$$

Let $\mathbf{1}_p$ denote the unit square matrix of order p.

Proposition 2.4.1. *Let A denote a pth square matrix or a pth square matrix valued function $A(z)$ in $z \in E$ $(\subset \mathbf{C}^n)$. Let B be another pth square matrix. Then the following hold:*

(i) $\|A+B\| \le \|A\| + \|B\|$.
(ii) $\|AB\| \le \|A\| \cdot \|B\|$; in particular, $\|A^k\| \le \|A\|^k$ $(k \in \mathbf{N})$.
(iii) If $\|A\|_E \le \varepsilon < 1$ for $A = A(z)$ $(z \in E)$, $(\mathbf{1}_p - A(z))$ is invertible and

$$(\mathbf{1}_p - A(z))^{-1} = \mathbf{1}_p + A(z) + A(z)^2 + \cdots,$$

where the right-hand side converges uniformly on E with respect to the operator norm, and $\|(\mathbf{1}_p - A)^{-1}\|_E \le \frac{1}{1-\varepsilon}$; in particular, $\|(\mathbf{1}_p - A)^{-1}\|_E \le 2$ with $\varepsilon = \frac{1}{2}$.
(iv) Assume that sequences of $0 < \varepsilon_k < 1$ $(k = 0,1,\ldots)$ and pth square matrix valued functions $A_k(z)$ in $z \in E$ $(\subset \mathbf{C}^n)$, are given and satisfy $\|A_k\|_E \le \varepsilon_k$ and $\sum_{k=0}^{\infty} \varepsilon_k < \infty$. Then the two infinite products

$$\lim_{k\to\infty} (\mathbf{1}_p - A_0(z)) \cdots (\mathbf{1}_p - A_k(z)),$$
$$\lim_{k\to\infty} (\mathbf{1}_p - A_k(z)) \cdots (\mathbf{1}_p - A_0(z))$$

converge uniformly on E, and the limits are both invertible matrix valued functions.

Proof. (i) and (ii) are immediate from definitions. (iii) follows from the following equations with $k \to \infty$:

$$(\mathbf{1}_p - A(z))(\mathbf{1}_p + A(z) + A(z)^2 + \cdots + A(z)^k) = \mathbf{1}_p - A(z)^{k+1},$$
$$\|\mathbf{1}_p + A(z) + A(z)^2 + \cdots + A(z)^k\|_E \le \sum_{j=0}^{k} \|A\|_E^j \le \sum_{j=0}^{k} \varepsilon^j = \frac{1-\varepsilon^{k+1}}{1-\varepsilon}.$$

(iv) The proofs of the two are similar, and so we prove the first. We set

$$G_k(z) = (\mathbf{1}_p - A_0(z)) \cdots (\mathbf{1}_p - A_k(z)) = \prod_{j=0}^{k} (\mathbf{1}_p - A_j(z)),$$
$$k = 0, 1, \ldots.$$

It suffices to show that $\{G_k\}_{k=0}^{\infty}$ is a uniform Cauchy sequence, and that $\{G_k^{-1}\}_{k=0}^{\infty}$ converges uniformly. Setting $C_0 = \exp(\sum_{k=0}^{\infty} \varepsilon_k)$, we have

$$\|G_k\|_E \le \prod_{j=0}^{k} \|\mathbf{1}_p - A_j\|_E \le \prod_{j=0}^{k} (1 + \|A_j\|_E) \le \prod_{j=0}^{k} (1 + \varepsilon_j)$$
$$= \exp\left(\sum_{j=0}^{k} \log(1+\varepsilon_j)\right) < \exp\left(\sum_{j=0}^{k} \varepsilon_j\right) < C_0.$$

Making use of the equations above, we have for $l > k > 0$

$$\begin{aligned}
\|G_l - G_k\|_E &\le \|G_k\|_E \cdot \left\|(\mathbf{1}_p - A_{k+1})(\mathbf{1}_p - A_{k+2}) \cdots (\mathbf{1}_p - A_l) - \mathbf{1}_p\right\|_E \\
&\le C_0 \| - A_{k+1} - A_{k+2} - \cdots - A_l + A_{k+1}A_{k+2} \\
&\qquad + \cdots + (-1)^{l-k} A_{k+1} \cdots A_l \|_E \\
&\le C_0 (\|A_{k+1}\|_E + \|A_{k+2}\|_E + \cdots + \|A_l\|_E + \|A_{k+1}\|_E \cdot \|A_{k+2}\|_E \\
&\qquad + \cdots + \|A_{k+1}\|_E \cdots \|A_l\|_E) \\
&= C_0 \left(\prod_{j=k+1}^{l} (1 + \|A_j\|_E) - 1 \right) \le C_0 \left(\prod_{j=k+1}^{l} (1+\varepsilon_j) - 1 \right) \\
&\le C_0 \left(\exp\left(\sum_{j=k+1}^{l} \varepsilon_j \right) - 1 \right) \longrightarrow 0 \quad (l > k \to \infty).
\end{aligned}$$

For $G_k^{-1} = \prod_{j=k}^{0} (\mathbf{1}_p - A_j)^{-1}$, we set $B_k = -A_k(\mathbf{1}_p - A_k)^{-1}$. It follows that

$$(\mathbf{1}_p - A_k)^{-1} = \mathbf{1}_p - B_k,$$

so that by (iii) above

$$\|B_k\|_E \le \|A_k\|_E \cdot \|(\mathbf{1}_p - A_k)^{-1}\|_E \le \frac{\varepsilon_k}{1-\varepsilon_k}.$$

Setting $0 < \theta := \max_k\{\varepsilon_k\} < 1$, we get

$$\|B_k\|_E \le \frac{\varepsilon_k}{1-\theta}.$$

Therefore there is some $k_0 \in \mathbf{N}$ such that

$$\frac{\varepsilon_k}{1-\theta} < 1, \quad \forall k \ge k_0.$$

Since for all $k \ge k_0$, B_k fulfill the conditions which A_k satisfies, $\{G_k^{-1}\}_{k=0}^{\infty}$ converges uniformly on E. □

Let S, T be pth square matrices. Assuming the existences of the inverses $(\mathbf{1}_p - S)^{-1}, (\mathbf{1}_p - T)^{-1}$, we put

$$\begin{aligned} M(S,T) &= (\mathbf{1}_p - S)^{-1}(\mathbf{1}_p - S - T)(\mathbf{1}_p - T)^{-1}, \\ N(S,T) &= \mathbf{1}_p - M(S,T). \end{aligned} \tag{2.4.2}$$

The next lemma is a key in the later arguments on convergences.

Lemma 2.4.3. *Assume that* $\max\{\|S\|, \|T\|\} \le \varepsilon$ *with* $0 < \varepsilon < 1$. *Then*

$$\|N(S,T)\| \le \left(\frac{1}{1-\varepsilon}\right)^2 (\max\{\|S\|, \|T\|\})^2.$$

In particular, with $\varepsilon = \frac{1}{2}$ *we have*

$$\|N(S,T)\| \le 2^2(\max\{\|S\|, \|T\|\})^2.$$

Proof. Noting that $(\mathbf{1}_p - T)^{-1} = \mathbf{1}_p + T(\mathbf{1}_p - T)^{-1} = \mathbf{1}_p + T + T^2(\mathbf{1}_p - T)^{-1}$, we have

$$\begin{aligned} M(S,T) &= (\mathbf{1}_p - S)^{-1}(\mathbf{1}_p - S - T)(\mathbf{1}_p - T)^{-1} \\ &= (\mathbf{1}_p - (\mathbf{1}_p - S)^{-1}T)(\mathbf{1}_p - T)^{-1} \\ &= \mathbf{1}_p + T + T^2(\mathbf{1}_p - T)^{-1} \\ &\quad - (\mathbf{1}_p + S(\mathbf{1}_p - S)^{-1})T(\mathbf{1}_p + T(\mathbf{1}_p - T)^{-1}) \\ &= \mathbf{1}_p + T + T^2(\mathbf{1}_p - T)^{-1} \\ &\quad - T - T^2(\mathbf{1}_p - T)^{-1} - S(\mathbf{1}_p - S)^{-1}T(\mathbf{1}_p - T)^{-1} \\ &= \mathbf{1}_p - S(\mathbf{1}_p - S)^{-1}T(\mathbf{1}_p - T)^{-1}, \\ N(S,T) &= S(\mathbf{1}_p - S)^{-1}T(\mathbf{1}_p - T)^{-1}. \end{aligned}$$

It follows from Proposition 2.4.1 (iii) that

$$\|N(S,T)\| \le \|S\| \cdot \left(\frac{1}{1-\varepsilon}\right) \cdot \|T\| \cdot \left(\frac{1}{1-\varepsilon}\right) \le \left(\frac{1}{1-\varepsilon}\right)^2 (\max\{\|S\|, \|T\|\})^2.$$ □

2.4.2 Cartan's Matrix Decomposition

We assume the following setting:

2.4.4 (**closed cuboid**). Here a closed *cuboid* or a closed rectangle is bounded and the edges are parallel to the real and imaginary axes, and the case of some degenerate edges with width 0 is included.

Let $E', E'' \Subset \Omega$ be closed cuboids as follows: There are a closed cuboid $F \Subset \mathbf{C}^{n-1}$ and two adjacent closed rectangles $E'_n, E''_n \Subset \mathbf{C}$ sharing an edge ℓ, such that

$$E' = F \times E'_n, \quad E'' = F \times E''_n, \quad \ell = E'_n \cap E''_n \quad \text{(cf. Fig. 2.1)}.$$

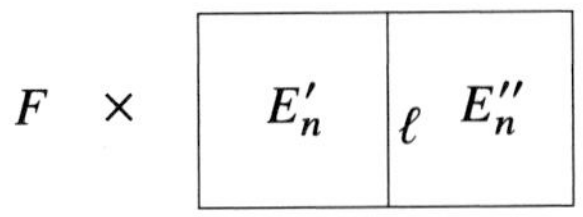

Fig. 2.1 Adjacent closed cuboids.

Let $GL(p;\mathbf{C})$ be the general linear group of all invertible pth square matrices.

Lemma 2.4.5 (Cartan's matrix decomposition). *Let the notation be as above. Then there is a neighborhood $V_0 \subset GL(p;\mathbf{C})$ of $\mathbf{1}_p$ such that for a matrix-valued holomorphic function $A : U \to V_0$ in a neighborhood U of $F \times \ell$, there is a matrix-valued holomorphic function $A' : U' \to GL(p;\mathbf{C})$ (resp. $A'' : U'' \to GL(p;\mathbf{C})$) in a neighborhood U' (resp. U'') of E' (resp. E'') satisfying $A = A' \cdot A''$ in a neighborhood of $F \times \ell$.*

Proof. We widen each edge of F, E'_n, E''_n by the same length $\delta > 0$ outward and denote the resulting closed cube and closed rectangles by $\tilde{F}$, $\tilde{E}'_{n(1)}$ and $\tilde{E}''_{n(1)}$, respectively. With sufficiently small $\delta > 0$ we have

$$F \times \ell \subset \tilde{F} \times (\tilde{E}'_{n(1)} \cap \tilde{E}''_{n(1)}) \Subset U.$$

We set the boundary of $\tilde{E}'_{n(1)} \cap \tilde{E}''_{n(1)}$ with positive orientation as in Fig. 2.2, divided by ℓ to the right-hand side and the left-hand:

(2.4.6) $$\partial\left(\tilde{E}'_{n(1)} \cap \tilde{E}''_{n(1)}\right) = \gamma_{(1)} = \gamma'_{(1)} + \gamma''_{(1)}.$$

Similarly, keeping the inner $\frac{\delta}{2}$ of the width δ as E'_n is widened to $\tilde{E}'_{n(1)}$, we successively shrink inward by dividing the outer $\frac{\delta}{2}$ in half. That is, $\tilde{E}'_{n(2)}$ denotes the closed cuboid shrunk inward by $\frac{\delta}{4}$ from $\tilde{E}'_{n(1)}$. Assuming $\tilde{E}'_{n(k)}$ determined, we

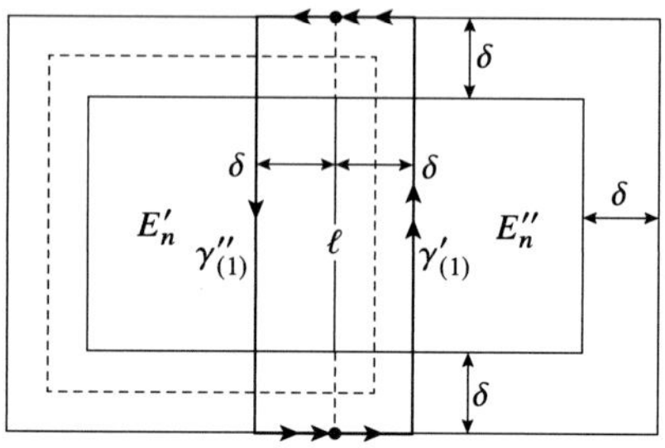

Fig. 2.2 δ-closed neighborhoods of adjacent closed rectangles.

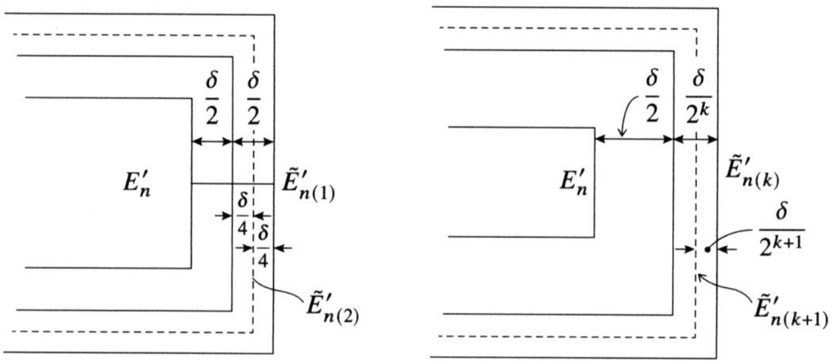

Fig. 2.3 $\frac{\delta}{2^k}$-closed neighborhood of $\tilde{E}'_{n(k)}$.

denote by $\tilde{E}'_{n(k+1)}$ the closed cuboid shrunk inward by $\frac{\delta}{2^{k+1}}$ from $\tilde{E}'_{n(k)}$ (cf. Fig. 2.3).

Since $\frac{\delta}{4} + \frac{\delta}{8} + \cdots = \frac{\delta}{2}$, we see that

$$\bigcap_{k=1}^{\infty} \tilde{E}'_{n(k)} = \text{the closed cuboid widened from } E'_n \text{ by } \frac{\delta}{2}.$$

We set $\tilde{E}''_{n(k)}$, similarly. As (2.4.6) we write

$$\partial\left(\tilde{E}'_{n(k)} \cap \tilde{E}''_{n(k)}\right) = \gamma_{(k)} = \gamma'_{(k)} + \gamma''_{(k)}. \tag{2.4.7}$$

Let

$$\tilde{E}'_{(k)} = \tilde{F} \times \tilde{E}'_{n(k)}, \qquad \tilde{E}''_{(k)} = \tilde{F} \times \tilde{E}''_{n(k)}$$

be the closed cuboid neighborhoods of E' and E'', respectively.

We set $B_1(z) = \mathbf{1}_p - A(z)$ for $z \in \tilde{E}'_{(1)} \cap \tilde{E}''_{(1)}$. Here we apply the idea of the Cousin integral discussed in §1.3.1. For $(z', z_n) \in \tilde{E}'_{(2)} \cap \tilde{E}''_{(2)}$, we have by Cauchy's integral formula

$$B_1(z', z_n) = \frac{1}{2\pi i} \int_{\gamma_{(1)}} \frac{B_1(z', \zeta)}{\zeta - z_n} d\zeta \tag{2.4.8}$$
$$= \frac{1}{2\pi i} \int_{\gamma'_{(1)}} \frac{B_1(z', \zeta)}{\zeta - z_n} d\zeta + \frac{1}{2\pi i} \int_{\gamma''_{(1)}} \frac{B_1(z', \zeta)}{\zeta - z_n} d\zeta$$
$$= B'_1(z', z_n) + B''_1(z', z_n).$$

Here, $B'_1(z', z_n)$ (resp. $B''_1(z', z_n)$) is holomorphic in $(z', z_n) \in \tilde{E}'_{(2)}$ (resp. $(z', z_n) \in \tilde{E}''_{(2)}$), and

$$|z_n - \zeta| \geq \frac{\delta}{4}, \qquad \forall (z', z_n) \in \tilde{E}'_{(2)}, \quad \forall \zeta \in \gamma'_{(1)}. \tag{2.4.9}$$

Letting L be the length of $\gamma'_{(1)}$, we get

$$L = \text{the length of } \gamma''_{(1)} \geq \text{the length of } \gamma'_{(k)} \text{ (also, the length of } \gamma''_{(k)}),$$
$$k = 1, 2, \ldots.$$

It follows from (2.4.8) and (2.4.9) that for $(z', z_n) \in \tilde{E}'_{(2)}$

$$\|B'_1(z', z_n)\| \leq \frac{1}{2\pi} \cdot \frac{4}{\delta} L \cdot \max_{\gamma_{(1)}} \|B_1(z', \zeta)\|.$$

Therefore

$$\|B'_1\|_{\tilde{E}'_{(2)}} \leq \frac{2L}{\pi\delta} \|B_1\|_{\tilde{E}'_{(1)} \cap \tilde{E}''_{(1)}}.$$

Similarly, we have

$$\|B''_1\|_{\tilde{E}''_{(2)}} \leq \frac{2L}{\pi\delta} \|B_1\|_{\tilde{E}'_{(1)} \cap \tilde{E}''_{(1)}}.$$

We set

$$\varepsilon_1 = \max \left\{ \|B'_1\|_{\tilde{E}'_{(2)}}, \|B''_1\|_{\tilde{E}''_{(2)}} \right\} \left(\leq \frac{2L}{\pi\delta} \|B_1\|_{\tilde{E}'_{(1)} \cap \tilde{E}''_{(1)}} \right). \tag{2.4.10}$$

Choose $\delta > 0$, smaller if necessary so that $\frac{\pi\delta}{2^5 L} \leq \frac{1}{2}$. Suppose that

$$\|B_1\|_{\tilde{E}'_{(1)} \cap \tilde{E}''_{(1)}} \leq \frac{\pi^2 \delta^2}{2^6 L^2}.$$

Then

$$\varepsilon_1 \leq \frac{\pi\delta}{2^5 L} \leq \frac{1}{2}, \tag{2.4.11}$$

(2.4.12)
$$A(z) = (\mathbf{1}_p - B_1(z)) = (\mathbf{1}_p - B'_1(z))(\mathbf{1}_p - N(B'_1(z), B''_1(z))) \cdot (\mathbf{1}_p - B''_1(z)), \qquad z \in \tilde{E}'_{(2)} \cap \tilde{E}''_{(2)}.$$

Here it follows from Lemma 2.4.3 that

$$\|N(B'_1(z), B''_1(z))\|_{\tilde{E}'_{(2)} \cap \tilde{E}''_{(2)}} \le 2^2 \varepsilon_1^2;$$

here the estimate "ε_1^2" is the point.

In what follows, we proceed inductively; for $j = 1, \ldots, k (\in \mathbf{N})$ we choose pth square matrix-valued holomorphic functions

$$B'_j(z)\ (z \in \tilde{E}'_{(j+1)}), \quad B''_j(z)\ (z \in \tilde{E}''_{(j+1)})$$

satisfying

(2.4.13)
$$\varepsilon_j := \max\left\{\|B'_j\|_{\tilde{E}'_{(j+1)}}, \|B''_j\|_{\tilde{E}''_{(j+1)}}\right\} \le \frac{\pi\delta}{2^{j+4}L} \left(\le \frac{1}{2^j}\right), \ 1 \le j \le k,$$

(2.4.14)
$$A(z) = (\mathbf{1}_p - B'_1(z)) \cdots (\mathbf{1}_p - B'_k(z)) \cdot (\mathbf{1}_p - N(B'_k(z), B''_k(z))) \cdot (\mathbf{1}_p - B''_k(z)) \cdots (\mathbf{1}_p - B''_1(z)), \qquad z \in \tilde{E}'_{(k+1)} \cap \tilde{E}''_{(k+1)}.$$

The case of $k = 1$ holds by (2.4.11) and (2.4.12).

We set $B_{k+1}(z) = N(B'_k(z), B''_k(z))$ for $z \in \tilde{E}'_{(k+2)} \cap \tilde{E}''_{(k+2)}$ (cf. (2.4.2)). With $\gamma'_{(k+1)}, \gamma''_{(k+1)}$ defined by (2.4.7) we set

$$B'_{k+1}(z', z_n) = \frac{1}{2\pi i} \int_{\gamma'_{(k+1)}} \frac{B_{k+1}(z', \zeta)}{\zeta - z_n} d\zeta, \quad (z', z_n) \in \tilde{E}'_{(k+2)},$$
$$B''_{k+1}(z', z_n) = \frac{1}{2\pi i} \int_{\gamma''_{(k+1)}} \frac{B_{k+1}(z', \zeta)}{\zeta - z_n} d\zeta, \quad (z', z_n) \in \tilde{E}''_{(k+2)}.$$

Since $|\zeta - z_n| \ge \frac{\delta}{2^{k+2}}$ in the integrand above, it follows from (2.4.13) and Lemma 2.4.3 that

$$\begin{aligned} \varepsilon_{k+1} &\le \frac{L}{2\pi} \frac{2^{k+2}}{\delta} \|N(B'_k, B''_k)\|_{\tilde{E}'_{(k+1)} \cap \tilde{E}''_{(k+1)}} \\ &\le \frac{L}{2\pi} \frac{2^{k+2}}{\delta} 2^2 \varepsilon_k^2 \le \frac{1}{2} \varepsilon_k \le \frac{\pi\delta}{2^{k+5}L}, \end{aligned}$$
$$\mathbf{1}_p - N(B'_k(z), B''_k(z)) = (\mathbf{1}_p - B'_{k+1}(z))(\mathbf{1}_p - N(B'_{k+1}(z), B''_{k+1}(z))) \cdot (\mathbf{1}_p - B''_{k+1}(z)), \qquad z \in \tilde{E}'_{(k+2)} \cap \tilde{E}''_{(k+2)}.$$

Thus we deduce (2.4.13) and (2.4.14) for $k + 1$.

By (2.4.13) and Proposition 2.4.1 (iv), the infinite products

$$A'(z) = \lim_{k \to \infty} (\mathbf{1}_p - B'_1(z)) \cdots (\mathbf{1}_p - B'_k(z)), \quad z \in \tilde{E}' := \bigcap_{k=1}^{\infty} \tilde{E}'_{(k)},$$

$$A''(z) = \lim_{k\to\infty} (\mathbf{1}_p - B''_k(z)) \cdots (\mathbf{1}_p - B''_1(z)), \quad z \in \tilde{E}'' := \bigcap_{k=1}^{\infty} \tilde{E}''_{(k)}$$

converge uniformly in each defining domain, and the limits are pth square invertible matrix-valued holomorphic functions. It follows from (2.4.13) and Lemma 2.4.3 that for $z \in \tilde{E}' \cap \tilde{E}''$

$$\|N(B'_k(z), B''_k(z))\| \le 2^2 \varepsilon_k^2 \le \frac{1}{2^{2k-2}} \longrightarrow 0 \qquad (k \to \infty),$$

and so $A(z) = A'(z)A''(z)$ by (2.4.14). □

2.4.3 Cartan's Merging Lemma

It is the aim to prove Cartan's Merging Lemma:

Lemma 2.4.15. *Let $E', E'' \Subset \Omega$ be as in Lemma 2.4.5, and let $\mathscr{F}$ be an analytic sheaf over Ω. Let $\left\{\underline{\sigma'_j} \in \Gamma(E', \mathscr{F}) : 1 \le j \le p'\right\}$ (resp. $\left\{\underline{\sigma''_k} \in \Gamma(E'', \mathscr{F}) : 1 \le k \le p''\right\}$) be a finite generator system of $\mathscr{F}$ over E' (resp. E''). Assume further that there are $a_{jk}, b_{kj} \in \mathscr{O}(E' \cap E'')$, $1 \le j \le p'$, $1 \le k \le p''$, satisfying*

$$\underline{\sigma'_j} = \sum_{k=1}^{p''} \underline{a_{jk}} \cdot \underline{\sigma''_k}, \quad \underline{\sigma''_k} = \sum_{j=1}^{p'} \underline{b_{kj}} \cdot \underline{\sigma'_j} \quad (\text{on } E' \cap E'').$$

Then, there exists a finite generator system $\left\{\underline{\sigma_l} \in \Gamma(E' \cup E'', \mathscr{F}) : 1 \le l \le p' + p''\right\}$ of $\mathscr{F}$ over $E' \cup E''$.

Proof. Set column vectors and matrices,

$$\sigma' = \begin{pmatrix} \underline{\sigma'_1} \\ \vdots \\ \underline{\sigma'_{p'}} \end{pmatrix}, \qquad \sigma'' = \begin{pmatrix} \underline{\sigma''_1} \\ \vdots \\ \underline{\sigma''_{p''}} \end{pmatrix},$$

and

$$A = (a_{jk}), \qquad B = (b_{kj}).$$

Unless confusion occurs, A (resp. B) stands for $(\underline{a_{jk}})$ (resp. $(\underline{b_{kj}})$):

$$\sigma' = A\sigma'', \qquad \sigma'' = B\sigma'. \tag{2.4.16}$$

Adding 0 to σ' and σ'' to have the same size, we set $p\,(= p' + p'')$th column vectors as follows:

$$\tilde{\sigma}' = \begin{pmatrix} \sigma'_1 \\ \vdots \\ \sigma'_{p'} \\ \hdashline 0 \\ \vdots \\ 0 \end{pmatrix}, \qquad \tilde{\sigma}'' = \begin{pmatrix} 0 \\ \vdots \\ 0 \\ \hdashline \sigma''_1 \\ \vdots \\ \sigma''_{p''} \end{pmatrix}.$$

Also we set

$$\tilde{A} = \left(\begin{array}{c:c} \mathbf{1}_{p'} & A \\ \hdashline -B & \mathbf{1}_{p''} - BA \end{array}\right).$$

By making use of $BA\,\sigma'' = \sigma''$ by (2.4.16), we get

$$\tilde{\sigma}' = \tilde{A}\,\tilde{\sigma}''. \tag{2.4.17}$$

We take the compositions of fundamental transformations,

$$\begin{aligned} P &= \left(\begin{array}{c:c} \mathbf{1}_{p'} & A \\ \hdashline 0 & \mathbf{1}_{p''} \end{array}\right), \quad P^{-1} = \left(\begin{array}{c:c} \mathbf{1}_{p'} & -A \\ \hdashline 0 & \mathbf{1}_{p''} \end{array}\right), \\ Q &= \left(\begin{array}{c:c} \mathbf{1}_{p'} & 0 \\ \hdashline B & \mathbf{1}_{p''} \end{array}\right), \quad Q^{-1} = \left(\begin{array}{c:c} \mathbf{1}_{p'} & 0 \\ \hdashline -B & \mathbf{1}_{p''} \end{array}\right). \end{aligned} \tag{2.4.18}$$

We transform $\tilde{A}$ from right and left: $Q\,\tilde{A}\,P^{-1} = \mathbf{1}_p$. Since $\tilde{A} = Q^{-1}P$, with $R := P^{-1}Q$ we get

$$R = \left(\begin{array}{c:c} \mathbf{1}_{p'} & -A \\ \hdashline 0 & \mathbf{1}_{p''} \end{array}\right)\left(\begin{array}{c:c} \mathbf{1}_{p'} & 0 \\ \hdashline B & \mathbf{1}_{p''} \end{array}\right), \quad \tilde{A}R = \mathbf{1}_p. \tag{2.4.19}$$

Because of the form, R is invertible for any choices of A and B. Since the elements a_{jk}, b_{kj} of A and B are holomorphic in a neighborhood of $E' \cap E'' = F \times \ell$, it follows from Theorem 1.3.2 that a_{jk}, b_{jk} are uniformly approximated by polynomials $\tilde{a}_{jk}, \tilde{b}_{kj}$ on a neighborhood $W_0 \Supset E' \cap E''$ with $W_0 \Subset U' \cap U''$. We define $\tilde{R}$ by (2.4.19) with those polynomials $\tilde{a}_{jk}, \tilde{b}_{kj}$. Taking the sufficiently enough approximations, we deduce that for the neighborhood V of $\mathbf{1}_p$ in Lemma 2.4.5,

$$\hat{A}(z) = \tilde{A}(z)\,\tilde{R}(z) \in V_0, \qquad z \in W_0. \tag{2.4.20}$$

Then by Lemma 2.4.5 there is a pth invertible matrix valued function $\hat{A}'$ (resp. $\hat{A}''$) on E' (resp. E''), such that

$$\hat{A} = \hat{A}'\,\hat{A}'' \tag{2.4.21}$$

on $E' \cap E''$. From this and (2.4.20) we deduce $\tilde{A} = \hat{A}'\hat{A}''\tilde{R}^{-1}$, and so from (2.4.17)

$$\hat{A}'^{-1}\tilde{\sigma}' = \hat{A}''\,\tilde{R}^{-1}\,\tilde{\sigma}'' \tag{2.4.22}$$

on $E' \cap E''$. Therefore $\underline{\tau_h} \in \Gamma(E' \cup E'', \mathscr{F})$, $1 \le h \le p$, are well-defined by

$$\begin{pmatrix} \underline{\tau_1} \\ \vdots \\ \underline{\tau_p} \end{pmatrix} = \begin{cases} \hat{A}'^{-1}\,\tilde{\sigma}', & \text{on } E', \\ \hat{A}''\,\tilde{R}^{-1}\,\tilde{\sigma}'', & \text{on } E''. \end{cases}$$

Since $\hat{A}'^{-1}$ and $\hat{A}''\,\tilde{R}^{-1}$ are invertible, the finite system $\{\underline{\tau_h} : 1 \le h \le p\}$ generates $\mathscr{F}$ over $E' \cup E''$. □

The above obtained $\{\underline{\tau_h}\}$ is called a *merged system* from $\{\underline{\sigma'_j}\}$ and $\{\underline{\sigma''_k}\}$.

2.5 Oka's Joku-Iko Principle

Let $S \subset \mathrm{P}\Delta$ denote a complex submanifold of a polydisk $\mathrm{P}\Delta \subset \mathbf{C}^N$. *Oka's Joku-Iko Principle* is a methodological principle to reduce a problem on S to that of a polydisk $\mathrm{P}\Delta$ of even higher dimension by embedding the original domain under the consideration onto $S \subset \mathrm{P}\Delta$ and extending the problem over $\mathrm{P}\Delta$, to solve the problem by virtue of the simple shape of $\mathrm{P}\Delta$, and then to get the solution on S by restriction.

It is the innovative point of Oka's idea to solve a problem caused by the increase of numbers of variables by increasing more the number of variables.

2.5.1 Oka Syzygy

Let $E \Subset \mathbf{C}^n$ be a closed cuboid (cf. 2.4.4). The number of the edges of E with positive length is called the *dimension of* E, and denoted by $\dim E$, where $0 \le \dim E \le 2n$.

We consider an analytic sheaf $\mathscr{F} \subset \mathscr{O}_U^q$ over a neighborhood U of E. The two claims of the next lemma are the most fundamental properties of coherent sheaves.

Lemma 2.5.1 (Oka Syzygy). *Let $E \Subset \mathbf{C}^n$ be a closed cuboid as above, and let $\mathscr{F}$ be a coherent sheaf over E. Then we have:*

(i) *$\mathscr{F}$ has a finite generator system over E.*

(ii) *Let $\{\underline{\sigma_j}\}_{1 \le j \le N}$ be a finite generator system of $\mathscr{F}$ over E. Then, for every $\underline{\sigma} \in \Gamma(E, \mathscr{F})$ there are holomorphic functions $a_j \in \mathscr{O}(E)$, $1 \le j \le N$, such that*

$$\underline{\sigma} = \sum_{j=1}^{N} \underline{a_j} \cdot \underline{\sigma_j} \quad (on\ E). \tag{2.5.2}$$

Proof. We prove (i) and (ii) simultaneously by the double induction on $\nu := \dim E$.

Writing respectively $(\mathrm{i})_\nu$ and $(\mathrm{ii})_\nu$ for the statements (i) and (ii) in the case of $\dim E = \nu$, we shall proceed as

$$[(\mathrm{i})_{\nu-1} + (\mathrm{ii})_{\nu-1}] \Longrightarrow (\mathrm{i})_\nu \Longrightarrow (\mathrm{ii})_\nu.$$

(a) The case of $\nu = 0$: Since E is a singleton, the both are immediate by the assumptions.

(b) Suppose $\nu \geq 1$, and that each of $(\mathrm{i})_{\nu-1}$ and $(\mathrm{ii})_{\nu-1}$ holds for any coherent sheaf over an arbitrary closed cuboid of dimension $\nu - 1$. (Here it is noted that the induction hypothesis is not claiming the validities of $(\mathrm{i})_{\nu-1}$ and $(\mathrm{ii})_{\nu-1}$ for the same coherent sheaf, but the validity of each of $(\mathrm{i})_{\nu-1}$ and $(\mathrm{ii})_{\nu-1}$ for arbitrary coherent sheaves: Either, E is not fixed.)

(i) Let E be a closed cuboid of dimension ν, and let $\mathscr{F}$ be a coherent sheaf over a neighborhood of E. Let $z = (z_j) \in \mathbf{C}^n$ be the coordinates. By parallel transitions, changes of the order of the coordinates and the multiplications to the coordinates by the imaginary unit i, we may assume without loss of generality that E is expressed as

$$\begin{aligned} E &= F \times \{z_n : 0 \leq \Re z_n \leq T, |\Im z_n| \leq \theta\}, \\ &\quad T > 0,\ \theta \geq 0. \end{aligned} \tag{2.5.3}$$

Here, in the case $\theta = 0$ (resp. $\theta > 0$), F is a closed cuboid of dimension $\nu - 1$ (resp. $\nu - 2$).

Take arbitrarily a point $t \in [0, T]$ and set $E_t := F \times \{z_n : \Re z_n = t, |\Im z_n| \leq \theta\}$. Since E_t is a closed cuboid of dimension $\nu - 1$, it follows from $(\mathrm{i})_{\nu-1}$ that $\mathscr{F}$ has a finite generator system over a neighborhood of E_t. By Heine–Borel's Theorem there is a finite partition

$$0 = t_0 < t_1 < \cdots < t_L = T \tag{2.5.4}$$

with finite generator systems $\{\underline{\sigma_{\alpha j}}\}_{j=1}^{N_\alpha}$ of $\mathscr{F}$ over

$$E_\alpha := F \times \{z_n : t_{\alpha-1} \leq \Re z_n \leq t_\alpha, |\Im z_n| \leq \theta\} \quad (1 \leq \alpha \leq L).$$

Since $E_\alpha \cap E_{\alpha+1} = E_{t_\alpha}$ is a closed cuboid of dimension $\nu - 1$, the induction hypothesis $(\mathrm{ii})_{\nu-1}$ implies the existence of holomorphic functions a_{jk}, b_{kj} on $E_\alpha \cap E_{\alpha+1}$ satisfying

$$\underline{\sigma_{\alpha j}} = \sum_k \underline{a_{jk}} \cdot \underline{\sigma_{\alpha+1 k}}, \quad \underline{\sigma_{\alpha+1 k}} = \sum_j \underline{b_{kj}} \cdot \underline{\sigma_{\alpha j}}.$$

Applying Cartan's Merging Lemma 2.4.15, we obtain a finite generator system of $\mathscr{F}$ over $E_\alpha \cup E_{\alpha+1}$.

Merging first the finite generator systems of $\mathscr{F}$ over E_1 and over E_2, we obtain a merged finite generator system of $\mathscr{F}$ over $E_1 \cup E_2$ (cf. Fig. 2.4). Similarly, from the finite generator system above over $E_1 \cup E_2$ and the one over E_3, we obtain a merged

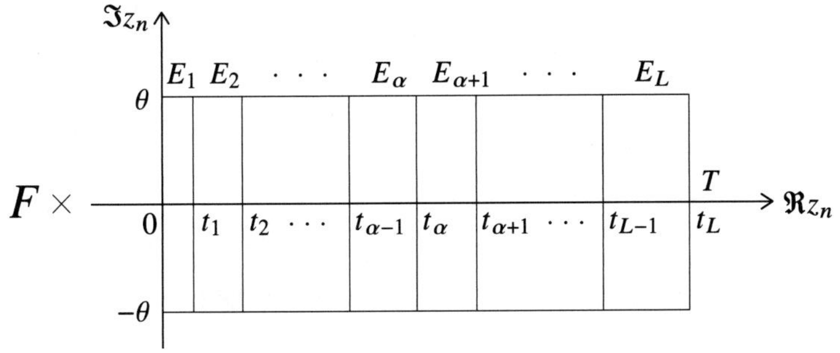

Fig. 2.4 $E_\alpha = F \times [t_{\alpha-1}, t_\alpha]$.

finite generator system of $\mathscr{F}$ over $\bigcup_{\alpha=1}^{3} E_\alpha$. Repeating the process, we have a finite generator system of $\mathscr{F}$ over $\bigcup_{\alpha=1}^{L} E_\alpha = E$.

(ii) Let $\{\underline{\sigma_j}\}_{1\le j\le N}$, and $\underline{\sigma}$ be the given ones. We take a closed cuboid E as in (2.5.3), and use the same notation there.

For every point $t \in [0,T]$, E_t is a closed cuboid of dimension $\nu - 1$, and hence it follows from (ii)$_{\nu-1}$ that there are holomorphic functions $a_{tj} \in \mathscr{O}(E_t)$ such that

$$\underline{\sigma} = \sum_{j=1}^{N} \underline{a_{tj}} \cdot \underline{\sigma_j} \quad (\text{on } E_t).$$

By the same arguments as in (i) there is a finite partition (2.5.4) (cf. Fig. 2.4) with holomorphic functions $a_{\alpha j}$ on E_α satisfying

$$\underline{\sigma} = \sum_{j} \underline{a_{\alpha j}} \cdot \underline{\sigma_j} \quad (\text{on } E_\alpha). \tag{2.5.5}$$

Let $\mathscr{R} := \mathscr{R}\left((\underline{\sigma_j})_j\right)$ denote the relation sheaf given by $(\underline{\sigma_j})_j$. By Corollary 2.3.38, $\mathscr{R}$ is coherent. If $E_{\alpha+1}$ is adjacent to E_α, we have

$$\sum_{j} \underline{(a_{\alpha j} - a_{\alpha+1 j})} \cdot \underline{\sigma_j} = 0 \tag{2.5.6}$$

over $E_\alpha \cap E_{\alpha+1}$. Therefore

$$\left(\underline{b_{\alpha j}}\right)_j := \left(\underline{a_{\alpha j} - a_{\alpha+1 j}}\right)_j \in \Gamma(E_\alpha \cap E_{\alpha+1}, \mathscr{R}). \tag{2.5.7}$$

Since $\mathscr{R}$ is coherent, it follows from the above result (i)$_\nu$ that $\mathscr{R}$ has a finite generator system $\{\tau_h\}$ over a neighborhood of E. Since $E_\alpha \cap E_{\alpha+1}$ is a closed cuboid of dimension $\nu - 1$, it follows again from (ii)$_{\nu-1}$ that there exist $c_{\alpha h} \in \mathscr{O}(E_\alpha \cap E_{\alpha+1})$ such that

(2.5.8) $$\left(\underline{b_{\alpha j}}\right)_j = \sum_h c_{\alpha h} \cdot \underline{\tau_h} \quad (\text{on } E_\alpha \cap E_{\alpha+1}).$$

Note that $E_\alpha \cap E_{\alpha+1} = F \times \{z_n : \Re z_n = t_\alpha, |\Im z_n| \leq \theta\}$. For a sufficiently small $\delta > 0$, (2.5.8) holds over

$$F \times \{z_n : \Re z_n = t_\alpha, |\Im z_n| \leq \theta + \delta\}.$$

We consider the Cousin integral (1.3.7) of $c_{\alpha h}$ along the line segment $\ell_\alpha = \{z_n : \Re z_n = t_\alpha, |\Im z_n| \leq \theta + \delta\}$ in z_n-plane with orientation as $\Im z_n$ increases. Let the Cousin decomposition be

(2.5.9) $$\begin{aligned} &d_{\alpha h} \in \mathscr{O}(E_\alpha), \quad d_{\alpha+1 h} \in \mathscr{O}(E_{\alpha+1}), \\ &c_{\alpha h} = d_{\alpha h} - d_{\alpha+1 h} \quad (\text{on } E_\alpha \cap E_{\alpha+1}). \end{aligned}$$

(Cf. Lemma 1.3.10.) It hence follows from (2.5.7)–(2.5.9) that

$$\left(\underline{a_{\alpha j}} - \underline{a_{\alpha+1 j}}\right)_j = \sum_h \left(\underline{d_{\alpha h}} - \underline{d_{\alpha+1 h}}\right) \underline{\tau_h}.$$

With $\underline{\tau_h} = (\underline{\tau_{hj}})$ expressed by the components $\tau_{hj} \in \mathscr{O}(E)$, we have

$$a_{\alpha j} - \sum_h d_{\alpha h} \tau_{hj} = a_{\alpha+1 j} - \sum_h d_{\alpha+1 h} \tau_{hj}$$

over $E_\alpha \cap E_{\alpha+1}$. The left-hand side is an element of $\mathscr{O}(E_\alpha)$, and the right-hand side is that of $\mathscr{O}(E_{\alpha+1})$, and so it defines an element $\tilde{a}_{\alpha j} \in \mathscr{O}(E_\alpha \cup E_{\alpha+1})$. Since $\underline{\tau_h}$ is a section of $\mathscr{R}$,

(2.5.10) $$\underline{\sigma} = \sum_j \underline{\tilde{a}_{\alpha j}}\, \underline{\sigma_j} \quad (\text{on } E_\alpha \cup E_{\alpha+1}).$$

We thus patched together the expression (2.5.5) of $\underline{\sigma}$ over E_α and that over $E_{\alpha+1}$ to obtain an expression (2.5.10) of $\underline{\sigma}$ over $E_\alpha \cup E_{\alpha+1}$.

Beginning with $\alpha = 1$, we patch together the expression of $\underline{\sigma}$ over E_1 and that over E_2 to obtain an expression of $\underline{\sigma}$ over $E_1 \cup E_2$. We then patch together it and that over E_3 to obtain an expression of $\underline{\sigma}$ over $E_1 \cup E_2 \cup E_3$. Repeating this procedure up to $\alpha = L - 1$, we obtain an expression (2.5.2) of $\underline{\sigma}$ over $E = E_1 \cup \cdots \cup E_L$.

Thus, the proof by induction is completed. □

Applying the Oka Syzygy Lemma 2.5.1 for a geometric ideal sheaf $\mathscr{F} = \mathscr{I}\langle S\rangle$ (cf. footnote at p. 38) with a complex submanifold S, we immediately have by Theorem 2.3.40:

Lemma 2.5.11 (Geometric Syzygy)**.** *Let S be a complex submanifold of a neighborhood of a closed cuboid $E \Subset \mathbf{C}^n$. Then we have:*

(i) *$\mathscr{I}\langle S\rangle$ has a finite generator system over E.*

(ii) *If* $\underline{\{\sigma_j\}}_{1\le j\le N}$ ($\sigma_j \in \mathscr{O}(E)$, $\sigma_j|_S = 0$) *is a finite generator system of* $\mathscr{I}\langle S\rangle$ *over* E, *then every section* $\underline{\sigma} \in \Gamma(E, \mathscr{I}\langle S\rangle)$ ($\sigma \in \mathscr{O}(E)$, $\sigma|_S = 0$) *is written as*

$$\sigma = \sum_{j=1}^{N} a_j \cdot \sigma_j \quad (on\ E), \tag{2.5.12}$$

where $a_j \in \mathscr{O}(E)$, $1 \le j \le N$.

Remark 2.5.13. (i) In the proof of the Oka Syzygy Lemma 2.5.1 the induction on the cuboid dimension combined with the Cousin integral worked very efficiently. Since the method will be used repeatedly henceforth, we call it the *induction on cuboid dimension* for convenience.

(ii) (Syzygy) Let $\underline{\{\sigma_j\}}_{j=1}^{l}$ be a finite generator system obtained in the Oka Syzygy Lemma 2.5.1 (i). Then the map

$$\phi : (\underline{a_j}_z) \in \mathscr{O}_z^l \subset \mathscr{O}_E^l \longrightarrow \sum_{j=1}^{l} \underline{a_j}_z \cdot \underline{\sigma_j}_z \in \mathscr{F}_z \subset \mathscr{F}|_E \quad (z \in E)$$

is surjective, and the condition of the image to be 0 is given by

$$\sum_{j=1}^{l} \underline{a_j}_z \cdot \underline{\sigma_j}_z = 0 \quad (z \in E),$$

which yields a relation sheaf $\mathscr{R}_V$ over a neighborhood V of E. With the inclusion map $\iota : \mathscr{R}_V \hookrightarrow \mathscr{O}_V^l$, we have $\iota(\mathscr{R}_V) = \operatorname{Ker}\phi := \phi^{-1}0$, called the kernel of the above surjection; the sequence of maps

$$\mathscr{R}_V \stackrel{\iota}{\longrightarrow} \mathscr{O}_V^l \stackrel{\phi}{\longrightarrow} \mathscr{F}|_V$$

is called an *exact sequence*. Let $\mathscr{F}|_V \to 0$ be the zero map. Since ϕ is surjective, the sequence of maps

$$\mathscr{O}_V^l \stackrel{\phi}{\longrightarrow} \mathscr{F}|_V \longrightarrow 0$$

is exact. Therefore we obtain the following composite of those two exact sequences:

$$\mathscr{R}_V \stackrel{\iota}{\longrightarrow} \mathscr{O}_V^l \stackrel{\phi}{\longrightarrow} \mathscr{F}|_V \longrightarrow 0,$$

which is also said to be exact: This exact sequence is called a *syzygy* of $\mathscr{F}$ (the term "syzygy" is due to D. Hilbert).

2.5.2 Oka Extension of the Joku-Iko Principle

The following is the key of Oka's Joku-Iko Principle.

Theorem 2.5.14 (Oka Extension). *Let $S \subset \mathrm{P}\Delta$ be a complex submanifold. Then, for every holomorphic function $f \in \mathcal{O}(S)$ there exists a holomorphic function $F \in \mathcal{O}(\mathrm{P}\Delta)$ with $F|_S = f$; i.e., the following is an exact sequence:*

$$\mathcal{O}(\mathrm{P}\Delta) \ni F \longmapsto F|_S \in \mathcal{O}(S) \longrightarrow 0.$$

Proof. By Riemann's mapping Theorem 1.4.9 we may assume that $\mathrm{P}\Delta$ is an open cuboid $\mathrm{P} \Subset \mathbf{C}^n$.

(1) We first consider the problem quasi-globally; that is, we are going to solve it in a neighborhood of an arbitrary compact subset of P. We will employ the induction on cuboid dimension.

Lemma 2.5.15 (Extension). *Let $E \Subset \mathrm{P}$ be an arbitrary closed cuboid. Then there is a neighborhood W of E such that for every $g \in \mathcal{O}(W \cap S)$ there exists $G \in \mathcal{O}(E)$ with the restriction to $E \cap S$ $(\Subset S)$ satisfying*

$$G|_{E \cap S} = g|_{E \cap S},$$

where the equality means the validity of the equation in a neighborhood of $E \cap S$ in S. We call G a solution *on E.*

$\because$) (a) When $\dim E = 0$, E consists of one point $a \in \mathrm{P}$, and so we choose a neighborhood $W \ni a$ and a local coordinate system such as (2.3.41). For a holomorphic function $g(w_{k+1}, \ldots, w_n) \in \mathcal{O}(W \cap S)$ we set

$$G(w_1, \ldots, w_{k+1}, \ldots, w_n) = g(w_{k+1}, \ldots, w_n) \in \mathcal{O}(W),$$

which satisfies the required property.

(b) Suppose that the claimed statement holds in the case of $\dim E = \nu - 1$ with $\nu \geq 1$. Let $\dim E = \nu$. By the Geometric Syzygy Lemma 2.5.11 (i) there is a finite generator system $\{\sigma_j\}_{j=1}^N$ of $\mathscr{I}\langle S \rangle$ over a neighborhood W of E.

Without loss of generality we may assume that E is of the form (2.5.3). In the sequel, the symbols such as E_t, E_α stand for the same as those used after (2.5.3). Because of $\dim E_t = \nu - 1$ the induction hypothesis implies the existence of a solution G_t on E_t, so that $G_t|_{S \cap E_t} = g|_{S \cap E_t}$. By the Heine–Borel Theorem there are a partition (2.5.4) and holomorphic functions $G_\alpha \in \mathcal{O}(E_\alpha)$ on each E_α such that

$$G_\alpha|_{S \cap E_\alpha} = g|_{S \cap E_\alpha}.$$

Therefore we deduce that $G_{\alpha+1} - G_\alpha \in \Gamma(E_\alpha \cap E_{\alpha+1}, \mathscr{I}\langle S \rangle)$. By the Geometric Syzygy Lemma 2.5.11 (ii) there are $a_{\alpha j} \in \mathcal{O}(E_\alpha \cap E_{\alpha+1})$ $(1 \leq j \leq N)$ such that

(2.5.16)
$$G_{\alpha+1} - G_\alpha = \sum_{j=1}^{N} a_{\alpha j}\sigma_j \quad (\text{on } E_\alpha \cap E_{\alpha+1}).$$

Taking the Cousin decomposition (1.3.11) of $a_{\alpha j}$, we write

(2.5.17)
$$a_{\alpha j} = b_{\alpha j} - b_{\alpha+1 j}, \quad b_{\alpha j} \in \mathscr{O}(E_\alpha),\ b_{\alpha+1 j} \in \mathscr{O}(E_{\alpha+1}).$$

Then it follows that

(2.5.18)
$$G_\alpha + \sum_{j=1}^{N} b_{\alpha j}\sigma_j = G_{\alpha+1} + \sum_{j=1}^{N} b_{\alpha+1 j}\sigma_j \quad (\text{on } E_\alpha \cap E_{\alpha+1}).$$

Since the left-hand side above is a solution on E_α and the right-hand side is a solution on $E_{\alpha+1}$, we obtain a solution on $E_\alpha \cup E_{\alpha+1}$, which we call a solution patched together with the solutions G_α on E_α and $G_{\alpha+1}$ on $E_{\alpha+1}$.

Beginning with $\alpha = 1$, we patch together G_1 and G_2 to obtain a solution H_2 on $E_1 \cup E_2$, and then we patch together H_2 and G_3 to obtain a solution H_3 on $E_1 \cup E_2 \cup E_3$. Repeating this up to $\alpha = L-1$, We have a solution H_L on $E = \bigcup_{\alpha=1}^{L} E_\alpha$. It suffices to set $G = H_L$. △

(2) We take a covering of P by increasing open cuboids

$$\mathrm{P}_1 \Subset \mathrm{P}_2 \Subset \cdots \Subset \mathrm{P}_\mu \Subset \cdots, \qquad \bigcup_{\mu=1}^{\infty} \mathrm{P}_\mu = \mathrm{P}.$$

By the Geometric Syzygy Lemma 2.5.11 (i) there is a finite generator system $\underline{\{\sigma_{\mu j}\}}_{j=1}^{N_\mu}$ of $\mathscr{I}\langle S\rangle$ on each closure $\overline{\mathrm{P}}_\mu$.

For a given $f \in \mathscr{O}(S)$ there is a solution G_μ on each $\overline{\mathrm{P}}_\mu$ by Extension Lemma 2.5.15.

Set $F_1 = G_1$ ($\in \mathscr{O}(\overline{\mathrm{P}}_1)$). Assume that solutions $F_j \in \mathscr{O}(\overline{\mathrm{P}}_j)$, $1 \le j \le \mu$, are determined so that

(2.5.19)
$$\|F_{j-1} - F_j\|_{\overline{\mathrm{P}}_{j-1}} < \frac{1}{2^{j-1}} \quad 2 \le j \le \mu.$$

Since $\underline{G_{\mu+1} - F_\mu} \in \Gamma(\overline{\mathrm{P}}_\mu, \mathscr{I}\langle S\rangle)$, it follows from the Geometric Syzygy Lemma 2.5.11 (ii) that there are holomorphic functions $h_j \in \mathscr{O}(\overline{\mathrm{P}}_\mu)$ $(1 \le j \le N_{\mu+1})$ satisfying

$$G_{\mu+1} - F_\mu = \sum_{j=1}^{N_{\mu+1}} h_j \sigma_{\mu+1 j} \quad (\text{on } \overline{\mathrm{P}}_\mu).$$

By Theorem 1.3.2 we approximate h_j by polynomials $\tilde{h}_j$ on $\overline{\mathrm{P}}_\mu$, so sufficiently that

$$\left\| \sum_{j=1}^{N_{\mu+1}} (h_j - \tilde{h}_j)\sigma_{\mu+1j} \right\|_{\overline{\mathrm{P}}_\mu} < \frac{1}{2^\mu}.$$

Set

$$F_{\mu+1} = G_{\mu+1} - \sum_{j=1}^{N_{\mu+1}} \tilde{h}_j \sigma_{\mu+1j} \in \mathcal{O}(\overline{\mathrm{P}}_{\mu+1}).$$

Since all $\sigma_{\mu+1j}$ vanish on S, $F_{\mu+1}$ is a solution on $\overline{\mathrm{P}}_{\mu+1}$, which satisfies (2.5.19) with $j = \mu + 1$.

Thus the limit

$$\begin{aligned} F = \lim_{\mu\to\infty} F_\mu &= F_1 + \sum_{j=1}^{\infty} (F_{j+1} - F_j) \\ &= F_\mu + \sum_{j=\mu}^{\infty} (F_{j+1} - F_j), \qquad \forall \mu \geq 1 \end{aligned}$$

converges uniformly on compact subsets of P, so that $F \in \mathcal{O}(\mathrm{P})$ and $F|_S = f$. □

Note. An element of $\mathcal{O}_a$ is defined on a domain which varies by the choice of each element; that is, we do not consider functions defined on a definite or determined domain. The idea of thinking analytic functions defined in undetermined domains was first introduced by K. Oka in his VII-th paper [48] written in 1948. And there is a special importance in an object carrying the algebraic structure of *ideal*, which he called "idéal de domaines indéterminés". Immediately, H. Cartan realized that the new notion of Oka is the same as that of "sheaves" defined by J. Leray around the same time (cf. Grauert–Remmert [25] Introduction).

Lemma 2.4.5 was proved by H. Cartan [9]. Since then historically, a number of various proofs have been given to this lemma. The present proof is based on the original proof of H. Cartan [9] with a simplification by Lemma 2.4.3.

The results of H. Cartan in §2.4 are due to [9] published in 1940, which was the last information K. Oka received from Europe. World War II had begun and the Great East Asian War (the Pacific War in the U.S.A.) followed; communications between Japan and countries overseas ceased. During that time Oka's research advanced considerably with the *complete affirmative solutions* of the Three Big Problems ((P1)–(P3) in the Preface) for unramified domains over $\mathbf{C}^n$ in a series of unpublished papers of 1943 ([44])) and the preliminary studies of the notion of "*coherence*" or *ideals of undetermined domains (idéal de domaines indéterminés)*, which was later completed and published in Oka VII, 1948/'50 and VIII, 1951 (cf. [48], [49], [50]; see also [39] Chap. 9, [42], [44]).

Exercises

1. For $t_j \in \mathbf{C}, 1 \le j \le n$, set

$$s_\nu = (-1)^\nu \sum_{1 \le j_1 < \cdots < j_\nu \le n} t_{j_1} \cdots t_{j_\nu}, \quad 1 \le \nu \le n,$$

$$\sigma_\mu = \sum_{j=1}^{n} t_i^\mu, \quad \mu \ge 1.$$

Show that

$$\sigma_m + \sigma_{m-1}s_1 + \cdots + \sigma_1 s_{m-1} + m s_m = 0, \quad m \le n,$$
$$\sigma_m + \sigma_{m-1}s_1 + \cdots + \sigma_{m-n} s_n = 0, \quad m > n.$$

(Hint: For the first, set $f(X) = \prod_{j=1}^n (1 - t_j X)$. Then, $\frac{f'(X)}{f(X)} = -\sum_{j=1}^n \frac{t_j}{1-t_j X} = -\sum_{\nu=0}^\infty \sigma_{\nu+1} X^\nu$. Then, use $f'(X) = -f(X)\sum_{\nu=0}^\infty \sigma_{\nu+1} X^\nu$. For the second, use $X^m + s_1 X^{m-1} + \cdots + s_m X^{m-n} = X^{m-n} \prod_{j=1}^n (X - t_j)$. Then, substitute $X = t_j$, $1 \le j \le n$.)

2. Show that in Lemma 2.4.5 there are positive constants η and C dependent on E', E'' and U, and a closed cuboid $\tilde{E}'$ (resp. $\tilde{E}''$) such that the interior of $\tilde{E}'$ (resp. $\tilde{E}''$) contains E' (resp. E'') and $\tilde{E}' \cap \tilde{E}'' \subset U$, and that if $\|B\|_U \le \eta$ with $A = \mathbf{1}_p - B$, we can choose $A' = \mathbf{1}_p - B'$ and $A'' = \mathbf{1}_p - B''$ to satisfy

$$A(z) = A'(z)A''(z), \quad z \in \tilde{E}' \cap \tilde{E}'',$$
$$\max\{\|B'\|_{\tilde{E}'}, \|B''\|_{\tilde{E}''}\} \le C\|B\|_U.$$

3. Show that $\mathscr{O}_{\mathbf{C}}$ of dimension one is coherent.
4. Show that if a vector-valued holomorphic function $(f_1(z), f_2(z))$ in $z \in \mathbf{C}$ satisfies

$$f_1(z)\sin z + f_2(z)\cos z = 1,$$

then there is a holomorphic function $g(z) \in \mathscr{O}(\mathbf{C})$ satisfying

$$(f_1(z), f_2(z)) = (\sin z, \cos z) + g(z)(\cos z, -\sin z).$$

5. We consider a bounded holomorphic function $f \in \mathscr{O}(\mathrm{P}\Delta(r))$, $|f| \le M$ with a constant $M \in \mathbf{R}_+$, where $\mathrm{P}\Delta(r)$ is a polydisk of polyradius $r = (r_1, \ldots, r_n)$ with center at the origin. Show that if $f(0) = 0$, there are holomorphic functions $g_j \in \mathscr{O}(\mathrm{P}\Delta(r))$ $(1 \le j \le n)$ such that $|g_j| \le 2M/r_j$ and

$$f(z) = \sum_{j=1}^n z_j g_j(z).$$

6. Assume that finitely many holomorphic functions $f_j, 1 \le j \le q$, in a domain $\Omega \subset \mathbf{C}^n$ are given so that for every $z \in \Omega$ there is an f_j with $f_j(z) \neq 0$. Then, show that the relation sheaf $\mathscr{R}(\underline{f_1}, \ldots, \underline{f_q})$ is generated over Ω by the trivial solutions

$$\underline{T_{ij}} = (0, \ldots, 0, \overset{i\,\text{th}}{-\underline{f_j}}, 0, \ldots, 0, \overset{j\,\text{th}}{\underline{f_i}}, 0, \ldots, 0), \quad 1 \le i < j \le q.$$

7. Let the notation be as in the previous exercise. Let $E \subset \Omega$ be a closed cuboid. Show that there exist holomorphic functions $c_j \in \mathscr{O}(E), 1 \le j \le q$, satisfying

$$c_1(z) f_1(z) + \cdots + c_q(z) f_q(z) = 1, \qquad z \in E.$$

8. Let the notation be the same as in the previous exercise, and furthermore let Ω be a polydisk $\mathrm{P}\Delta$. Then, show that there are $b_j \in \mathscr{O}(\mathrm{P}\Delta), 1 \le j \le q$, satisfying

$$b_1(z) f_1(z) + \cdots + b_q(z) f_q(z) = 1, \qquad z \in \mathrm{P}\Delta.$$

(This was proved in H. Cartan [9] §7.)

9. Let $\mathrm{P}\Delta \subset \mathbf{C}^n$ be a polydisk, and let $f \in \mathscr{O}(\mathrm{P}\Delta)$.
Show that there are holomorphic functions $g_j(x, w) \in \mathscr{O}(\mathrm{P}\Delta \times \mathrm{P}\Delta))$ $(1 \le j \le n)$ satisfying

$$f(z) - f(w) = \sum_{j=1}^{n} g_j(z, w)(z_j - w_j).$$

(Cf. Weil's condition (3.8.4).)

Chapter 3
Domains of Holomorphy

The purpose of this chapter is to solve affirmatively the first two of the Three Big Problems. We first define the notion of domains of holomorphy, and then introduce holomorphically convex domains. We prove the Cartan–Thullen Theorem claiming the equivalence of the two domains in $\mathbf{C}^n$. We solve the Approximation Problem by Oka's Joku-Iko Principle. We then formulate the Continuous Cousin Problem and prove that it is solvable on domains of holomorphy. We give the solutions of the Cousin Problem (I and II) with the Oka principle. Similarly, we deal with the $\bar{\partial}$-equation, and the interpolation problem. Lastly, we introduce unramified domains over $\mathbf{C}^n$, on which we discuss the obtained results.

3.1 Definitions and Elementary Properties

In §1.2 we saw that if $n \geq 2$ the Hartogs phenomenon takes place according to the shape of a domain Ω of $\mathbf{C}^n$. It is necessary to think of the maximal domain from the viewpoint of analytic continuation.

3.1.1 Relatively Compact Hull

We begin with a definition. Let $\Omega \subset \mathbf{C}^n$ be a domain.

Definition 3.1.1. We define the *relatively compact hull* $K^{\star}_{\Omega}$[1] of a compact set $K \Subset \Omega$ by

$$K^{\star}_{\Omega} = K \cup \left(\bigcup U \right), \tag{3.1.2}$$

where U runs over all connected components of $\Omega \setminus K$ that are relatively compact in Ω.

[1] The term and the symbol are not standard, but convenient.

J. Noguchi, *Basic Oka Theory in Several Complex Variables*, Universitext,
https://doi.org/10.1007/978-981-97-2056-9_3

The relatively compact hull $K^{\star}_{\Omega}$ is purely topologically defined, and will play an interesting role in the approximation theory in complex analysis; in particular, it distinguishes the case of $n = 1$ and that of $n \geq 2$. We know the convex hull in $\mathbf{R}^n$, and will define later several notions of convex hulls such as the holomorphically convex hull; among them, $K^{\star}_{\Omega}$ is the samllest.

We here investigate a preliminary properties of $K^{\star}_{\Omega}$ with respect to the euclidean metric (distance). Let $A \subset \mathbf{C}^n$ be a subset in general. The distance from a point $z \in \mathbf{C}^n$ to A is defined by

$$(3.1.3) \qquad d(z, A) = \inf\{\|z - w\| : w \in A\}.$$

If A is closed,

$$d(z, A) = \min\{\|z - w\| : w \in A\}.$$

By a simple computation we have

$$(3.1.4) \qquad |d(z, A) - d(w, A)| \leq \|z - w\|,$$

provided that $A \neq \emptyset$ (cf. Exercise 7 at the end of this chapter). Therefore, $d(z, A)$ is Lipschitz continuous in z. The distance $d(A, B)$ between A and another subset $B \subset \mathbf{C}^n$ is defined by

$$(3.1.5) \qquad d(A, B) = \inf\{d(z, B) : z \in A\}.$$

Clearly, $d(A, B) = d(B, A)$, and if A and B are closed,

$$d(A, B) = \min\{\|z - w\| : z \in A, w \in B\}.$$

It is noticed that if U is an open subset, then

$$(3.1.6) \qquad d(U, B) = d(\bar{U}, B) = d(\partial U, B).$$

For a domain $\Omega \subset \mathbf{C}^n$, $d(z, \partial\Omega)$ $(z \in \Omega)$ is called the *boundary distance function*.

Proposition 3.1.7. *Let K be a compact subset of a domain Ω $(\subset \mathbf{C}^n)$. Then:*

(i) *The relatively compact hull $K^{\star}_{\Omega}$ is compact.*
(ii) $d(K, \partial\Omega) = d(K^{\star}_{\Omega}, \partial\Omega)$.
(iii) $(K^{\star}_{\Omega})^{\star}_{\Omega} = K^{\star}_{\Omega}$.

Proof. (i), (ii) By definition, $\Omega \setminus K^{\star}_{\Omega}$ is the intersection of the complements of all connected components of $\Omega \setminus K$ which are not relatively compact in Ω; hence, $\Omega \setminus K^{\star}_{\Omega}$ is closed in Ω.

Let $U \subset \Omega \setminus K$ be a connected component. Suppose that $U \Subset \Omega$. Then, $\partial U \subset K$. We see that $K^{\star}_{\Omega}$ is bounded in $\mathbf{C}^n$, and by (3.1.6) that

$$d(U, \partial\Omega) = d(\partial U, \partial\Omega) \geq d(K, \partial\Omega).$$

Therefore, $d(K^\star_\Omega, \partial\Omega) \geq d(K, \partial\Omega)$; Since the opposite inequality is trivial,

$$d(K^\star_\Omega, \partial\Omega) = d(K, \partial\Omega) > 0.$$

Thus, $K^\star_\Omega$ is compact.

(iii) This is immediate, since $\Omega \setminus K^\star_\Omega$ has no relatively compact connected component in Ω. □

3.1.2 Domain of Holomorphy and Holomorphic Convexity

Now, we consider a maximal domain from the viewpoint of analytic continuation:

Definition 3.1.8 (Domain of holomorphy). A domain Ω ($\subset \mathbf{C}^n$) is called a *domain of holomorphy* if there occurs no Hartogs' phenomenon for Ω (cf. Definition 1.2.4).

The notion of a domain of holomorphy takes account of all $f \in \mathcal{O}(\Omega)$, but Ω can be maximal for one $f \in \mathcal{O}(\Omega)$ with respect to the analytic continuation. In such a case Ω is called the *domain of existence* of f (cf. §3.6 for multivalued functions).

In general, because of the definition it is difficult to check if a given domain is a domain of holomorphy or not. The following notion provides a criterion for it by the property of only the interior points.

Let $\Omega \subset \mathbf{C}^n$ be domain, and let $\mathscr{F} \subset \mathcal{O}(\Omega)$ be a family. For a subset $A \subset \Omega$ we define the *$\mathscr{F}$-convex hull* $\hat{A}_{\mathscr{F}}$ of A in Ω by

$$\hat{A}_{\mathscr{F}} = \{z \in \Omega : |f(z)| \leq \|f\|_A, \ \forall f \in \mathscr{F}\}. \tag{3.1.9}$$

If $A = \hat{A}_{\mathscr{F}}$, we say that A is an *$\mathscr{F}$-convex set* in Ω.

In the case of $\mathscr{F} = \mathcal{O}(\Omega)$ (resp. $\mathscr{F} = \mathbf{C}[(z_j)]$) we write

$$\hat{A}_\Omega = \hat{A}_{\mathcal{O}(\Omega)} \quad (\text{resp. } \hat{A}_{\mathrm{poly}} = \hat{A}_{\mathbf{C}[(z_j)]}) \tag{3.1.10}$$

and call it the *holomorphically* (resp. *polynomially*) *convex hull* of A. If $A = \hat{A}_\Omega$ (resp. $A = \hat{A}_{\mathrm{poly}}$), we say that A is a *holomorphically* (resp. *polynomially*) *convex set* in Ω.

Definition 3.1.11. Ω is called an $\mathscr{F}$-(resp. *holomorphically*, resp. *polynomially*) *convex domain* if $\widehat{K}_{\mathscr{F}} \Subset \Omega$ (resp. $\widehat{K}_\Omega \Subset \Omega$, resp. $\widehat{K}_{\mathrm{poly}} \Subset \Omega$) for all $K \Subset \Omega$.

Example 3.1.12. The following are immediate from the Maximum Principle (Theorem 1.1.48):

(i) $\mathbf{C}^n$ and a polydisk $\mathrm{P}\Delta(a; (r_j))$ are polynomially convex domains and are typical examples of holomorphically convex domains. If $A \subset \mathbf{C}^n$ is bounded, then so is $\hat{A}_{\mathrm{poly}}$.

(ii) A Hartogs domain Ω_{H} (see (1.2.11)) is *not* holomorphically convex (cf. Theorem 1.2.12).

(iii) With $\mathbf{R}^n = \{(z_j) \in \mathbf{C}^n : \Im z_j = 0, 1 \le j \le n\}$, $\mathbf{C}^n \setminus \mathbf{R}^n$ or $\mathrm{P}\Delta((r_j)) \setminus \mathbf{R}^n$ is *not* holomorphically convex (cf. Theorem 1.2.13).

(iv) With the notation of Theorem 1.2.7, $\mathbf{C}^n \setminus S$ or $\mathrm{P}\Delta((r_j)) \setminus S$ is *not* holomorphically convex.

Proposition 3.1.13. (i) *For a compact subset $K \Subset \Omega$ of a domain Ω $(\subset \mathbf{C}^n)$, $\widehat{(\widehat{K}_\Omega)}_\Omega = \widehat{K}_\Omega$.*

(ii) *For a compact subset $K \Subset \mathbf{C}^n$, $\widehat{K}_{\mathrm{poly}} = \widehat{K}_{\mathbf{C}^n}$.*

(iii) *A compact convex subset of $\mathbf{C}^n$ is polynomially convex.*

(iv) *A convex domain is polynomially convex.*

(v) *If domains $\Omega_j \subset \mathbf{C}^{n_j}$ $(1 \le j \le l)$ are polynomially (resp. holomorphically) convex, then the product $\prod_{j=1}^{l} \Omega_j$ is a polynomially (resp. holomorphically) convex domain.*

(vi) *Let K be a compact subset of a domain Ω $(\subset \mathbf{C}^n)$. Then,*

$$K^\star_\Omega \subset \widehat{K}_\Omega \subset \widehat{K}_{\mathrm{poly}}. \tag{3.1.14}$$

In particular, if Ω is holomorphically convex, $\Omega \setminus \widehat{K}_\Omega$ has no connected component which is relatively compact in Ω.

(vii) *For any planar domain $\Omega \subset \mathbf{C}$ and a compact subset $K \Subset \Omega$ we have*

$$d(K, \partial\Omega) = d(K^\star_\Omega, \partial\Omega) = d(\widehat{K}_\Omega, \partial\Omega);$$

hence, Ω is holomorphically convex.

(viii) *A cylinder domain $\prod_{i=1}^{n} \Omega_i$ with $\Omega_i \subset \mathbf{C}$ is holomorphically convex.*

Proof. (i) This is clear by definition.

(ii) By definition $\widehat{K}_{\mathrm{poly}} \supset \widehat{K}_{\mathbf{C}^n}$. Every $f \in \mathcal{O}(\mathbf{C}^n)$ is expanded to a power series

$$f(z) = \sum_{\alpha \in \mathbf{Z}^n_+} c_\alpha z^\alpha,$$

which converges uniformly on compact subsets of $\mathbf{C}^n$. It follows that $\widehat{K}_{\mathrm{poly}} \subset \widehat{K}_{\mathbf{C}^n}$.

(iii) Let $A \Subset \mathbf{C}^n$ be a compact convex subset. Take a point $b \in \mathbf{C}^n \setminus A$. By (i) it suffices to show that there is a function $f \in \mathcal{O}(\mathbf{C}^n)$ with $|f(b)| > \|f\|_A$. Let $z_j = x_j + iy_j$ $(1 \le j \le n)$ be the natural coordinates. By the convexity assumption there is a real linear functional $L(z)$ such that

$$L(z) = L(x_1, y_1, \ldots, x_n, y_n) = \sum_{j=1}^{n} (a_j x_j + b_j y_j), \quad a_j, b_j \in \mathbf{R},$$

$$L(b) > \max\{L(z) : z \in A\}. \tag{3.1.15}$$

Then $L_0(z) := \sum_{j=1}^{n} (a_j - ib_j) z_j$ is holomorphic and satisfies

$$L(z) = \Re L_0(z), \quad |e^{L_0(z)}| = e^{L(z)}.$$

With $f(z) = e^{L_0(z)} \in \mathcal{O}(\mathbf{C}^n)$ it follows from (3.1.15) that $|f(b)| > \|f\|_A$.

(iv) This is immediate from (ii).

(v) This is clear, too.

(vi) Let $U \subset \Omega \setminus K$ be a connected component. Suppose that $U \Subset \Omega$. Then, $\partial U \subset K$. It follows from the Maximum Principle (Theorem 1.1.48) that for any point $a \in U$ and any function $f \in \mathcal{O}(\Omega)$

$$|f(a)| \le \|f\|_{\partial U} \le \|f\|_K.$$

Thus $a \in \widehat{K}_\Omega$, and so $U \subset \widehat{K}_\Omega$; consequently, $K^\star_\Omega \subset \widehat{K}_\Omega$.

The second inclusion relation is clear.

(vii) Let $\Omega \subset \mathbf{C}$ be a domain. If $\Omega = \mathbf{C}$, $\partial\Omega = \emptyset$ and Ω is holomorphically convex (see Example 3.1.12). So, we assume $\Omega \ne \mathbf{C}$. Let $K \Subset \Omega$ be a compact subset. With $b \in \partial\Omega$ we consider $f(z) = 1/(z-b) \in \mathcal{O}(\Omega)$, and then deduce that

$$|f(z)| \le \|f\|_K, \quad \forall z \in \widehat{K}_\Omega,$$
$$|z-b| \ge d(b,K), \ \forall z \in \widehat{K}_\Omega.$$

Therefore, $d(\widehat{K}_\Omega, \partial\Omega\} \ge d(K,\Omega)$. Because $K \subset K^\star_\Omega \subset \widehat{K}_\Omega$ by (3.1.14),

$$d(\widehat{K}_\Omega, \Omega) = d(K^\star_\Omega, \partial\Omega) = d(K, \partial\Omega) > 0;$$

therefore, $\widehat{K}_\Omega \Subset \Omega$.

(viii) This follows from (vii) and (v). □

Corollary 3.1.16. *Let K be a compact subset of a domain $\Omega \subset \mathbf{C}^n$. Then there is no connected component of $\Omega \setminus \widehat{K}_\Omega$ which is relatively compact in Ω.*

Remark 3.1.17. If $n = 1$, the equality $K^\star_\Omega = \widehat{K}_\Omega$ holds in Proposition 3.1.13 (vi) (see Lemma 3.3.20).

Example 3.1.18. (i) An open ball $\mathrm{B}(a;r)$ is polynomially convex (Proposition 3.1.13 (iii)).

(ii) Let $\Omega = \Delta(1)$ be the unit disk, and let $K = \{z \in \Delta(1) : |z| = r\}$ with $0 < r < 1$. Then by the Maximum Principle, it is easy to have

$$K^\star_{\Delta(1)} = \bar{\Delta}(r) = \widehat{K}_\Delta = \hat{\Delta}_{\mathrm{poly}}.$$

(iii) Let $\Omega = \mathrm{P}\Delta = \Delta(1) \times \Delta(1) \subset \mathbf{C}^2$ be the 2-dimensional unit polydisk, and let $K = \{(z,w) \in \mathrm{P}\Delta : |z| = |w| = r\}$ with $0 < r < 1$. Then $\mathrm{P}\Delta \setminus K$ is connected and so $K^\star_{\mathrm{P}\Delta} = K$, but $\widehat{K}_{\mathrm{P}\Delta} = \widehat{K}_{\mathrm{poly}} = \bar{\Delta}(r) \times \bar{\Delta}(r)$ by the Maximum Principle; hence,

$$K = K^\star_{\mathrm{P}\Delta} \subsetneqq \widehat{K}_{\mathrm{P}\Delta} = \widehat{K}_{\mathrm{poly}},$$
$$d(K, \partial\mathrm{P}\Delta) = d(K^\star_{\mathrm{P}\Delta}, \partial\mathrm{P}\Delta) = d(\widehat{K}_{\mathrm{P}\Delta}, \partial\mathrm{P}\Delta) = d(\widehat{K}_{\mathrm{poly}}, \partial\mathrm{P}\Delta).$$

3.2 Cartan–Thullen Theorem

We prove that a domain of holomorphy and a holomorphically convex domain are the same. This result was proved by a paper of Cartan and Thullen in 1932 before K. Oka started to work on the Three Big Problems.[2] The first two problems of approximation (development of functions) and Cousin had been posed for domains of holomorphy. Oka, in fact, proved them for holomorphically convex domains. Therefore the contents of this section are not necessary, if Oka's solutions of the first two problems are stated for holomorphically convex domains. It is the third Pseudoconvexity Problem where the results of the present section are used in full (cf. Chaps. 4 and 5).

Let $\Omega \subset \mathbf{C}^n$ be a domain, and let $\mathrm{B}(1)$ be the unit ball of $\mathbf{C}^n$. Then the boundary distance function $d(z, \partial\Omega)$ defined by (3.1.3) is written as

$$d(z, \partial\Omega) = \delta_{\mathrm{B}(1)}(z, \partial\Omega) := \sup\{s > 0 : z + s\mathrm{B}(1) \subset \Omega\},$$

where $s\mathrm{B}(1) = \{sz : z \in \mathrm{B}(1)\}$. With $\mathrm{B}(R)$ $(R > 0)$ we define $\delta_{\mathrm{B}(R)}(z, \partial\Omega)$ similarly to $\delta_{\mathrm{B}(1)}(z, \partial\Omega)$, and then have

$$\delta_{\mathrm{B}(R)}(z, \partial\Omega) = \frac{1}{R}\delta_{\mathrm{B}(1)}(z, \partial\Omega).$$

Let $r = (r_j) = (r_1, \ldots, r_n)$ with $r_j > 0$ be a polyradius, and fix a polydisk $\mathrm{P}\Delta = \mathrm{P}\Delta((r_j))$ of polyradius r with center at the origin. For $s > 0$ we put

$$s\mathrm{P}\Delta = s \cdot \mathrm{P}\Delta = \mathrm{P}\Delta((sr_j)).$$

Similarly to $\delta_{\mathrm{B}(R)}(z, \partial\Omega)$, we define:

$$(3.2.1) \qquad \begin{aligned} \delta_{\mathrm{P}\Delta}(z, \partial\Omega) &= \sup\{s > 0 : z + s\mathrm{P}\Delta \subset \Omega\} (> 0), \quad z \in \Omega, \\ \|z\|_{\mathrm{P}\Delta} &= \inf\{s > 0 : z \in s\mathrm{P}\Delta\} \geq 0, \quad z \in \mathbf{C}^n. \end{aligned}$$

An easy computation yields that $\|z\|_{\mathrm{P}\Delta}$ satisfies the axioms of a norm (cf. Exercise 2 at the end of this chapter) and

$$(3.2.2) \qquad |\delta_{\mathrm{P}\Delta}(z, \partial\Omega) - \delta_{\mathrm{P}\Delta}(z', \partial\Omega)| \leq \|z - z'\|_{\mathrm{P}\Delta}, \quad z, z' \in \Omega.$$

We call $\delta_{\mathrm{P}\Delta}(z, \partial\Omega)$ the *boundary distance function* of Ω with respect to $\mathrm{P}\Delta$.

Let $\|z\|$ be the euclidean norm defined by (1.1.1). Then there is a constant $C > 0$ such that

$$C^{-1}\|z\| \leq \|z\|_{\mathrm{P}\Delta} \leq C\|z\|.$$

Thus $\delta_{\mathrm{P}\Delta}(z, \partial\Omega)$ is a Lipschitz continuous function by (3.2.2).

---

[2] Cf. the beginning of Preface.

For a subset $A \subset \Omega$ we put

$$\delta_{\mathrm{P}\Delta}(A, \partial\Omega) = \inf\{\delta_{\mathrm{P}\Delta}(z, \partial\Omega) : z \in A\}. \tag{3.2.3}$$

If $A \Subset \Omega$, then $\delta_{\mathrm{P}\Delta}(A, \partial\Omega) > 0$.

Lemma 3.2.4. *Let $f \in \mathcal{O}(\Omega)$ and let $K \Subset \Omega$ be compact. Assume that*

$$|f(z)| \le \delta_{\mathrm{P}\Delta}(z, \partial\Omega), \quad z \in K. \tag{3.2.5}$$

Expand an arbitrary $u \in \mathcal{O}(\Omega)$ to a power series about a point $\xi \in \widehat{K}_\Omega$:

$$u(z) = \sum_{\alpha} \frac{\partial^\alpha u(\xi)}{\alpha!}(z-\xi)^\alpha. \tag{3.2.6}$$

Then this power series converges at $z \in \xi + |f(\xi)| \cdot \mathrm{P}\Delta$.

Proof. Fix $0 < t < 1$ and set

$$\begin{aligned}
\Omega_t &= \{(z_j) : \exists w \in K, |z_j - w_j| \le t r_j |f(w)|, \ 1 \le j \le n\} \\
&\subset \bigcup_{w \in K} \left\{(z_j) : (z_j) \in (w_j) + t\delta_{\mathrm{P}\Delta}(w, \partial\Omega) \cdot \overline{\mathrm{P}\Delta}\right\}.
\end{aligned}$$

Then Ω_t is compact in Ω. There is a constant $M > 0$ with $|u(z)| \le M$, $z \in \Omega_t$. From this we obtain the estimates of partial derivatives: Let $\alpha = (\alpha_j) \in \mathbf{Z}_+^n$ and $w \in K$. Keeping $\rho_j > 0$ so that the appearing variables stay in Ω, we have

$$\begin{aligned}
u(z) &= \left(\frac{1}{2\pi i}\right)^n \int \cdots \int_{|\xi_j - w_j| = \rho_j} \frac{u(\xi)}{\prod_j (\xi_j - z_j)} d\xi_1 \cdots d\xi_n, \\
\partial^\alpha u(z) &= \left(\frac{1}{2\pi i}\right)^n \alpha! \int \cdots \int_{|\xi_j - w_j| = \rho_j} \frac{u(\xi)}{(\xi_1 - z_1)^{\alpha_1 + 1} \cdots (\xi_n - z_n)^{\alpha_n + 1}} d\xi_1 \cdots d\xi_n.
\end{aligned}$$

Suppose that $f(w) \neq 0$. Set $z = w$, $\rho_j = t r_j |f(w)|$ and $\rho = (\rho_j)$. Then

$$\begin{aligned}
&|\partial^\alpha u(w)| \le \frac{\alpha! M}{\rho^\alpha} = \frac{\alpha! M}{t^{|\alpha|} |f(w)|^{|\alpha|} r^\alpha}, \\
&\frac{|\partial^\alpha u(w)| t^{|\alpha|} |f(w)|^{|\alpha|} r^\alpha}{\alpha!} \le M, \quad w \in K.
\end{aligned}$$

The last inequality holds trivially in the case $f(w) = 0$. Therefore we get

$$\left| f(w)^{|\alpha|} \partial^\alpha u(w) \right| \le \frac{\alpha! \cdot M}{t^{|\alpha|} r^\alpha}, \quad w \in K.$$

Since $f(w)^{|\alpha|} \partial^\alpha u(w) \in \mathcal{O}(\Omega)$, we deduce from the definition of $\widehat{K}_\Omega$ that

$$\left|f(w)^{|\alpha|}\partial^\alpha u(w)\right| \le \frac{\alpha! M}{t^{|\alpha|} r^\alpha}, \quad w \in \widehat{K}_\Omega.$$

With $w = \xi \in \widehat{K}_\Omega$, (3.2.6) converges at $z \in \xi + |f(\xi)| t \cdot \mathrm{P}\Delta$. As $t \nearrow 1$, we see that (3.2.6) converges at $z \in \xi + |f(\xi)| \cdot \mathrm{P}\Delta$. □

Lemma 3.2.7 (Cartan–Thullen). *Let $\Omega \subset \mathbf{C}^n$ be a domain of holomorphy, let $f \in \mathcal{O}(\Omega)$, and let $K \Subset \Omega$ be compact. By $\delta_*(z, \partial\Omega)$ we denote either $\delta_{\mathrm{P}\Delta}(z, \partial\Omega)$ or $\delta_{\mathrm{B}(R)}(z, \partial\Omega)$. If*

$$|f(z)| \le \delta_*(z, \partial\Omega), \quad z \in K, \tag{3.2.8}$$

then

$$|f(z)| \le \delta_*(z, \partial\Omega), \quad z \in \widehat{K}_\Omega. \tag{3.2.9}$$

In particular, with $f \equiv \delta_*(K, \partial\Omega)$ (constant) we get

$$\delta_*(K, \partial\Omega) = \delta_*\left(\widehat{K}_\Omega, \partial\Omega\right). \tag{3.2.10}$$

Proof. **(a)** Case of $\delta_{\mathrm{P}\Delta}(z, \partial\Omega)$: It follows from Lemma 3.2.4 that for arbitrary $u \in \mathcal{O}(\Omega)$ and $z \in \widehat{K}_\Omega$, u is holomorphic in $z + |f(z)| \cdot \mathrm{P}\Delta$. The assumption of Ω being a domain of holomorphy implies that $z + |f(z)| \cdot \mathrm{P}\Delta \subset \Omega$. Therefore

$$|f(z)| \le \delta_{\mathrm{P}\Delta}(z, \partial\Omega), \quad z \in \widehat{K}_\Omega.$$

In particular, we take $f \equiv \delta_{\mathrm{P}\Delta}(K, \partial\Omega)$. Then

$$\delta_{\mathrm{P}\Delta}(K, \partial\Omega) \le \delta_{\mathrm{P}\Delta}\left(\widehat{K}_\Omega, \partial\Omega\right).$$

The opposite inequality follows from $K \subset \widehat{K}_\Omega$, and so the equality holds.

(b) Case of $\delta_{\mathrm{B}(R)}(z, \partial\Omega)$: By the multiplication of a positive constant to coordinates we may assume $R = 1$, and write $\mathrm{B}(1) = \mathrm{B}$. We take a special polydisk $\mathrm{P}\Delta = \{(z_j) \in \mathbf{C}^n : |z_j| \le 1/\sqrt{n}\}$. Then, $\mathrm{P}\Delta \subset \mathrm{B}$. We denote by $U(n)$ all unitary transformations of the coordinates. Then

$$\mathrm{B} = \bigcup_{A \in U(n)} A\mathrm{P}\Delta.$$

Note that what was proved in (a) above remains valid for the changes of coordinates, in particular after replacing $\mathrm{P}\Delta$ with $A\mathrm{P}\Delta$ ($A \in U(n)$). Suppose that

$$|f(z)| \le \delta_{\mathrm{B}}(z, \partial\Omega), \quad z \in K.$$

Since $\delta_{\mathrm{B}}(z, \partial\Omega) \le \delta_{A\mathrm{P}\Delta}(z, \partial\Omega)$,

$$|f(z)| \le \delta_{A\mathrm{P}\Delta}(z, \partial\Omega), \quad z \in \widehat{K}_\Omega, \ \forall A \in U(n).$$

Therefore

$$|f(z)| \le \delta_{\mathrm{B}}(z, \partial\Omega), \quad z \in \widehat{K}_\Omega. \qquad \square$$

Theorem 3.2.11 (Cartan–Thullen). *The following three conditions are equivalent for a domain $\Omega \subset \mathbf{C}^n$:*

(i) *Ω is a domain of holomorphy.*
(ii) *There is a holomorphic function $f \in \mathcal{O}(\Omega)$ whose domain of existence is Ω.*
(iii) *Ω is a holomorphically convex domain.*

Proof. (i)⇒(iii): We take an arbitrary compact subset $K \Subset \Omega$. By the definition, $\widehat{K}_\Omega$ is bounded and closed in Ω. Since Ω is a domain of holomorphy, (3.2.10) implies

$$\delta_{\mathrm{P}\Delta}\left(\widehat{K}_\Omega, \partial\Omega\right) = \delta_{\mathrm{P}\Delta}(K, \partial\Omega) > 0.$$

Therefore, $\widehat{K}_\Omega \Subset \Omega$.

(iii)⇒(ii): We take a discrete sequence $\{a_j\}_{j=1}^\infty$ of points of Ω such that it has no accumulation point in Ω and all points of $\partial\Omega$ are the accumulation points of it. Set

$$D_j = a_j + \delta_{\mathrm{P}\Delta}(a_j, \partial\Omega) \cdot \mathrm{P}\Delta \subset \Omega.$$

Let K_j, $j = 1, 2, \ldots$, be an increasing sequence of compact subsets of Ω such that

$$K_j \Subset K_{j+1}^\circ, \qquad \bigcup_{j=1}^\infty K_j^\circ = \Omega,$$

where K_j° denotes the interior of K_j. By the choice, $D_j \cap (\Omega \setminus \widehat{K_j}_\Omega) \neq \emptyset$ for all $j \ge 1$. It follows that for a point $z_j \in D_j \setminus \widehat{K_j}_\Omega$ there is an $f_j \in \mathcal{O}(\Omega)$ satisfying

$$\max_{K_j} |f_j| < |f_j(z_j)|.$$

Dividing f_j by $f_j(z_j)$, we may assume $f_j(z_j) = 1$ and that

$$\max_{K_j} |f_j| < |f_j(z_j)| = 1.$$

Taking a power f_j^ν with sufficiently large ν and denoting it again by f_j, we have

$$\max_{K_j} |f_j| < \frac{1}{2^j}, \quad f_j(z_j) = 1.$$

Since $\sum_j \frac{j}{2^j} < \infty$, the infinite product

$$f(z) = \prod_{j=1}^\infty (1 - f_j(z))^j$$

converges uniformly on compact subsets of Ω (cf. [38] Chap. 2 §6). Necessarily, $f \not\equiv 0$.

We show that Ω is a domain of existence of $f(z)$. If it were not the case, there should exist a domain $V \not\subset \Omega$ with $V \cap \Omega \neq \emptyset$, an element $g \in \mathcal{O}(V)$ and a connected component $W \subset V \cap \Omega$ such that $f|_W = g|_W$. For a boundary point $b \in \partial\Omega \cap \partial W \cap V$ there is a subsequence $\{a_{j_\nu}\}$ converging to b. Since $\delta_{\mathrm{P}\Delta}(a_{j_\nu}, \partial\Omega) \to 0$ ($\nu \to \infty$), $\{z_{j_\nu}\}$ also converges to b. Note that $f(z)$ has a zero of order j_ν at $z = z_{j_\nu}$. Thus for an arbitrary partial derivative ∂^α with $|\alpha| \leq j_\nu$

$$\partial^\alpha f(z_{j_\nu}) = 0.$$

With an arbitrary fixed ∂^α ($|\alpha| > 0$) $\partial^\alpha f(z_{j_\nu}) = 0$ for $\nu \gg 1$, and then

$$\partial^\alpha f(z_{j_\nu}) \to \partial^\alpha f(b), \quad \nu \to \infty.$$

It thus follows that

$$\partial^\alpha f(b) = 0, \quad \forall \alpha.$$

The Identity Theorem 1.1.46 implies $f \equiv 0$, which is absurd.

(ii)⇒(i): Since f itself cannot be analytically continued over a properly larger domain than Ω, Ω is a domain of holomorphy.

Thus the proof is completed. □

Corollary 3.2.12. *Let $\{\Omega_\gamma\}_{\gamma \in \Gamma}$ be a family of domains of holomorphy. Then every connected component of the interior points of $\bigcap_{\gamma \in \Gamma} \Omega_\gamma$ is a domain of holomorphy.*

Proof. Let Ω be such a component. Let $K \Subset \Omega$ be a compact subset. Then, $K \subset \widehat{K}_\Omega \subset \widehat{K}_{\Omega_\gamma}$. Because of Ω_γ being a domain of holomorphy, (3.2.10) implies

$$\delta_0 := \delta_{\mathrm{P}\Delta}(K, \partial\Omega) \leq \delta_{\mathrm{P}\Delta}(K, \partial\Omega_\gamma) = \delta_{\mathrm{P}\Delta}(\widehat{K}_{\Omega_\gamma}, \partial\Omega_\gamma).$$

The inclusion relations lead to

$$\delta_{\mathrm{P}\Delta}(K, \partial\Omega_\gamma) \geq \delta_{\mathrm{P}\Delta}(\widehat{K}_\Omega, \partial\Omega_\gamma) \geq \delta_{\mathrm{P}\Delta}(\widehat{K}_{\Omega_\gamma}, \partial\Omega_\gamma).$$

It follows that $\delta_{\mathrm{P}\Delta}(\widehat{K}_\Omega, \partial\Omega_\gamma) = \delta_{\mathrm{P}\Delta}(K, \partial\Omega_\gamma) \geq \delta_0 > 0$. Thus, for every $a \in \widehat{K}_\Omega$ we see

$$a + \delta_0 \mathrm{P}\Delta \subset \Omega_\gamma, \quad \forall \gamma \in \Gamma.$$

Since $a + \delta_0 \mathrm{P}\Delta$ is connected, $a + \delta_0 \mathrm{P}\Delta \subset \Omega$. It follows that $\widehat{K}_\Omega \Subset \Omega$; hence, Ω is holomorphically convex. By Theorem 3.2.11 Ω is a domain of holomorphy. □

Remark 3.2.13. In the proof of Theorem 3.2.11, Lemmata 3.2.4 and 3.2.7 were used only with conditions (3.2.5) and (3.2.8) restricted to $f =$ constant. There may be some readers who wonder why general f is considered. In fact, the matter remains the same until the solution of the Cousin Problem. It will make sense to deal with a general holomorphic function f in those conditions after going into the Pseudoconvexity

Problem; it plays an essential role in the proof of Oka's Theorem 4.3.1 of Boundary Distance Functions. It leads to the solution of the Pseudoconvexity Problem (see Chap. 5).

3.3 Analytic Polyhedron and Oka–Weil Approximation

3.3.1 Analytic Polyhedron

Let $\Omega \subset \mathbf{C}^n$ be a domain.

Definition 3.3.1 (Analytic polyhedron). With holomorphic functions $\varphi_j \in \mathcal{O}(\Omega)$ and positive numbers $\rho_j > 0$ $(1 \le j \le l)$ given, a union P $(\Subset \Omega)$ of a finite number of connected components of the open set

$$\{z \in \Omega : |\varphi_j(z)| < \rho_j,\ 1 \le j \le l\}, \tag{3.3.2}$$

which are relatively compact in Ω, is called an $\mathcal{O}(\Omega)$-*analytic polyhedron* or simply *analytic polyhedron* of Ω. We say that P is defined by φ_j $(1 \le j \le N)$ of (3.3.2).

In particular, if $\varphi_j \in \mathcal{O}(\Omega)$ $(1 \le j \le l)$ are polynomials, P is called a *polynomial polyhedron*. A connected analytic (resp. polynomial) polyhedron is called an *analytic* (resp. *polynomial*) *polyhedral domain*.

Let $\mathrm{P} \Subset \Omega$ be an analytic polyhedron defined by (3.3.2). Since P is bounded, there is a polydisk $\mathrm{P}\Delta((r_j)) \supset \mathrm{P}$. We consider a holomorphic map

$$\begin{aligned} &\varphi : z \in \mathrm{P} \to (z, \varphi_1(z), \ldots, \varphi_l(z)) \in \mathrm{P}\Delta, \\ &\mathrm{P}\Delta := \mathrm{P}\Delta((r_j)) \times \mathrm{P}\Delta((\rho_j)) \ (\subset \mathbf{C}^n \times \mathbf{C}^l). \end{aligned} \tag{3.3.3}$$

We call φ the *Oka map* of P. We denote by $\mathbf{C}[z, \varphi]$ $(\subset \mathcal{O}(\Omega))$ the set of all polynomials in $z_1, \ldots, z_n, \varphi_1, \ldots, \varphi_l$ with complex coefficients.

Lemma 3.3.4. *The Oka map* $\varphi : \mathrm{P} \to \mathrm{P}\Delta$ *is proper, and the image* $\varphi(\mathrm{P})$ *is a complex submanifold of* $\mathrm{P}\Delta$.

Proof. As z $(\in \mathrm{P}) \to \partial\mathrm{P}$ (z approaches $\partial\mathrm{P}$), $\varphi(z) \to \partial\mathrm{P}\Delta$; therefore, φ is proper, and the image $\varphi(\mathrm{P})$ is a closed subset of $\mathrm{P}\Delta$. The rest is clear. □

Lemma 3.3.5. *Let* $\mathrm{P} \Subset \Omega$ *be an analytic polyhedron defined by* $\varphi_j \in \mathcal{O}(\Omega)$ $(1 \le j \le l)$. *Let* $K \Subset \mathrm{P}$ *be compact and let* $g \in \mathcal{O}(\mathrm{P})$. *Then* g *is approximated uniformly on* K *by elements of* $\mathbf{C}[z, \varphi]$.

Proof. Let $\varphi : \mathrm{P} \to \mathrm{P}\Delta$ be the Oka map and set $S = \varphi(\mathrm{P})$. We may consider g as a holomorphic function on S and $K \Subset S$. It follows from the Oka Extension Theorem 2.5.14 that there is an element $G \in \mathcal{O}(\mathrm{P}\Delta)$ with $G|_S = g$. We expand G to a power series:

$$G(w) = \sum_{\beta \in \mathbf{Z}_+^{n+l}} c_\beta w^\beta, \quad w \in \mathrm{P}\Delta.$$

For every $\varepsilon > 0$ there is a sufficiently large $N \in \mathbf{N}$ such that with $G_N(w) := \sum_{|\beta| \le N} c_\beta w^\beta$,

$$(3.3.6) \qquad |G(w) - G_N(w)| < \varepsilon, \quad w \in K.$$

Setting $g_N(z) := G_N(\varphi(z)) \in \mathbf{C}[z,\varphi] \subset \mathcal{O}(\Omega)$, we have by (3.3.6) that

$$\|g - g_N\|_K < \varepsilon.$$

□

Remark 3.3.7. If $\varphi_j(z)$ $(1 \le j \le l)$ in Lemma 3.3.5 are bounded, then elements of $\mathbf{C}[z,\varphi]$ are also bounded on P.

Proposition 3.3.8. *Let $\Omega \subset \mathbf{C}^n$ be a holomorphically convex domain, and let $\mathrm{P} \Subset \Omega$ be an analytic polyhedron. Then the closure $\overline{\mathrm{P}}$ is holomorphically convex.*

Proof. Assume that $\widehat{\overline{\mathrm{P}}}_\Omega \supsetneq \overline{\mathrm{P}}$. By the assumption,

$$(3.3.9) \qquad \widehat{\overline{\mathrm{P}}}_\Omega \Subset \Omega.$$

We take a point $b \in \widehat{\overline{\mathrm{P}}}_\Omega \setminus \overline{\mathrm{P}}$. Let P be defined by (3.3.2). With sufficiently small $\varepsilon > 0$ we denote by P_ε the finite union of connected components of

$$(3.3.10) \qquad \{z \in \Omega : |\varphi_j(z)| < \rho_j + \varepsilon,\ 1 \le j \le l\},$$

which are relatively compact in Ω and contain points of $\overline{\mathrm{P}}$; P_ε is an analytic polyhedron and satisfies $b \notin \overline{\mathrm{P}}_\varepsilon$. With more φ_j if necessary, it follows from (3.3.9) that the connected component Q of (3.3.10) containing b is relatively compact in Ω. We consider the analytic polyhedron

$$\widetilde{\mathrm{P}}_\varepsilon = \mathrm{P}_\varepsilon \cup Q \Supset \overline{\mathrm{P}} \cup \{b\}, \quad \mathrm{P}_\varepsilon \cap Q = \emptyset.$$

We define $g \in \mathcal{O}(\widetilde{\mathrm{P}}_\varepsilon)$ by

$$g(z) = \begin{cases} 0, & z \in \mathrm{P}_\varepsilon, \\ 1, & z \in Q. \end{cases}$$

By Lemma 3.3.5 $g(z)$ is uniformly approximated on $\overline{\mathrm{P}} \cup \{b\}$ by elements of $\mathcal{O}(\Omega)$. Therefore there is an element $f \in \mathcal{O}(\Omega)$ such that

$$\|f\|_{\overline{\mathrm{P}}} < \frac{1}{2} < |f(b)|.$$

However, this contradicts $b \in \widehat{\overline{\mathrm{P}}}_\Omega$. □

Lemma 3.3.11. *A holomorphically convex compact subset $K \Subset \Omega$ has a fundamental neighborhood system of analytic polyhedra; that is, for an arbitrary neighborhood $U \supset K$ there is an $\mathscr{O}(\Omega)$-analytic polyhedron* P *such that* $K \Subset \mathrm{P} \Subset U$.

Proof. We may assume $U \Subset \Omega$. Since $K = \widehat{K}_\Omega$, for each $b \in \partial U$ there is a function $f \in \mathscr{O}(\Omega)$ such that

$$\|f\|_K < 1 < |f(b)|.$$

Then $V_f(b) = \{z \in \Omega : |f(z)| > 1\}$ is a neighborhood of b. Since ∂U is compact, we can take finitely many such $V_{f_j}(b_j), 1 \le j \le l$, so that

$$\partial U \subset \bigcup_{j=1}^{l} V_{f_j}(b_j).$$

Let P be the finite union of connected components of $W = \{z \in \Omega : |f_j(z)| < 1, 1 \le j \le l\}$ which contain the points of K. Then $\mathrm{P} \supset K$; since $\overline{W} \cap \partial U = \emptyset$, $\mathrm{P} \Subset U$. □

3.3.2 Oka–Weil Approximation Theorem

Now we solve the Approximation Problem (P1).[3] Approximation Theorem 1.3.2 of Runge type is generalized to:

Theorem 3.3.12 (Oka–Weil Approximation). *Let $K \Subset \Omega$ be a holomorphically convex set in Ω. Then a holomorphic function on K is uniformly approximated on K by elements of $\mathscr{O}(\Omega)$.*

Proof. This follows from Lemmata 3.3.5 and 3.3.11. □

Corollary 3.3.13 (Weil). *If $K = \widehat{K}_{\mathbf{C}^n} (= \widehat{K}_{\mathrm{poly}})$, a holomorphic function on K is approximated uniformly on K by polynomials.*

Proof. It is immediate from Theorem 3.3.12 with $\Omega = \mathbf{C}^n$, since holomorphic functions in $\mathbf{C}^n$ are expanded to power series. □

Corollary 3.3.14. *A holomorphic function on a compact convex subset K of $\mathbf{C}^n$ is approximated uniformly on K by polynomials.*

Proof. Use Proposition 3.1.13 (iii) and Corollary 3.3.13. □

Remark 3.3.15. It is noted that Oka's Joku-Iko Principle enables us to get rid of the "cylinder" condition relative to the coordinate system for the domain in Corollary 3.3.14.

Definition 3.3.16. We consider a pair of open sets $\Omega_1 \subset \Omega_2 \subset \mathbf{C}^n$. Assume that every connected component of Ω_j $(j = 1, 2)$ is holomorphically convex. If every

[3] Cf. Preface.

holomorphic function $f \in \mathcal{O}(\Omega_1)$ is approximated uniformly on compact subsets of Ω_1 by elements of $\mathcal{O}(\Omega_2)$, (Ω_1, Ω_2) is called a *Runge pair*.

Lemma 3.3.17. *Let Ω be a holomorphically convex domain and let* P *be an analytic polyhedron of Ω. Then* (P, Ω) *is a Runge pair.*

Proof. Use Lemma 3.3.5. □

Theorem 3.3.18. *Let $\Omega_1 \subset \Omega_2$ $(\subset \mathbf{C}^n)$ be a pair of holomorphically convex domains. The following four conditions are equivalent:*

(i) (Ω_1, Ω_2) *is a Runge pair.*
(ii) *For every compact subset $K \Subset \Omega_1$, $\widehat{K}_{\Omega_1} = \widehat{K}_{\Omega_2}$.*
(iii) *For every compact subset $K \Subset \Omega_1$, $\widehat{K}_{\Omega_2} \Subset \Omega_1$.*
(iv) *For every compact subset $K \Subset \Omega_1$, $\widehat{K}_{\Omega_2} \cap \Omega_1 \Subset \Omega_1$.*

Proof. (i)⇒(ii): For K given, $\widehat{K}_{\Omega_1} \Subset \Omega_1$. We take arbitrarily its neighborhood U with $\widehat{K}_{\Omega_1} \Subset U \Subset \Omega_1$. It follows from Lemma 3.3.11 that there is an $\mathcal{O}(\Omega_1)$-analytic polyhedron P_1 satisfying

$$K \Subset \mathrm{P}_1 \Subset U.$$

Let P_1 be defined by $|\varphi_j| < 1$, $\varphi_j \in \mathcal{O}(\Omega_1)$ $(1 \le j \le l)$. By the assumption, φ_j can be approximated uniformly on $\bar{\mathrm{P}}_1$ by functions $\psi_j \in \mathcal{O}(\Omega_2)$, so that

$$\mathrm{Q} := \{z \in \Omega_2 : |\psi_j(z)| < 1,\ 1 \le j \le l\} \supset K, \quad \mathrm{P}_2 \Subset U,$$

where P_2 denotes the finite union of connected components of Q which contain points of K. From Proposition 3.3.8 it follows that

$$\widehat{K}_{\Omega_1} \subset \widehat{K}_{\Omega_2} \subset \widehat{\bar{\mathrm{P}}}_{2\Omega_2} = \bar{\mathrm{P}}_2 \subset U.$$

Since $U \Supset \widehat{K}_{\Omega_1}$ is arbitrary, $\widehat{K}_{\Omega_1} = \widehat{K}_{\Omega_2}$.

(ii)⇒(iii)⇒(iv): These are clear.

(iv)⇒(i): Take an element $f \in \mathcal{O}(\Omega_1)$ and a compact subset $K \Subset \Omega_1$. By the assumption there is an $\mathcal{O}(\Omega_2)$-analytic polyhedron P such that $K \Subset \mathrm{P} \Subset \Omega_1$. It follows from Lemma 3.3.17 that f can be approximated uniformly on K by elements of $\mathcal{O}(\Omega_2)$. Therefore, (Ω_1, Ω_2) is a Runge pair. □

Proposition 3.3.19. *A holomorphically convex domain Ω carries an increasing open covering by $\mathcal{O}(\Omega)$-analytic polyhedral domains P_ν $(\nu \in \mathbf{N})$ such that*

$$\mathrm{P}_1 \Subset \mathrm{P}_2 \Subset \cdots \Subset \mathrm{P}_\nu \Subset \cdots, \qquad \bigcup_{\nu=1}^{\infty} \mathrm{P}_\nu = \Omega.$$

Proof. If $\Omega = \mathbf{C}^n$, it suffices to take polydisks P_ν $(\nu = 1, 2, \ldots)$ of monotone increasing polyradii diverging to infinity.

Suppose that $\Omega \neq \mathbf{C}^n$, $\partial\Omega \neq \emptyset$. Fix arbitrarily a point $a_0 \in \Omega$. With the boundary distance function $d(z, \partial\Omega)$ we have

$$U_1 = \left\{ z \in \Omega : \|z\| < r_0, d(z, \partial\Omega) > \frac{1}{r_0} \right\} \ni a_0$$

for a large $r_0 > 0$. Let V_1 be the connected component of U_1 containing a_0. Similarly, let V_ν be the connected components of

$$U_\nu = \left\{ z \in \Omega : \|z\| < \nu r_0, d(z, \partial\Omega) > \frac{1}{\nu r_0} \right\}, \quad \nu = 1, 2, \ldots,$$

which contain a_0. It follows that $\bigcup_{\nu=1}^{\infty} U_\nu = \Omega$.

For every point $z \in \Omega$ there is a curve connecting a_0 and z. Since C is compact and $C \subset \bigcup_{\nu=1}^{\infty} U_\nu$, there is a number ν_0 with $C \subset U_{\nu_0}$. Because of the connectedness, $C \subset V_{\nu_0}$, so that $z \in V_{\nu_0}$. Thus, $\bigcup_{\nu=1}^{\infty} V_\nu = \Omega$. We obtain an increasing open covering of Ω by

$$V_\nu \Subset V_{\nu+1}, \qquad \bigcup_{\nu=1}^{\infty} V_\nu = \Omega.$$

Let $\widehat{\bar{V}}_{1\Omega}$ be the holomorphically convex hull of $\bar{V}_1$. Since Ω is holomorphically convex, $\widehat{\bar{V}}_{1\Omega}$ is compact. It follows from Lemma 3.3.11 that there is an $\mathcal{O}(\Omega)$-analytic polyhedron P_1 with $\widehat{\bar{V}}_{1\Omega} \Subset \mathrm{P}_1 \Subset \Omega$. We rewrite P_1 for the connected component of P_1 containing V_1. We next take V_{ν_2} such that $V_{\nu_2} \supset \bar{\mathrm{P}}_1 \cup \bar{V}_2$. Repeating the same argument as above again to $\bar{V}_{\nu_2}$, we obtain an $\mathcal{O}(\Omega)$-analytic polyhedral domain P_2 such that $V_{\nu_2} \Subset \mathrm{P}_2 \Subset \Omega$. Repeating this, we obtain the required analytic polyhedral domains P_ν, $\nu = 1, 2, \ldots$. □

3.3.3 Runge Approximation Theorem (One Variable)

In this subsection we prove the classical Runge Approximation Theorem on a domain of $\mathbf{C}$. The Oka–Weil Approximation Theorem 3.3.12 restricted to the case $n = 1$ is very close to Runge's, but not exactly the same. In one variable we will see that the holomorphic convex hull is determined by the topology.

Let $n = 1$ and let $\Omega \subset \mathbf{C}$ be a domain. Let $K \Subset \Omega$ be a compact subset. Let $K_\Omega^\star$ be the relatively compact hull defined by (3.1.2). In general, $K_\Omega^\star$ is compact (Proposition 3.1.7). In one variable the first inclusion relation in (3.1.14) holds in fact with equality:

Lemma 3.3.20. *With the notation above,* $K_\Omega^\star = \widehat{K}_\Omega$.

Proof. Take a point $a \in \Omega \setminus K_\Omega^\star$. It suffices to obtain an element $f \in \mathcal{O}(\Omega)$ such that $|f(a)| > \|f\|_{K_\Omega^\star}$.

Let ω be the connected component of $\Omega \setminus K_\Omega^\star$ containing a. In the case when ω is bounded in $\mathbf{C}$, we take a point $c \in \partial\omega \cap \partial\Omega$ ($\neq \emptyset$): By the linear transform $w = 1/(z - c)$, we can reduce ω to an unbounded domain. Henceforth, ω is unbounded.

Let $\Delta(R)$ be a large disk such that $\Delta(R) \Supset K^{\star}_{\Omega} \cup \{a\}$. Taking a point $b \in \omega \setminus \Delta(R)$, we connect a (the initial point) and b (the terminal point) by a curve $C \subset \omega$. Set

$$\delta_0 := \min\{|\zeta - z| : \zeta \in C,\ z \in K^{\star}_{\Omega}\} > 0.$$

If $b \in \overline{\Delta(a;\delta_0/2)}$, we let $b_1 = b$. Otherwise, we consider a moving point ζ on C starting from a; then, ζ crosses the boundary $\partial\Delta(a;\delta_0/2)$. We take the last crossing point denoted by $b_1 \in C \cap \partial\Delta(a;\delta_0/2)$. We set $g(z) = 1/(z - b_1)$, which is a rational function with a pole only at b_1, and

$$(3.3.21) \qquad |g(a)| \geq \frac{2}{\delta_0} > \frac{1}{\delta_0} = \|g\|_{K^{\star}_{\Omega}}.$$

We next consider $\overline{\Delta(b_1;\delta_0/4)}$. If $b \in \overline{\Delta(b_1;\delta_0/4)}$, we set $b_2 = b$. Otherwise, the moving point $\zeta \in C$ crosses $\partial\Delta(b_1;\delta_0/4)$, and the last point is denoted by $b_2 \in C \cap \partial\Delta(b_1;\delta_0/4)$. Repeating this procedure, we can take a finite sequence of points on C such that

$$(3.3.22) \qquad \begin{aligned} &C \ni b_1,\ b_2,\ \ldots,\ b_l = b, \quad |b_j - b_{j+1}| \leq \frac{\delta_0}{4},\ 1 \leq j \leq l-1, \\ &|z - b_j| \geq \frac{\delta_0}{2}, \quad z \in K^{\star}_{\Omega} \cup \{a\},\ 1 \leq j \leq l. \end{aligned}$$

If $|z - b_2| > \delta_0/4$, then $\left|\frac{b_1 - b_2}{z - b_2}\right| < 1$, and

$$g(z) = \frac{1}{z - b_2} \cdot \frac{1}{1 - \frac{b_1 - b_2}{z - b_2}} = \sum_{\nu=0}^{\infty} \frac{(b_1 - b_2)^{\nu}}{(z - b_2)^{\nu+1}}.$$

Here the convergence is uniform on compact subsets of $|z - b_2| > \delta_0/4$ and in particular, by (3.3.22) it is uniform on $K^{\star}_{\Omega} \cup \{a\}$. That is, $g(z)$ can be approximated uniformly on $K^{\star}_{\Omega} \cup \{a\}$ by rational functions with only one pole at b_2.

Applying the same argument to $\frac{1}{z - b_2}$, we have

$$\frac{1}{z - b_2} = \sum_{\nu=0}^{\infty} \frac{(b_2 - b_3)^{\nu}}{(z - b_3)^{\nu+1}},$$

where the convergence is uniform on compact subsets of $|z - b_3| > \delta_0/4$. Thus $\frac{1}{z - b_2}$ can be approximated uniformly on $K^{\star}_{\Omega} \cup \{a\}$ by rational functions with a pole only at b_3.

Combining the above two steps, we see that $g(z)$ can be approximated uniformly on $K^{\star}_{\Omega} \cup \{a\}$ by rational functions with a pole only at b_3. Repeating the process up to $b_l = b$, we see that $g(z)$ can be approximated uniformly on $K^{\star}_{\Omega} \cup \{a\}$ by rational functions with a pole only at b.

Now, since $K^{\star}_{\Omega} \cup \{a\} \subset \Delta(R)$ and $|b| > R$, we have an expansion

$$\frac{1}{z-b} = \sum_{\nu=0}^{\infty} -\frac{z^\nu}{b^{\nu+1}}, \qquad z \in \Delta(R),$$

where the convergence is uniform on $K^\star_\Omega \cup \{a\}$. Therefore $g(z)$ can be approximated uniformly on $K^\star_\Omega \cup \{a\}$ by polynomials. It follows from (3.3.21) that there is a polynomial $f(z)$ satisfying $|f(a)| > \|f\|_{K^\star_\Omega}$; here, needless to say, $f \in \mathscr{O}(\Omega)$. □

Theorem 3.3.23 (Runge Approximation). *Let $\Omega \subset \mathbf{C}$ be a domain and let $K \Subset \Omega$ be a compact subset such that $K^\star_\Omega = K$. Then a holomorphic function on K is approximated uniformly on K by functions of $\mathscr{O}(\Omega)$.*

Proof. Use Lemma 3.3.20 and the Oka–Weil Approximation Theorem 3.3.12. □

Remark 3.3.24. It is a specialty of one variable that the possibility of the uniform approximation on K can be characterized by the topological property. There is a more elementary proof of Theorem 3.3.23 by means of only the Cauchy integral formula and Lemma 3.3.20 (see Exercise 5 at the end of this chapter).

3.4 Cousin Problem

We solve the Cousin Problem (P2).[4] The Cousin Problem consists of I and II; we shall formulate a Continuous Cousin Problem and solve I and II simultaneously.

We begin with the definition of meromorphic functions. Let $\Omega \subset \mathbf{C}^n$ be an open set.

Definition 3.4.1. A pair (f, Ω') of $f \in \mathscr{O}(\Omega')$ and an open dense subset $\Omega' \subset \Omega$ is said to be *meromorphic* in Ω if for every point $a \in \Omega$ there are a connected neighborhood U of a and $g, h \in \mathscr{O}(U)$ with $g \not\equiv 0$ satisfying

(3.4.2) $$f(z) = \frac{h(z)}{g(z)}, \qquad z \in \Omega' \cap U \setminus \{g = 0\}.$$

Two meromorphic pairs (f_j, Ω'_j) $(j = 1, 2)$ in Ω are equivalent if there is an open dense subset $\Omega'' \subset \Omega$ such that

(3.4.3) $$\begin{aligned} &\Omega'' \subset \Omega'_1 \cap \Omega'_2, \quad f_j|_{\Omega''} \in \mathscr{O}(\Omega''), \\ &f_1(z) = f_2(z), \qquad z \in \Omega''. \end{aligned}$$

It is easy to check that this relation is in fact an equivalence relation (cf. Exercise 4 at the end of this chapter); the equivalence class of (f, Ω') is called a *meromorphic function* in Ω, denoted simply by f or $f(z)$. Let $\mathscr{M}(\Omega)$ denote the set of all meromorphic functions in Ω. If an element $f \in \mathscr{M}(\Omega)$ has a representative (f_0, Ω) with $f_0 \in \mathscr{O}(\Omega)$, then f_0 is uniquely determined, so that we have the natural inclusion $\mathscr{O}(\Omega) \subset \mathscr{M}(\Omega)$.

[4] Cf. Preface.

If $g(z)$ in (3.4.2) can be chosen to satisfy $g(a) \neq 0$, after shrinking U smaller if necessary, we have $1/g \in \mathcal{O}(U)$. In this case, $f \in \mathcal{O}(U)$, and so f is said to be holomorphic at a. The set Ω_0 of all points where f is holomorphic is an open set, and $Z_\infty := \Omega \setminus \Omega_0$ is a closed set; A point of Z_∞ is called a *pole* of f, and Z_∞ is called the *polar set* of f.

Let $f \in \mathcal{M}(\Omega)$. For a connected component Ω' of Ω we have either $f|_{\Omega'} \equiv 0$ or $f|_{\Omega'} \not\equiv 0$. In the first case, the zero set of f in Ω' is Ω' itself. In the latter case, $1/(f|_{\Omega'}) \in \mathcal{M}(\Omega')$, and the *zero set* of f in Ω' is defined to be the polar set of $1/(f|_{\Omega'})$ in Ω'; the *zero set* of f is the union Z_0 of all those zero sets in the connected components of Ω.

Theorem 3.4.4. *Let Ω be a domain and let $f \in \mathcal{M}(\Omega)$ be a non-zero meromorphic function. Let $Z = Z_\infty \cup Z_0$ with the polar set Z_∞ and the zero set Z_0 of f.*

(i) *Z is a nowhere dense closed subset.*
(ii) *$\Omega \setminus Z$ is a domain.*
(iii) *If $n \geq 2$, Z contains no isolated point.*
(iv) *About every point $a \in Z_\infty$, f is unbounded; i.e., for any neighborhood U of a in Ω, $f|_{U \setminus Z_\infty}$ is unbounded; equivalently, if f is bounded in a dense open subset of a neighborhood of a point $a \in \Omega$, then f is holomorphic about a.*

Proof. (i), (ii) By definition there are a connected neighborhood U of every $a \in \Omega$ and a proper analytic subset Y of U such that $Z \cap U \subset Y$. By Theorem 1.5.2 $U \setminus Y$ is a connected open dense subset of U, and hence so is $U \setminus Z$ in U. Therefore Z is a nowhere dense closed subset, and $\Omega \setminus Z$ is connected if so is Ω.

(iii) This follows from Theorem 1.2.7.

(iv) Suppose that for a point $a \in \Omega$ there is a neighborhood $U \ni a$ in Ω such that $f|_{U \setminus Z_\infty}$ is bounded. By definition, with U chosen smaller if necessary, there are $g, h \in \mathcal{O}(U)$ such that $g \not\equiv 0$ and

$$f(z) = \frac{h(z)}{g(z)}, \quad z \in U \setminus \{g = 0\} \subset U \setminus Z_\infty.$$

Theorem 1.5.11 implies $f \in \mathcal{O}(U)$, and so $a \notin Z_\infty$. □

Assume that Ω is a domain. It follows from the above theorem that if $h(z)$ of (3.4.2) satisfies $h \neq 0$ (i.e., $h(z) \not\equiv 0$), then this holds for any other point of Ω. In that case, setting $1/f(z) = g(z)/h(z)$ locally in U, we may obtain a unique inverse element of f. Thus $\mathcal{M}(\Omega)$ naturally carries a structure of a field.

Remark 3.4.5. It may be helpful for understanding the notion of meromorphic functions to learn that the ring $\mathcal{O}_{n,a}$ is a unique factorization domain (see, e.g., [39] Theorem 2.2.12). Keeping the above notation, we suppose that $f \not\equiv 0$ and $Z_\infty \neq \emptyset$. The following is known for Z_∞: For every pole $a \in Z_\infty$ there are a neighborhood U of a and functions $g, h \in \mathcal{O}(U)$ satisfying

$$f = \frac{h}{g}, \qquad Z_\infty \cap U = \{g = 0\}, \quad Z_0 \cap U = \{h = 0\}$$

(cf. [39] §2.2.2). In particular, Z_∞ and Z_0 are complex hypersurfaces of Ω. But, we will not use this fact.

3.4.1 Cousin I Problem

Problem 3.4.6 (Cousin I). Let $\{U_\lambda\}_{\lambda\in\Lambda}$ be an open covering of a domain $\Omega \subset \mathbf{C}^n$ and let $f_\lambda \in \mathscr{M}(U_\lambda)$ be given so that

$$f_\lambda - f_\mu \in \mathscr{O}(U_\lambda \cap U_\mu) \quad \text{on } U_\lambda \cap U_\mu, \quad \forall \lambda, \mu \in \Lambda. \tag{3.4.7}$$

Here, if $U_\lambda \cap U_\mu = \emptyset$, we consider it to hold.

Then, find $F \in \mathscr{M}(\Omega)$ such that

$$F - f_\lambda \in \mathscr{O}(U_\lambda), \qquad \forall \lambda \in \Lambda. \tag{3.4.8}$$

A family $\mathscr{C}_I := \{(U_\lambda, f_\lambda)\}_{\lambda\in\Lambda}$ of pairs (U_λ, f_λ) satisfying (3.4.7) is called *Cousin I data* (also, a Cousin I distribution) on Ω. The above F, if it exists, is called a *solution* of $\mathscr{C}_I$. Also in this case, the polar sets Z_λ of f_λ in U_λ satisfy

$$Z_\lambda \cap (U_\lambda \cap U_\mu) = Z_\mu \cap (U_\lambda \cap U_\mu)$$

for all $\nu, \mu \in \Lambda$. Therefore the *polar set* Z of $\mathscr{C}_I$ is well-defined in Ω. By Theorem 3.4.4, Z is a nowhere dense closed subset of Ω, and if $n \geq 2$, Z contains no isolated point.

Remark 3.4.9. In one variable ($n = 1$) the Cousin I Problem is known to hold as the Mittag-Leffler Theorem ([38] Theorem (7.2.3)). The Cousin I Problem is a version of it in several variables.

Remark 3.4.10 (Non-solvable example). In the case of $n \geq 2$ there is a non-solvable example of $\mathscr{C}_I$ due to the shape of Ω. For example, we consider a Hartogs domain Ω_{H} in the space of two variables $(z, w) \in \mathbf{C}^2$:

$$\begin{aligned} \Omega_1 &= \{(z,w) \in \mathbf{C}^2 : |z| < 3, |w| < 1\}, \\ \Omega_2 &= \{(z,w) \in \mathbf{C}^2 : 2 < |z| < 3, |w| < 3\}, \\ \Omega_{\mathrm{H}} &= \Omega_1 \cup \Omega_2. \end{aligned} \tag{3.4.11}$$

We consider Cousin I data $\mathscr{C}_I$ defined by $f_1 = 0$ on Ω_1 and $f_2 = 1/(z-w)$ on Ω_2 (cf. Fig. 3.1). Suppose that $\mathscr{C}_I$ has a solution $F \in \mathscr{M}(\Omega_{\mathrm{H}})$. If we set $g(z,w) = (z-w)F(z,w) \in \mathscr{O}(\Omega_{\mathrm{H}})$, we deduce from Hartogs' Phenomenon (Theorem 1.2.12) that $g \in \mathscr{O}(\mathrm{P}\Delta((3,3)))$. Then, $g(z,z)$ is holomorphic in $\Delta(3)$, and $g(z,z) = 0$ in $|z| < 1$, but $g(z,z) = 1$ in $2 < |z| < 3$; this contradicts the Identity Theorem 1.1.46.

Therefore, for the solvability of the Cousin I Problem in $n \geq 2$, the domain Ω cannot be arbitrary, and it was asked for Ω domain of holomorphy.

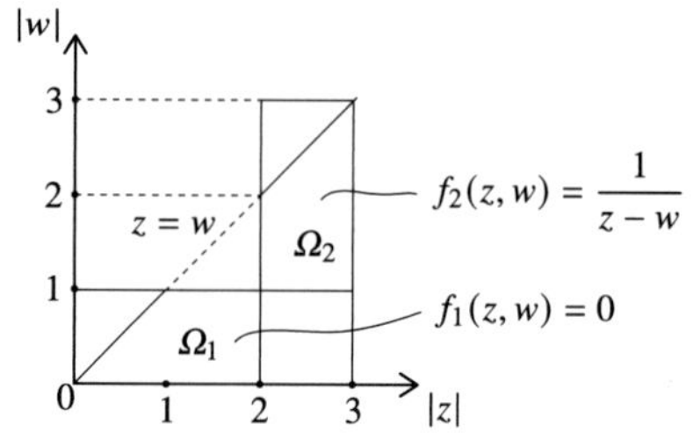

Fig. 3.1 Non-solvable Cousin I data.

3.4.2 Continuous Cousin Problem

There is the Cousin II Problem, but before going into it we would like to prepare some topological setting and formulate the Continuous Cousin Problem in order to make the consideration easier.

Let $\Omega \subset \mathbf{C}^n$ be an open set. We set

$$W_j = \left\{ z \in \Omega : \|z\| < j,\ d(z, \partial\Omega) > \frac{1}{j} \right\}, \quad j = 1, 2, \ldots. \tag{3.4.12}$$

Then,

$$W_j \Subset W_{j+1}, \qquad \bigcup_{j=1}^{\infty} W_j = \Omega.$$

Definition 3.4.13 (Locally finite). Let $\{U_\lambda\}_{\lambda \in \Lambda}$ be an arbitrary open covering of Ω. We say that $\{U_\lambda\}_{\lambda \in \Lambda}$ is *locally finite*, if one of the following equivalent conditions holds:

(i) For every point $a \in \Omega$ there is a neighborhood V of a such that there are only finitely many U_λ with $U_\lambda \cap V \neq \emptyset$.
(ii) For every compact subset $K \Subset \Omega$ there are only finitely many U_λ such that $U_\lambda \cap K \neq \emptyset$.

Definition 3.4.14 (Refinement). Let $\mathscr{U} = \{U_\lambda\}_{\lambda \in \Lambda}$ and $\mathscr{V} = \{V_\gamma\}_{\gamma \in \Gamma}$ be two open coverings of Ω. We say that $\mathscr{V}$ is a *refinement* of $\mathscr{U}$ (denoted by $\mathscr{U} \prec \mathscr{V}$) if for every V_γ there is a corresponding U_λ with $V_\gamma \subset U_\lambda$. If necessary, the correspondence is written as $\phi : \Gamma \to \Lambda$ (or the map between the families is denoted by $\phi : \mathscr{V} \to \mathscr{U}$), so that $V_\gamma \subset U_{\phi(\gamma)}$ (or $V_\gamma \subset \phi(V_\gamma)$).

Proposition 3.4.15. *For any open covering $\{U_\lambda\}_{\lambda \in \Lambda}$ of Ω there is its refinement $\{V_\mu\}_{\mu=1}^N$ ($N \leq \infty$) which is at most countable and locally finite.*

Proof. If Ω is covered by finitely many U_λ ($\lambda \in \Lambda$), it is sufficient to put $V_\lambda := U_\lambda$ for those U_λ. In the sequel we assume that Ω cannot be covered by finitely many members of U_λ ($\lambda \in \Lambda$).

Let W_j be as defined in (3.4.12). The closure $\bar{W}_j$ is compact. Since $\bar{W}_1$ is covered by finitely many U_λ, we denote them by $U_1, U_2, \ldots, U_{\nu_1}$. In the same way as above, we take finitely many $U_{\nu_1+1}, U_{\nu_1+2}, \ldots, U_{\nu_2}$ which cover $\bar{W}_2 \setminus W_1$. Repeating this we take a countable open covering $\{U_\mu\}_{\nu=1}^\infty$ out of $\{U_\lambda\}_{\lambda\in\Lambda}$.

Next, we set

$$V_\mu = U_\mu, \quad 1 \le \mu \le \nu_1,$$

which covers $\bar{W}_1$. We then set

$$V_\mu = U_\mu \setminus \bar{W}_1, \quad \nu_1 + 1 \le \mu \le \nu_2.$$

It follows that $\bar{W}_2 \subset \bigcup_{\mu=1}^{\nu_2} V_\mu$. Repeating this procedure, we obtain $\{V_\mu\}_{\mu=1}^\infty$. By the construction $\{V_\mu\}_{\mu=1}^\infty$ is a locally finite open covering of Ω, and a refinement of $\{U_\lambda\}_{\lambda\in\Lambda}$. □

Proposition 3.4.16 (Partition of unity). *Let $\Omega = \bigcup_{j=1}^N U_j$ $(N \le \infty)$ be an open covering of an open set $\Omega \subset \mathbf{C}^n$ which is at most countable and locally finite. Then there are continuous functions $\chi_j \in \mathscr{C}^0(\Omega)$ such that:*

(i) $0 \le \chi_j \le 1, \quad 1 \le j \le N,$
(ii) $\mathrm{Supp}\,\chi_j \subset U_j, \quad 1 \le j \le N,$
(iii) $\sum_{j=1}^N \chi_j(x) \equiv 1, \quad x \in \Omega.$

Proof. We take an open covering $\{V_j\}_{j=1}^N$ of Ω, so that the closures $\bar{V}_j$ in Ω are contained in U_j for all j. Then, since $\bar{V}_j \cap (\Omega \setminus U_j) = \emptyset$ for each j, there is a function $\rho_j \in \mathscr{C}^0(\Omega)$ (cf. Exercise 7 at the end of this chapter) such that:

(i) $\rho_j(x) \ge 0, \quad x \in \Omega;$
(ii) $\rho_j(x) = 1, \quad x \in V_j;$
(iii) $\mathrm{Supp}\,\rho_j \subset U_j.$

Since $\sum_{j=1}^N \rho_j(x) > 0$ $(\forall x \in \Omega)$, it suffices to put

$$\chi_j(x) = \frac{\rho_j(x)}{\sum_{j=1}^N \rho_j(x)}, \quad x \in \Omega.$$

□

Definition 3.4.17. The above $\{\chi_j\}$ is called a *(continuous or C^0-) partition of unity* subordinated to the open covering $\{U_j\}$.

Definition 3.4.18 (Continuous Cousin data). Let $\{U_\lambda\}_{\lambda\in\Lambda}$ be an open covering of Ω. A family $\mathscr{C} = \{(U_\lambda, f_\lambda)\}_{\lambda\in\Lambda}$ of pairs of U_λ and $f_\lambda \in \mathscr{C}^0(U_\lambda)$ is called *Continuous Cousin data* (or a continuous Cousin distribution) if for all $\lambda, \mu \in \Lambda$,

$$f_\lambda - f_\mu \in \mathscr{O}(U_\lambda \cap U_\mu) \quad \text{on } U_\lambda \cap U_\mu.$$

Here, if $U_\lambda \cap U_\mu = \emptyset$, we consider it to hold.

Problem 3.4.19 (Continuous Cousin). Let $\mathscr{C} = \{(U_\lambda, f_\lambda)\}_{\lambda\in\Lambda}$ be continuous Cousin data on Ω. Then, find a continuous function F on Ω, called a *solution of $\mathscr{C}$* such that

(3.4.20) $$F - f_\lambda \in \mathscr{O}(U_\lambda), \quad \forall \lambda \in \Lambda.$$

Remark 3.4.21. (i) In the above Continuous Cousin Problem, according to Proposition 3.4.15 we take a refinement $\{V_\mu\}_{\mu=1}^N$ of $\{U_\lambda\}_{\lambda\in\Lambda}$, which is at most countable and locally finite. Set

$$g_\mu = f_{\lambda(\mu)}|_{V_\mu} \in \mathscr{C}^0(V_\mu), \quad 1 \le \mu \le N.$$

Then $\mathscr{C}' := \{(V_\mu, g_\mu)\}_{\mu=1}^N$ is continuous Cousin data, which is called the *induced continuous Cousin data* of $\mathscr{C}$ by the refinement $\{V_\mu\}$. A solution $G \in \mathscr{C}^0(\Omega)$ of $\mathscr{C}'$ is clearly a solution of $\mathscr{C}$.

(ii) Therefore any Continuous Cousin Problem is reduced to the one with respect to an open covering which is at most countable and locally finite; we assume this henceforth.

Theorem 3.4.22. *If Ω is a holomorphically convex domain (or equivalently a domain of holomorphy), then every Continuous Cousin data on Ω has a solution.*

Proof. Suppose that continuous Cousin data $\mathscr{C} = \{(U_\lambda, f_\lambda)\}_{\lambda\in\Lambda}$ is given on Ω.

(1) Let $\mathrm{P} \Subset \Omega$ be an analytic polyhedron. We first obtain a solution of $\mathscr{C}$ on $\overline{\mathrm{P}}$. In what follows, a solution is that of $\mathscr{C}$. Let

(3.4.23) $$\varphi : \mathrm{P} \longrightarrow \mathrm{P}\Delta \subset \mathbf{C}^N$$

be the Oka map. We assume that φ is defined on a slightly larger analytic polyhedron $\widetilde{\mathrm{P}} \Supset \overline{\mathrm{P}}$ with the corresponding polydisk $\widetilde{\mathrm{P}\Delta} \Supset \mathrm{P}\Delta$:

(3.4.24) $$\varphi : \widetilde{\mathrm{P}} \longrightarrow \widetilde{\mathrm{P}\Delta}.$$

We set $\tilde{S} = \varphi(\widetilde{\mathrm{P}})$. By Riemann's Mapping Theorem 1.4.9 we may assume that $\widetilde{\mathrm{P}\Delta}$ is an open cuboid with edges parallel to coordinate axes (it is the same below). We take a closed cuboid E_0 in $\widetilde{\mathrm{P}\Delta}$, containing $\varphi(\overline{\mathrm{P}})$, and set

$$S = \varphi(\widetilde{\mathrm{P}}) \cap E_0.$$

We identify $\varphi^{-1}S (\supset \overline{\mathrm{P}})$ with S. In order to obtain a solution on S we consider:

Claim 3.4.25. *For an arbitrary closed cuboid $E \subset E_0$ there are a neighborhood W $(\supset E \cap S)$ in $\tilde{S}$ and a continuous function G on W (cf. Fig. 3.2) such that*

$$G - f_\lambda|_{E\cap S\cap U_\lambda} \in \mathscr{O}(E \cap S \cap U_\lambda), \quad \forall \lambda \in \Lambda;$$

that is, $G|_{W\cap U_\lambda} - f_\lambda|_{W\cap U_\lambda} \in \mathscr{O}(W \cap U_\lambda)$ for every $\lambda \in \Lambda$.

$\because$) For the proof we modify the induction on the cuboid dimension $\dim E$ used in §2.5.

(a) The case of $\dim E = 0$. Let $E = \{a\}$. If $a \notin S$, then $E \cap S = \emptyset$ and the claim holds. If $a \in S$, there is a $U_\lambda \ni a$, on which we set $G = f_\lambda$.

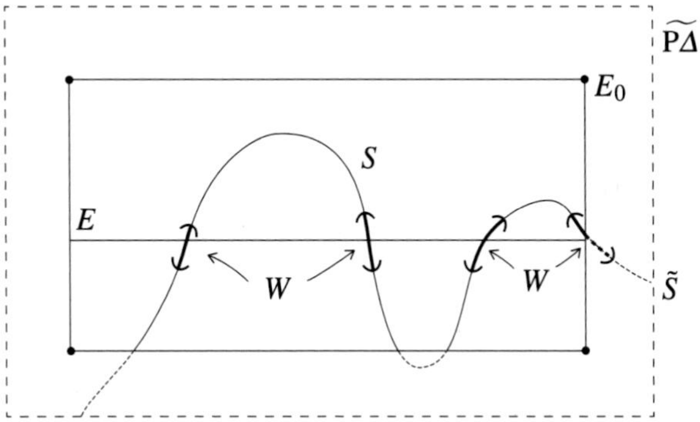

Fig. 3.2 $E \cap S \subset W \subset \tilde{S}$.

(b) Suppose that the case of $\dim E = \nu - 1$ ($\nu \geq 1$) holds. Let $\dim E = \nu$.
As in (2.5.3), E may be assumed to be of the form:

(3.4.26)
$$E = F \times \{z_N : 0 \leq \Re z_N \leq T, |\Im z_N| \leq \theta\},\quad T > 0, \quad \theta \geq 0.$$

For every $t \in [0, T]$

$$E_t := F \times \{z_N : \Re z_N = t, |\Im z_N| \leq \theta\}$$

is a closed cuboid of dimension $\nu - 1$. By the induction hypothesis there is a solution on $E_t \cap S$. By the Heine–Borel Theorem there is a finite partition

(3.4.27)
$$0 = t_0 < t_1 < \cdots < t_L = T$$

such that with

$$E_\alpha = F \times \{z_N : t_{\alpha-1} \leq \Re z_N \leq t_\alpha, |\Im z_N| \leq \theta\}, \quad 1 \leq \alpha \leq L,$$

either $E_\alpha \cap S = \emptyset$ or there is a solution G_α on $E_\alpha \cap S$ $(\neq \emptyset)$.

It is sufficient to consider those E_α with $E_\alpha \cap S \neq \emptyset$. We say that the adjacent E_α and $E_{\alpha+1}$ is connected through S if $E_\alpha \cap E_{\alpha+1} \cap S \neq \emptyset$. We take a maximal sequence of E_α mutually connected through S:

$$E' = E_{\alpha_0} \cup E_{\alpha_0+1} \cup \cdots \cup E_{\alpha_1}.$$

We are going to construct a solution G' on $E' \cap S (\subset \tilde{S})$. For the simplicity of indices we assume $\alpha_0 = 1$. By the choices of G_α $(1 \leq \alpha \leq \alpha_1)$ we see that

(3.4.28)
$$h := G_1|_{E_1 \cap E_2 \cap S} - G_2|_{E_1 \cap E_2 \cap S} \in \mathscr{O}(E_1 \cap E_2 \cap S).$$

With small $\delta > 0$ and a small open cuboid neighborhood $V' \subset \mathbf{C}^{N-1}$ of F we have

$$V := V' \times \{z_N : t_1 - \delta < \Re z_N < t_1 + \delta, |\Im z_N| < \theta + \delta\},$$
$$h \in \mathcal{O}(V \cap \tilde{S}).$$

Note that V is biholomorphic to a polydisk and $V \cap \tilde{S}$ is a submanifold of V. By the Oka Extension Theorem 2.5.14 there is a function $H \in \mathcal{O}(V)$ such that $H|_{V\cap\tilde{S}} = h$ (cf. Fig. 3.3). We consider the Cousin integral of H along the oriented line segment

$$\ell = \left\{ z_N : \Re z_N = t_1,\ -\theta - \frac{\delta}{2} \le \Im z_N \le \theta + \frac{\delta}{2} \right\}$$

and obtain the Cousin decomposition of H:

$$H = H_1 - H_2 \quad (\text{on } E_1 \cap E_2), \quad H_1 \in \mathcal{O}(E_1),\ H_2 \in \mathcal{O}(E_2).$$

This together with (3.4.28) implies

$$(G_1 - H_1|_{E_1\cap S})|_{E_1\cap E_2\cap S} = (G_2 - H_2|_{E_2\cap S})|_{E_1\cap E_2\cap S}. \tag{3.4.29}$$

The left-hand side of the above equation is a solution on $E_1 \cap S$, and the right-hand side is a solution on $E_2 \cap S$; hence we obtain a solution G'_2 on $(E_1 \cup E_2) \cap S$. We call G'_2 the merged solution of the solution G_1 (on $E_1 \cap S$) and G_2 (on $E_2 \cap S$).

We then merge G'_2 and the solution G_3 on $E_3 \cap S$ to obtain a solution G'_3 on $(E_1 \cup E_2 \cup E_3) \cap S$. Repeating this, we obtain a solution $G' := G'_{\alpha_1}$ on $(E_1 \cup \cdots \cup E_{\alpha_1}) \cap S = E' \cap S$. △

(2) We obtain a solution on Ω. It follows from Proposition 3.3.19 that there is an increasing sequence of analytic polyhedral domains

$$\mathrm{P}_1 \Subset \mathrm{P}_2 \Subset \cdots \Subset \mathrm{P}_\mu \Subset \cdots, \qquad \bigcup_{\mu=1}^{\infty} \mathrm{P}_\mu = \Omega.$$

By the result of the above (1) we have a solution G_μ on each $\overline{\mathrm{P}}_\mu$. Note that

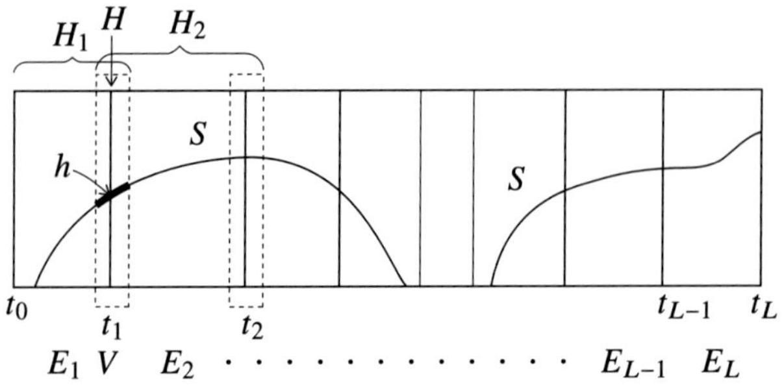

Fig. 3.3 $E_1, \ldots, E_L$.

$$(G_{\mu+1} - G_\mu)|_{\overline{\mathrm{P}}_\mu} \in \mathscr{O}(\overline{\mathrm{P}}_\mu).$$

Set $F_1 = G_1$. Since $(G_2 - F_1)|_{\overline{\mathrm{P}}_1} \in \mathscr{O}(\overline{\mathrm{P}}_1)$, the Oka–Weil Approximation Theorem 3.3.12 implies the existence of a function $h_2 \in \mathscr{O}(\Omega)$ such that

$$\|G_2 - F_1 - h_2\|_{\overline{\mathrm{P}}_1} < \frac{1}{2}.$$

We set $F_2 = G_2 - h_2$, which is a solution on $\overline{\mathrm{P}}_2$ and satisfies

$$\|F_2 - F_1\|_{\overline{\mathrm{P}}_1} < \frac{1}{2}.$$

Repeating this inductively, we construct a solution F_μ on $\overline{\mathrm{P}}_\mu$ such that

$$\|F_\mu - F_{\mu-1}\|_{\overline{\mathrm{P}}_{\mu-1}} < \frac{1}{2^{\mu-1}}, \quad \mu = 2, 3, \ldots.$$

We finally set

$$\begin{aligned} F &= F_1 + \sum_{\mu=1}^{\infty} (F_{\mu+1} - F_\mu) \\ &= F_\nu + \sum_{\mu=\nu}^{\infty} (F_{\mu+1} - F_\mu), \quad \forall \nu \geq 1. \end{aligned} \tag{3.4.30}$$

The second term of the right-hand side of (3.4.30) is a series of holomorphic functions on P_ν, and converges uniformly on P_ν (in fact, by majorant series) to a holomorphic function on P_ν. Therefore, F is a solution on every P_ν, and hence on Ω. □

3.4.3 Cousin I Problem — continued

Now we consider Cousin I Problem 3.4.6. We keep the notation used there. It is noticed that the covering $\{U_\lambda\}_{\lambda \in \Lambda}$ of the Cousin I data $\mathscr{C}_I = \{(U_\lambda, f_\lambda)\}_{\lambda \in \Lambda}$ may be assumed by the same reasoning as in Remark 3.4.21 to be at most countable and locally finite. Let

$$\Lambda = \{1, 2, \ldots, M\}, \quad M \leq \infty.$$

By Proposition 3.4.16 there is a partition of unity $\{\chi_\nu\}_{\nu=1}^{M}$ subordinated to $\{U_\nu\}_{\nu=1}^{M}$. We set

$$g_\nu = \sum_{\lambda=1}^{M} \chi_\lambda (f_\nu - f_\lambda).$$

Here, if $U_\nu \cap U_\lambda \neq \emptyset$, $f_\nu - f_\lambda \in \mathscr{O}(U_\nu \cap U_\lambda)$, so that $\chi_\lambda (f_\nu - f_\lambda)$ is extended as 0 in $U_\nu \setminus U_\lambda$ to be a continuous function on U_ν. By convention, we set $f_\nu - f_\lambda = 0$ when

$U_\nu \cap U_\lambda = \emptyset$. We have $g_\nu \in \mathscr{C}^0(U_\nu)$. If $U_\nu \cap U_\mu \neq \emptyset$,

$$\begin{aligned} g_\nu - g_\mu &= \sum_{\lambda=1}^{M} \chi_\lambda (f_\nu - f_\lambda) - \sum_{\lambda=1}^{M} \chi_\lambda (f_\mu - f_\lambda) \\ &= \sum_{\lambda=1}^{M} \chi_\lambda (f_\nu - f_\mu) = f_\nu - f_\mu \in \mathscr{O}(U_\nu \cap U_\mu). \end{aligned}$$

Thus $\{(U_\nu, g_\nu)\}_{\nu=1}^M$ is continuous Cousin data. If Ω is a holomorphically convex domain, it follows from Theorem 3.4.22 that there is a solution G of it:

$$g_\nu - G - (g_\mu - G) = f_\nu - f_\mu;$$

with $h_\nu = g_\nu - G \in \mathscr{O}(U_\nu)$ $(1 \le \nu \le M)$ we obtain

$$f_\nu - h_\nu = f_\mu - h_\mu.$$

Therefore, F defined by $F := f_\nu - h_\nu$ on each U_ν, gives a solution of $\mathscr{C}_I$ on Ω.

Theorem 3.4.31 (Oka). *If Ω is a holomorphically convex domain (or equivalently a domain of holomorphy), every Cousin I data on Ω has a solution.*

3.4.4 Hartogs Extension over a Compact Subset

Applying the solution of the Continuous Cousin Problem (on $\mathbf{C}^n$), we generalize Theorem 1.2.8 for domains of $\mathbf{C}^n$ to the following famous theorem of F. Hartogs, which is sometimes referred to as the Hartogs–Osgood Theorem, for the case of bounded holomorphic functions was obtained by W.F. Osgood.

Theorem 3.4.32 (Hartogs). *Let $n \ge 2$ and let $\Omega \subset \mathbf{C}^n$ be a domain.*

(i) *Let $K \Subset \Omega$ be a compact subset such that $\Omega \setminus K$ is connected. Then every holomorphic function f in $\Omega \setminus K$ extends uniquely to a holomorphic function in Ω.*

(ii) *Assume that Ω is a bounded domain of holomorphy. Then the boundary $\partial\Omega$ is connected.*

Proof. (i) Let $\Omega \Supset K$ be as given, and let $f \in \mathscr{O}(\Omega \setminus K)$. We set an open covering $\{U_0, U_1\}$ of $\mathbf{C}^n$ so that $U_0 = \Omega$ and $U_1 = \mathbf{C}^n \setminus K$. Note that $U_0 \cap U_1 = \Omega \setminus K$ and U_1 is connected. Let $\{\rho_0, \rho_1\} \subset \mathscr{C}^0(\mathbf{C}^n)$ be a partition of unity subordinated to $\{U_0, U_1\}$ (see Proposition 3.4.16). In U_0 we set $g_0 = \rho_1 f$ $(= 0$ in $U_0 \setminus U_1)$, and in U_1, $g_1 = -\rho_0 f$ $(= 0$ in $U_1 \setminus U_0)$. Then $\{(U_j, g_j)\}_{j=0,1}$ is continuous Cousin data on $\mathbf{C}^n$ such that

$$g_0 - g_1 = (\rho_0 + \rho_1) f = f \in \mathscr{O}(U_0 \cap U_1). \tag{3.4.33}$$

By Theorem 3.4.22 there is a continuous function G on $\mathbf{C}^n$ such that $G - g_j$ are holomorphic in U_j. Set $f_j = g_j - G \in \mathcal{O}(U_j)$. It follows from (3.4.33) that

$$f = f_0 - f_1. \tag{3.4.34}$$

By Theorem 1.2.8, f_1 extends uniquely and holomorphically on $\mathbf{C}^n$. Hence (3.4.34) implies that f extends uniquely and holomorphically on Ω.

(ii) Set $U = \mathbf{C}^n \setminus \bar{\Omega}$, and take a large open ball $\mathrm{B}(R) \supset \bar{\Omega}$. Then there is a connected component U_0 of U such that $U_0 \supset \mathbf{C}^n \setminus \bar{\mathrm{B}}(R)$.

We show that $U_0 = U$. Otherwise, there is a connected component ω of U other than U_0. Since $\omega \subset B(R)$, ω is bounded, and $\partial\omega \subset \partial\Omega$. Then $\bar{\omega}$ is compact and $\Omega' := \Omega \cup \bar{\omega}$ is a domain such that $\Omega' \setminus \bar{\omega} = \Omega$ is connected. Since Ω is a domain of holomorphy, it follows from (i) shown above that $\Omega' = \Omega$; this is a contradiction. Therefore, U is connected and $\partial U = \partial\Omega$. It is noticed that $\Gamma := \partial\Omega$ is compact and contains no isolated point.

We next show the connectedness of Γ. If otherwise, there would be a disjoint decomposition $\Gamma = \Gamma_1 \sqcup \Gamma_2$ with two compact subsets Γ_j $(j = 1, 2)$ such that $\Gamma_1 \cap \Gamma_2 = \emptyset$. Let $a_j \in \Gamma_j$ $(j = 1, 2)$ and take a small ball neighborhood $V_j = \mathrm{B}(a_j; s_j)$ $(s_j > 0)$ such that $V_1 \cap \Gamma_2 = \emptyset$ and $V_2 \cap \Gamma_1 = \emptyset$. We then choose points $b_j \in V_j \cap U$ and $c_j \in V_j \cap \Omega$ $(j = 1, 2)$. Let C_1 be an oriented curve in U from b_1 to b_2, and let C_2 be an oriented curve in Ω from c_2 to c_1. Let $L_{b_2c_2}$ (resp. $L_{c_1b_1}$) denote the oriented line segment form b_2 (resp. c_1) to c_2 (resp. b_1) in V_2 (resp. V_1). We obtain an oriented closed curve in $\mathbf{C}^n$,

$$C = C_1 + L_{b_2c_2} + C_2 + L_{c_1b_1}.$$

We define a continuous function

$$\tau(z) = \frac{d(z, \Gamma_1)}{d(z, \Gamma_1) + d(z, \Gamma_2)}, \qquad z \in \mathbf{C}^n \setminus (\Gamma_1 \sqcup \Gamma_2) = U \sqcup \Omega.$$

We set

$$\phi(z) = \begin{cases} e^{i\tau(z)\pi} \in S^1 := \{w \in \mathbf{C} : |w| = 1\}, & z \in U, \\ e^{i(2-\tau(z))\pi} \in S^1, & z \in \Omega. \end{cases} \tag{3.4.35}$$

With $\phi := 1$ (resp. -1) on Γ_1 (resp. Γ_2), $\phi(z)$ extends continuously to $\phi : \mathbf{C}^n \to S^1$. The image $\phi(C)$ gives rise to a homotopically non-zero cycle on S^1, but C is homotopically 0 ($\mathbf{C}^n$ is contractible to a point): This is a contradiction. □

Remark 3.4.36. (i) The boundedness assumption in (ii) above cannot be removed. For example, let $\Omega = (\Delta(R_1) \setminus \bar{\Delta}(r_1)) \times \mathbf{C} \subset \mathbf{C}^2$ with $0 < r_1 < R_1$. Then Ω is a domain of holomorphy and $\partial\Omega = (\{|z_1| = r_1\} \times \mathbf{C}) \sqcup (\{|z_1| = R_1\} \times \mathbf{C})$ consists of two connected components.

If we take $\Omega = (\Delta(R_1) \setminus \bar{\Delta}(r_1)) \times \Delta(R_2) \Subset \mathbf{C}^2$, which is a bounded domain of holomorphy, then $\partial\Omega = (\{|z_1| = r_1\} \times \bar{\Delta}(R_2)) \cup (\{|z_1| = R_1\} \times \bar{\Delta}(R_2)) \cup ((\bar{\Delta}(R_1) \setminus \Delta(r_1) \times \{|z_2| = R_2\})$, which is connected.

(ii) Because of Hartogs' domain Ω_{H} (see (1.2.11)), the converse of (ii) of the above theorem does not hold clearly.
(iii) Theorem 3.4.32 says that the shape of the boundary of a bounded domain of holomorphy in $\mathbf{C}^n$ with $n \geq 2$ is not arbitrary at all, contrary to the case of one variable. This fact motivated significantly the study of complex analytic functions of several variables.
(iv) The proof with ϕ in (3.4.35) is due to Czarnecki–Kulczycki–Lubawski [13].

3.4.5 Mittag-Leffler Theorem (One Variable)

In the present subsection we assume $n = 1$. Let $f(z)$ be a meromorphic function defined about $a \in \mathbf{C}$ with Laurent series expansion,

(3.4.37)
$$f(z) = \frac{c_{-m}}{(z-a)^m} + \frac{c_{-m+1}}{(z-a)^{m-1}} + \cdots + \frac{c_{-1}}{z-a} + c_0 + c_1(z-a) + \cdots \quad (m \in \mathbf{N}).$$

Then the term

(3.4.38)
$$Q(z) := \frac{c_{-m}}{(z-a)^m} + \frac{c_{-m+1}}{(z-a)^{m-1}} + \cdots + \frac{c_{-1}}{z-a}$$

is called the *main part* of $f(z)$ at a.

Theorem 3.4.39 (Mittag-Leffler). *Let $\{a_\nu\}_{\nu=1}^{\infty}$ be a discrete closed subset of a domain Ω ($\subset \mathbf{C}$) with distinct a_ν, at each of which a rational function*

(3.4.40)
$$Q_\nu(z) = \frac{c_{(\nu)-m_\nu}}{(z-a_\nu)^{m_\nu}} + \frac{c_{(\nu)-m_\nu+1}}{(z-a_\nu)^{m_\nu-1}} + \cdots + \frac{c_{(\nu)-1}}{z-a_\nu}$$

is given. Then there is a meromorphic function $f(z)$ in Ω such that $f(z)$ has the main part $Q_\nu(z)$ at each a_ν, and has no other pole.

Proof. We take a neighborhood $U_\nu \subset \Omega$ of each a_ν so that $U_\nu \cap U_\mu = \emptyset$ for $\nu \neq \mu$. With $U_0 := \Omega \setminus \{a_\nu\}_{\nu=1}^{\infty}$ we have a locally finite open covering, $\Omega = \bigcup_{\nu=0}^{\infty} U_\nu$, and give Cousin I data on Ω as follows:

$$f_0(z) = 0, \qquad z \in U_0,$$
$$f_\nu(z) = Q_\nu(z),\ z \in U_\nu, \quad \nu = 1, 2 \ldots.$$

Since Ω is a domain of holomorphy, it follows from Theorem 3.4.31 that there is a solution $f \in \mathscr{M}(\Omega)$ for the above Cousin I data. By the construction $f(z)$ has the main part $Q_\nu(z)$ at each a_ν, and has no other pole. □

Historically, Mittag-Leffler Theorem 3.4.39 gave a motivation of the Cousin I Problem, which Cousin solved in the case of cylinder domains.

3.4.6 Cousin II Problem and Oka Principle

Let $U \subset \mathbf{C}^n$ be an open set. We denote by $\mathscr{O}^*(U)$ (resp. $\mathscr{C}^*(U)$) the set of all holomorphic (resp. continuous) functions in U without zeros. We also denote by $\mathscr{M}^*(U)$ the set of all meromorphic functions in U which are nowhere locally 0 (identically).

Problem 3.4.41 (Cousin II). Let $\Omega \subset \mathbf{C}^n$ be a domain. Let $\{U_\lambda\}_{\lambda \in \Lambda}$ be an open covering of Ω and let $f_\lambda \in \mathscr{M}^*(U_\lambda)$ $(\lambda \in \Lambda)$. The family $\mathscr{C}_{II} := \{(U_\lambda, f_\lambda)\}_{\lambda \in \Lambda}$ of the pairs (U_λ, f_λ) is called *Cousin II data* (also, a Cousin II distribution) if

$$\frac{f_\lambda}{f_\mu} \in \mathscr{O}^*(U_\lambda \cap U_\mu), \qquad \forall \lambda, \mu \in \Lambda. \tag{3.4.42}$$

The *Cousin II Problem* is to ask if there exists an element $F \in \mathscr{M}^*(\Omega)$ satisfying

$$\frac{F}{f_\lambda} \in \mathscr{O}^*(U_\lambda), \qquad \forall \lambda \in \Lambda. \tag{3.4.43}$$

The above F, if it exists, is called an "*(analytic) solution*" of $\mathscr{C}_{II}$. $\mathscr{C}_{II}$ is said to be (*analytically*) *solvable* if F exists.

Remark 3.4.44.
(i) By (3.4.42) the polar set of f_λ and that of f_μ coincide on $U_\nu \cap U_\mu$, so that the polar set Z_∞ of $\mathscr{C}_{II}$ is well-defined in Ω: Similarly, the zero set Z_0 of $\mathscr{C}_{II}$ is well-defined in Ω. The both are closed and nowhere dense and f_λ are holomorphic, non-vanishing on open dense subsets $U_\lambda \setminus (Z_0 \cup Z_\infty) \subset U_\lambda$.
(ii) As we will see later (§3.4.7), in the case of one variable ($n = 1$) the Cousin II Problem always holds as Weierstrass' Theorem.

Definition 3.4.45 (Topological solution). Let $\mathscr{C}_{II} = \{(U_\lambda, f_\lambda)\}_{\lambda \in \Lambda}$ be Cousin II data on a domain Ω. A *topological solution* of $\mathscr{C}_{II}$ is a continuous function $G \in \mathscr{C}^*(\Omega')$ defined on an open dense subset $\Omega' \subset \Omega$, such that for every $\lambda \in \Lambda$ the function $\frac{G}{f_\lambda}$ defined on an open dense subset of U_λ continuously extends to a function of $\mathscr{C}^*(U_\lambda)$, for which we write

$$\frac{G}{f_\lambda} \in \mathscr{C}^*(U_\lambda). \tag{3.4.46}$$

If G exists, we say that $\mathscr{C}_{II}$ is *topologically solvable*.

Now, let $\Omega \subset \mathbf{C}^n$ be a domain.

Proposition 3.4.47. *Let $\mathscr{C}_{II} = \{(U_\lambda, f_\lambda)\}_{\lambda \in \Lambda}$ be Cousin II data on Ω. Then the following are equivalent:*

(i) *$\mathscr{C}_{II}$ is analytically (resp. topologically) solvable.*
(ii) *There are $g_\lambda \in \mathscr{O}^*(U_\lambda)$ (resp. $g_\lambda \in \mathscr{C}^*(U_\lambda)$) $(\lambda \in \Lambda)$ such that*

$$f_\lambda \cdot f_\mu^{-1} = g_\mu \cdot g_\lambda^{-1} \quad \text{on } U_\lambda \cap U_\mu (\neq \emptyset), \quad \forall \lambda, \mu \in \Lambda. \tag{3.4.48}$$

Proof. Let $F \in \mathscr{M}^*(\Omega)$ be an analytic solution of $\mathscr{C}_{II}$. Setting

$$g_\lambda := \frac{F|_{U_\lambda}}{f_\lambda} \in \mathscr{O}^*(U_\lambda), \qquad \lambda \in \Lambda,$$

we see that $\{g_\lambda\}_{\lambda \in \Lambda}$ satisfies (3.4.48).

Conversely, we assume the existence of $\{g_\lambda\}_{\lambda \in \Lambda}$ satisfying (3.4.48). Then

$$f_\lambda \cdot g_\lambda = f_\mu \cdot g_\mu \quad \text{on } U_\lambda \cap U_\mu (\neq \emptyset), \quad \forall \lambda, \mu \in \Lambda. \tag{3.4.49}$$

Defining $F := f_\lambda \cdot g_\lambda$ on each U_λ, we obtain an analytic solution $F \in \mathscr{M}^*(\Omega)$ of $\mathscr{C}_{II}$.

The case of the topological solution is just similar to the above. □

Remark 3.4.50. Let $\{V_\mu\}$ be an open covering of Ω which is a refinement of $\{U_\lambda\}$. Then the above $\mathscr{C}_{II}$ is analytically (resp. topologically) solvable if and only if the Cousin II data naturally induced from $\mathscr{C}_{II}$ (cf. Remark 3.4.21) is analytically (resp. topologically) solvable.

It is, of course, necessary for $\mathscr{C}_{II}$ to be analytically solvable for it to be topologically solvable. The next theorem asserts the sufficiency if the domain is of holomorphy or equivalently holomorphically convex.

Theorem 3.4.51 (Oka Principle). *On a holomorphically convex domain (or equivalently, a domain of holomorphy), Cousin II data has an analytic solution if and only if it has a topological solution.*

Proof. It suffices to show the "if" part. Let $\mathscr{C}_{II} = \{(U_\lambda, f_\lambda)\}_{\lambda \in \Lambda}$ be topologically solvable Cousin II data on a holomorphically convex domain Ω. Then there are $\varphi_\lambda \in \mathscr{C}^*(U_\lambda)$ $(\lambda \in \Lambda)$ satisfying (3.4.48). Every point $a \in U_\lambda$ has a simply connected neighborhood (e.g., a polydisk neighborhood); hence, taking a refinement of $\{U_\lambda\}$, we may assume without loss of generality that all U_λ are simply connected (cf. Remark 3.4.50).

In each U_λ we define a continuous function by $\psi_\lambda := \log \varphi_\lambda$ (1-valued branch). On $U_\lambda \cap U_\mu \neq \emptyset$ we see that

$$\psi_\lambda - \psi_\mu = \log \frac{\varphi_\lambda}{\varphi_\mu} = \log \frac{f_\mu}{f_\lambda} \in \mathscr{O}(U_\lambda \cap U_\mu).$$

Therefore $\{(U_\lambda, \psi_\lambda)\}$ is continuous Cousin data. By Theorem 3.4.22 there is a solution Ψ of it. It follows that

$$\log \frac{f_\mu}{f_\lambda} = \psi_\lambda - \Psi - (\psi_\mu - \Psi),$$

$$h_\lambda := \psi_\lambda - \Psi \in \mathscr{O}(U_\lambda),$$

$$\frac{f_\mu}{f_\lambda} = \frac{e^{h_\lambda}}{e^{h_\mu}}, \quad f_\lambda e^{h_\lambda} = f_\mu e^{h_\mu}.$$

By setting $F = f_\lambda e^{h_\lambda}$ on each U_λ, we obtain an analytic solution $F \in \mathscr{M}^*(\Omega)$ of $\mathscr{C}_{II}$. □

Example 3.4.52 (Non-solvable example 1). Similarly to the case of the Cousin I Problem (see Remark 3.4.10), there is an example of non-solvable Cousin II data on a Hartogs domain. Let $\Omega_{\mathrm{H}} = \Omega_1 \cup \Omega_2$ be a domain defined by (3.4.11). We define Cousin II data by setting $f_1 = 1$ on Ω_1 and $f_2 = z - w$ on Ω_2. Suppose that it has a solution $F \in \mathscr{O}(\Omega_{\mathrm{H}})$. By Theorem 1.2.12 F extends to $F \in \mathscr{O}(\mathrm{P}\Delta((3,3)))$. Restricting F to $z = w$, we have a holomorphic function $g(z) = F(z,z)$ in $\Delta(3)$ such that $g(z) \neq 0$ in $|z| < 1$, but $g(z) = 0$ in $2 < |z| < 3$; this contradicts the Identity Theorem 1.1.46.

Example 3.4.53 (Non-solvable example 2). When $n \geq 2$, even if Ω is a domain of holomorphy, there is non-solvable Cousin II data. We recall such a counter-example due to Oka, following Nishino [37] §3.5.

In the space $(z,w) \in \mathbf{C}^2$ we consider a cylinder domain given by

$$\Omega = \left\{ (z,w) : \frac{2}{3} < |z| < 1,\ \frac{2}{3} < |w| < 1 \right\};$$

Ω is a domain of holomorphy. Let $f(z,w) = w - z + 1$ and set a complex hypersurface $S = \{f(z,w) = 0\}$ $(\subset \mathbf{C}^2)$. It follows that

$$S \cap \{(z,w) \in \bar{\Omega} : \Im z = 0\} = \emptyset.$$

For, if $(z,w) \in \mathbf{R}^2$ of the set of the above left-hand side exists, then it is deduced from $w = z - 1$ and $z \leq 1$ that $w \leq 0$; hence, $-1 \leq w \leq -2/3$, i.e.,

$$0 \leq z \leq \frac{1}{3}.$$

This is a contradiction. Therefore, for a small $\delta > 0$,

$$S \cap \{(z,w) \in \bar{\Omega} : |\Im z| \leq \delta\} = \emptyset.$$

Set as follows (cf. Fig. 3.4):

$$\begin{aligned}
U_1 &= \{(z,w) \in \Omega : \Im z > -\delta\}, \\
U_2 &= \{(z,w) \in \Omega : \Im z < \delta\}, \\
\Omega &= U_1 \cup U_2, \\
S_1 &= S \cap U_1.
\end{aligned}$$

Then S_1 is a complex hypersurface of Ω. We take Cousin II data $\mathscr{C}_{II}$ with zero-set S_1 defined by

$$\begin{aligned}
f_1(z,w) &= f(z,w) = w - z + 1, & (z,w) &\in U_1, \\
f_2(z,w) &= 1, & (z,w) &\in U_2.
\end{aligned}$$

Claim 3.4.54. *$\mathscr{C}_{II}$ has no solution.*

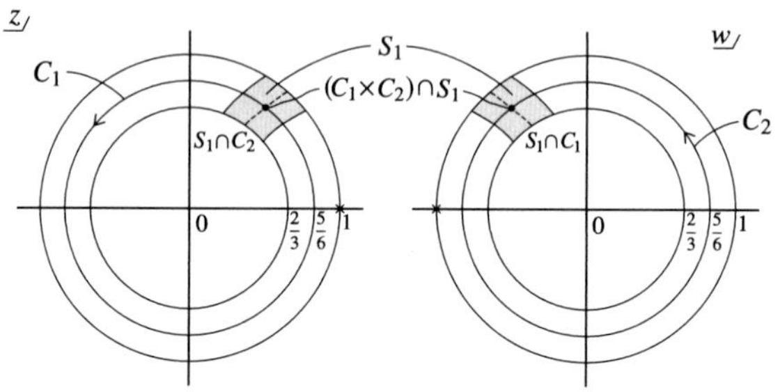

Fig. 3.4 Non-solvable Cousin II data $S_1 = \{w = z - 1\} \cap U_1 \subset \Omega$.

$\because$) Suppose that $\mathscr{C}_{II}$ has an analytic solution $F(z,w) \in \mathscr{O}(\Omega)$. We take circles with anti-clockwise orientation in the z-plane and the w-plane as follows:

$$C_1 = \left\{|z| = \frac{5}{6}\right\}, \quad C_2 = \left\{|w| = \frac{5}{6}\right\}.$$

We then see that (cf. Fig. 3.4)

$$C_1 \times C_2 \subset \Omega, \quad (C_1 \times C_2) \cap S_1 = \left\{\left(\frac{1}{2} + \frac{2}{3}i, -\frac{1}{2} + \frac{2}{3}i\right)\right\}.$$

For $z \in C_1$ with $(\{z\} \times C_2) \cap S_1 = \emptyset$ we set

$$\Theta(z) = \int_{w \in C_2} \partial_w \arg F(z,w)dw = \frac{1}{i}\int_{w \in C_2} \frac{\partial_w F(z,w)}{F(z,w)} dw \in 2\pi\mathbf{Z},$$

where $\partial_w = \partial / \partial w$. On U_1 we can write

$$F(z,w) = (w - z + 1)\phi(z,w), \quad \phi \in \mathscr{O}^*(U_1).$$

Therefore on U_1 we have

$$\Theta(z) = \int_{w \in C_2} \partial_w \arg(w - z + 1)dw + \int_{w \in C_2} \partial_w \arg \phi(z,w)dw,$$

$$\tau(z) := \int_{w \in C_2} \partial_w \arg \phi(z,w)dw.$$

We first consider $\tau(z)$. As z moves on $C_1 \cap \{\Im z \geq 0\}$ from $\frac{5}{6}$ to $-\frac{5}{6}$, $\tau(z)$ moves continuously; the value is in $2\pi\mathbf{Z}$. Hence, $\tau(z) = \tau_0$ must be a constant. On the other hand, we see that

$$\int_{w \in C_2} \partial_w \arg\left(w - \frac{5}{6} + 1\right) dw = 2\pi,$$

$$\int_{w\in C_2} \partial_w \arg\left(w+\frac{5}{6}+1\right) dw = 0.$$

It follows that

$$\Theta\left(\frac{5}{6}\right) = 2\pi + \tau_0, \quad \Theta\left(-\frac{5}{6}\right) = \tau_0,$$

(3.4.55) $$\Theta\left(\frac{5}{6}\right) - \Theta\left(-\frac{5}{6}\right) = 2\pi.$$

Now, as z moves on $C_1 \cap \{\Im z \leq 0\}$ from $\frac{5}{6}$ to $-\frac{5}{6}$, $F(z,w)$ has no zero there. Similarly to the arguments on $\tau(z)$, $\Theta(z)$ is a constant. Then

$$\Theta\left(\frac{5}{6}\right) - \Theta\left(-\frac{5}{6}\right) = 0 :$$

This contradicts (3.4.55) △

3.4.7 Weierstrass' Theorem (One Variable)

In the present subsection we assume $n = 1$. Let $\Omega \subset \mathbf{C}$ be a domain.

Let $\Gamma \subset \Omega$ be a closed discrete subset; i.e., it is discrete and has no accumulation point in Ω. Let $\{m_a \in \mathbf{Z} \setminus \{0\} : a \in \Gamma\}$ be given arbitrarily. We consider a problem to construct a meromorphic function on Ω such that it has a zero (in the case $m_a > 0$) or a pole (in the case $m_a < 0$) of order $|m_a|$ at every $a \in \Gamma$ and has no other zero or pole.

If Γ is finite, it suffices to set

$$f(z) = \prod_{a\in\Gamma} (z-a)^{m_a}.$$

The case of infinite Γ is the problem.

Theorem 3.4.56 (Weierstrass). *Let $\Omega \subset \mathbf{C}$ be a domain. Let $\{a_\nu\}_{\nu=1}^{\infty} \subset \Omega$ be a closed discrete subset with distinct a_ν and let $\{m_\nu\}_{\nu=1}^{\infty}$ be an arbitrary sequence in $\mathbf{Z} \setminus \{0\}$. Then there is a meromorphic function on Ω which has a zero (in the case $m_\nu > 0$) or a pole (in the case $m_\nu < 0$) of order $|m_\nu|$ at every a_ν and has no zero nor pole at other points.*

Proof. Just for a convenience, we assume by a parallel transform

(3.4.57) $$0 \in \Omega.$$

We take a sequence of subdomains P_μ, $\mu = 1, 2, \ldots$, such that $\mathrm{P}_\mu \Subset \mathrm{P}_{\mu+1}$ and $\Omega \setminus \bar{\mathrm{P}}_\mu$ has no relatively compact connected component in Ω. Set

$$Q_\mu(z) = \prod_{a_\nu \in \overline{\mathrm{P}}_\mu} (z - a_\nu)^{m_\nu}, \quad \mu = 1, 2, \ldots$$

We are going to modify $Q_\mu(z)$ to obtain a topological solution of the Cousin II Problem with the given zeros and poles.

$\mu = 1$: We set $F_1(z) = Q_1(z)$.

$\mu = 2$: We first consider

$$\prod_{a_\nu \in \overline{\mathrm{P}}_2 \setminus \overline{\mathrm{P}}_1} (z - a_\nu)^{m_\nu}. \tag{3.4.58}$$

For each $a_\nu \in \overline{\mathrm{P}}_2 \setminus \overline{\mathrm{P}}_1$ we take the connected component W_ν of $\Omega \setminus \overline{\mathrm{P}}_1$ containing a_ν. In the Riemann sphere $\hat{\mathbf{C}}$ we consider the boundary $\partial W_\nu = (\partial W_\nu \cap \partial \Omega) \cup (\partial W_\nu \cap \partial \mathrm{P}_1)$. Note that

$$(\partial W_\nu \cap \partial \Omega) \cap (\partial W_\nu \cap \partial \mathrm{P}_1) = \emptyset.$$

Since W_ν is not relatively compact in Ω,

$$\partial W_\nu \cap \partial \Omega \neq \emptyset.$$

Therefore there is a point $b_\nu \in \partial W_\nu \cap \partial \Omega$ such that there is a piecewise linear Jordan curve C_ν connecting a_ν and b_ν, and satisfying $C_\nu \setminus \{b_\nu\} \subset W_\nu$ (cf. Fig. 3.5).

When $b_\nu = \infty$, a line segment connecting a point c_ν of W_ν and ∞ is a half-line in $\mathbf{C}$ with one end at c_ν. Note that $b_\nu \neq 0$ by (3.4.57). Since $\hat{\mathbf{C}} \setminus C_\nu$ is simply connected, we may take a one-valued branch of

$$\log \frac{z - a_\nu}{-\frac{z}{b_\nu} + 1}, \quad z \in \hat{\mathbf{C}} \setminus C_\nu, \tag{3.4.59}$$

where, in the case $b_\nu = \infty$, $-z/b_\nu = 0$. We then set

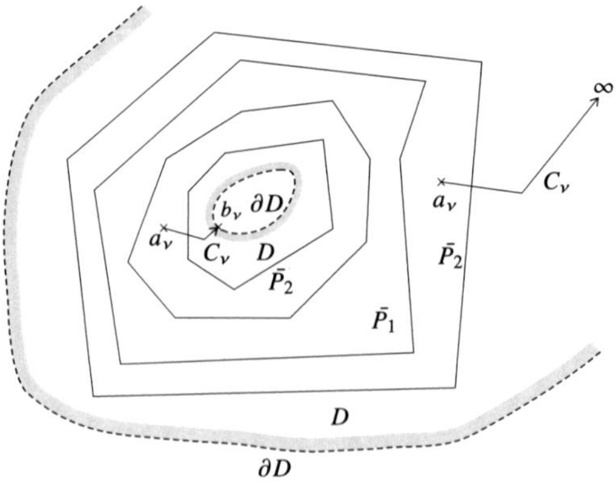

Fig. 3.5 Jordan curve C_ν.

$$G_2(z) = F_1(z) \prod_{a_\nu \in \overline{\mathrm{P}}_2 \setminus \overline{\mathrm{P}}_1} \left(\frac{z - a_\nu}{-\frac{z}{b_\nu} + 1} \right)^{m_\nu} \in \mathscr{M}(\Omega). \tag{3.4.60}$$

The zeros and poles of $G_2(z)$ in Ω are the same as those of $Q_2(z)$ in Ω. Let U_1 be a neighborhood of $\overline{\mathrm{P}}_1$ sufficiently close to $\overline{\mathrm{P}}_1$ such that

$$\overline{\mathrm{P}}_1 \Subset U_1 \Subset \mathrm{P}_2, \quad U_1 \cap \left(\bigcup_{a_\nu \in \overline{\mathrm{P}}_2 \setminus \overline{\mathrm{P}}_1} C_\nu \right) = \emptyset.$$

On U_1 we have by (3.4.59) that

$$\begin{aligned} G_2(z) &= F_1(z) e^{\sum_{a_\nu \in \overline{\mathrm{P}}_2 \setminus \overline{\mathrm{P}}_1} m_\nu \log \frac{z - a_\nu}{-\frac{z}{b_\nu} + 1}} = F_1(z) e^{h_2(z)}, \\ h_2(z) &:= \sum_{a_\nu \in \overline{\mathrm{P}}_2 \setminus \overline{\mathrm{P}}_1} m_\nu \log \frac{z - a_\nu}{-\frac{z}{b_\nu} + 1} \in \mathscr{O}(U_1). \end{aligned} \tag{3.4.61}$$

Let $\lambda_2(z) \in \mathscr{C}^0(\Omega)$ be such that $\lambda_2(z) = 0$ on $\overline{\mathrm{P}}_1$ and $\lambda_2(z) = 1$ on $\Omega \setminus U_1$ (cf. Exercise 7 at the end of this chapter). Set

$$F_2(z) = \begin{cases} F_1(z) e^{\lambda_2(z) h_2(z)}, & z \in U_1, \\ G_2(z), & z \in \overline{\mathrm{P}}_2 \setminus U_1. \end{cases}$$

Then $F_2/Q_2 \in \mathscr{C}^*(\overline{\mathrm{P}}_2)$; that is, $F_2(z)$ is a topological solution of the Cousin II Problem on $\overline{\mathrm{P}}_2$.

For $a_\nu \in \overline{\mathrm{P}}_3 \setminus \overline{\mathrm{P}}_2$ we take $b_\nu \in \partial\Omega$ in the same way as above, and set

$$G_3(z) = F_2(z) \prod_{a_\nu \in \overline{\mathrm{P}}_3 \setminus \overline{\mathrm{P}}_2} \left(\frac{z - a_\nu}{-\frac{z}{b_\nu} + 1} \right)^{m_\nu}. \tag{3.4.62}$$

The zeros and poles of $G_3(z)$ are the same as those of $Q_3(z)$. Let $\overline{\mathrm{P}}_2 \Subset U_2 \Subset \mathrm{P}_3$ be as in the case of U_1. On U_2 we have

$$\begin{aligned} G_3(z) &= F_2(z) e^{\sum_{a_\nu \in \overline{\mathrm{P}}_3 \setminus \overline{\mathrm{P}}_2} m_\nu \log \frac{z - a_\nu}{-\frac{z}{b_\nu} + 1}} = F_2(z) e^{h_3(z)}, \\ h_3(z) &:= \sum_{a_\nu \in \overline{\mathrm{P}}_3 \setminus \overline{\mathrm{P}}_2} m_\nu \log \frac{z - a_\nu}{-\frac{z}{b_\nu} + 1} \in \mathscr{O}(U_2). \end{aligned} \tag{3.4.63}$$

Let $\lambda_3(z) \in \mathscr{C}^0(\Omega)$ be such that $\lambda_3(z) = 0$ on $\overline{\mathrm{P}}_2$ and $\lambda_3(z) = 1$ on $\Omega \setminus U_2$. Set

$$F_3(z) = \begin{cases} F_2(z) e^{\lambda_3(z) h_3(z)}, & z \in U_2, \\ G_3(z), & z \in \overline{\mathrm{P}}_3 \setminus U_2. \end{cases}$$

Then $F_3(z)$ is a topological solution of the Cousin II Problem on $\overline{\mathrm{P}}_3$ such that $F_3|_{\overline{\mathrm{P}}_2} = F_2$ on $\overline{\mathrm{P}}_2$.

Inductively, we construct a topological solution F_μ of the Cousin II Problem on $\overline{\mathrm{P}}_\mu$ ($\mu = 1,2,3,\ldots$) such that $F_{\mu+1}|_{\overline{\mathrm{P}}_\mu} = F_\mu$. Therefore we have a topological solution $F(z) = \lim_{\mu\to\infty} F_\mu(z)$ of the Cousin II Problem on Ω.

Since Ω is holomorphically convex, by Oka's Theorem 3.4.51 there is a meromorphic function with given zeros and poles on Ω. □

3.4.8 $\bar{\partial}$-Equation

We briefly recall the notion of differential forms on complex domains (cf., e.g., [39] §3.5.3 for more).

Let $z = (z^1, \ldots, z^n)$ be the standard complex coordinate system of $\mathbf{C}^n$ (for tensor calculus it is better and convenient to use upper indices). Let x^j, y^j be respectively the real and imaginary parts of z^j, and let $\frac{\partial}{\partial x^j}$ and $\frac{\partial}{\partial y^j}$ be the constant vector fields defined respectively by x^j, y^j $(1 \le j \le n)$ with their dual differential forms dx^j, dy^j $(1 \le j \le n)$. We then define complex vector fields of type $(0,1)$ on $\mathbf{C}^n$ and their dual differential forms of type $(0,1)$ by

(3.4.64)
$$\frac{\partial}{\partial \bar{z}^j} = \frac{1}{2}\left(\frac{\partial}{\partial x^j} - \frac{1}{i}\frac{\partial}{\partial y^j}\right), \qquad 1 \le j \le n,$$
$$d\bar{z}^j = dx^j - i\,dy^j, \qquad 1 \le j \le n.$$

We consider a differential form of type $(0,1)$ with C^∞ function coefficients in an open set U of $\mathbf{C}^n$ (here, coefficient functions are always of C^∞-class):

(3.4.65)
$$f = f_1 d\bar{z}^1 + \cdots + f_n d\bar{z}^n, \quad f_j \in \mathscr{C}^\infty(U).$$

For a function $u \in \mathscr{C}^\infty(U)$ we define the $\bar{\partial}$-operator by

$$\bar{\partial}u = \frac{\partial u}{\partial \bar{z}^1} d\bar{z}^1 + \cdots + \frac{\partial u}{\partial \bar{z}^n} d\bar{z}^n.$$

For a given differential from f of type $(0,1)$, a differential equation with an unknown function u is defined by

(3.4.66)
$$\bar{\partial}u = f:$$

This is called the *$\bar{\partial}$-equation*. The Cauchy–Riemann condition (1.1.10) is equivalent to

(3.4.67)
$$\bar{\partial}u = 0.$$

If there is a solution $u \in C^{\infty}(U)$ of (3.4.66), then

$$\frac{\partial u}{\partial \bar{z}^j} = f_j, \quad \frac{\partial^2 u}{\partial \bar{z}^k \partial \bar{z}^j} = \frac{\partial f_j}{\partial \bar{z}^k}, \quad 1 \le j, k \le n.$$

Since $\frac{\partial^2 u}{\partial \bar{z}^k \partial \bar{z}^j} = \frac{\partial^2 u}{\partial \bar{z}^j \partial \bar{z}^k}$, we have

$$\frac{\partial f_j}{\partial \bar{z}^k} = \frac{\partial f_k}{\partial \bar{z}^j}, \quad 1 \le j, k \le n. \tag{3.4.68}$$

This is a necessary condition for $\bar{\partial}$-equation (3.4.66) to have a solution u, which is called the *integrable condition* of (3.4.66). A differential form of type $(0,2)$ is defined by

$$\bar{\partial} f = \sum_{1 \le j < k \le n} \left(\frac{\partial f_k}{\partial \bar{z}^j} - \frac{\partial f_j}{\partial \bar{z}^k} \right) d\bar{z}^j \wedge d\bar{z}^k.$$

The integrable condition (3.4.68) is equivalent to

$$\bar{\partial} f = 0. \tag{3.4.69}$$

In this case, f is said to be *$\bar{\partial}$-closed.* It follows from the above definitions that

$$\bar{\partial}\bar{\partial} u = 0, \quad u \in \mathscr{C}^{\infty}(U). \tag{3.4.70}$$

In what follows we would like to solve $\bar{\partial}$-equation (3.4.66) on a holomorphically convex domain as an application of Theorem 3.4.22 of the Continuous Cousin Problem. We prepare the local solvability.

We define the *Cauchy integral transform* $\mathrm{T}\psi(z)$ of a function ψ on $\mathbf{C}$ by

$$\mathrm{T}\psi(z) = \frac{1}{2\pi i} \int_{\mathbf{C}} \frac{\psi(\zeta)}{\zeta - z} d\zeta \wedge d\bar{\zeta}, \tag{3.4.71}$$

where the integral is considered under the condition of the existence.

Lemma 3.4.72. *Assume that $\psi \in \mathscr{C}^{\infty}(\mathbf{C})$ and* $\mathrm{Supp}\,\psi \Subset \mathbf{C}$. *Then $\mathrm{T}\psi$ is of C^{∞}-class and satisfies*

$$\frac{\partial \mathrm{T}\psi}{\partial \bar{z}} = \psi.$$

Proof. By a change of variable we have

$$\mathrm{T}\psi(z) = \frac{1}{2\pi i} \int_{\mathbf{C}} \frac{\psi(z+\zeta)}{\zeta} d\zeta \wedge d\bar{\zeta}.$$

From this we calculate by making use of Stokes' Theorem (as for the calculation of differential forms, cf. [39] §3.5.3):

$$
\begin{aligned}
\frac{\partial \mathrm{T}\psi}{\partial \bar{z}}(z) &= \frac{1}{2\pi i}\int_{\mathbf{C}} \frac{1}{\zeta}\frac{\partial \psi(z+\zeta)}{\partial \bar{z}} d\zeta \wedge d\bar{\zeta} = \frac{1}{2\pi i}\int_{\mathbf{C}} \frac{\frac{\partial}{\partial \bar{\zeta}}\psi(z+\zeta)}{\zeta} d\zeta \wedge d\bar{\zeta} \\
&= \frac{-1}{2\pi i}\int_{\mathbf{C}} d_{\zeta}\left(\frac{\psi(z+\zeta)}{\zeta} d\zeta\right) = \lim_{\varepsilon\to+0}\frac{1}{2\pi i}\int_{|\zeta|=\varepsilon} \frac{\psi(z+\zeta)}{\zeta} d\zeta \\
&= \lim_{\varepsilon\to+0}\frac{1}{2\pi}\int_0^{2\pi} \psi(z+\varepsilon e^{i\theta}) d\theta = \psi(z). \qquad \square
\end{aligned}
$$

Lemma 3.4.73 (Dolbeault)**.** *Let $\Omega = \prod_{j=1}^{n} \Omega_j \Subset \mathbf{C}^n$ be a bounded convex cylinder. For a $\bar{\partial}$-closed differential form f of type $(0,1)$ defined in a neighborhood U of the closure $\bar{\Omega}$, there is a differentiable function u on $\bar{\Omega}$ (i.e., in a neighborhood of it) such that*

$$\bar{\partial} u = f \quad on\ \Omega.$$

Proof. We take a C^∞ function $\chi(z) \geq 0$ on $\mathbf{C}^n$ such that $\chi|_\Omega = 1$ and $\operatorname{Supp}\chi \Subset U$. We set $\hat{f} := \chi f$ with $\hat{f}_j = \chi f_j$, and 0 outside U. Then $\hat{f}$ is a differential form of type $(0,1)$ on $\mathbf{C}^n$ with compact support. We write

$$\hat{f} = \hat{f}_q d\bar{z}^q + \cdots + \hat{f}_n d\bar{z}^n, \quad 1 \leq q \leq n.$$

We employ an induction on $q = n, n-1, \ldots, 1$.

The case of $q = n$: We have $\hat{f} = \hat{f}_n d\bar{z}^n$. We take the Cauchy integral transform with respect to the variable z^n:

$$\mathrm{T}_n \hat{f}_n(z^1, \ldots, z^n) = \frac{1}{2\pi i}\int_{\zeta^n \in \mathbf{C}} \frac{\hat{f}_n(z^1, \ldots, z^{n-1}, \zeta^n)}{\zeta^n - z^n} d\zeta^n \wedge d\bar{\zeta}^n.$$

Since $\chi|_\Omega = 1$,

$$\hat{f}(z) = f, \quad \bar{\partial}\hat{f}(z) = 0, \qquad z \in \Omega.$$

On Ω (noting $d\bar{z}^n \wedge d\bar{z}^n = 0$) we have

$$
\begin{aligned}
\bar{\partial}\hat{f} &= \sum_{h=1}^{n-1} \frac{\partial \hat{f}_n}{\partial \bar{z}^h} d\bar{z}^h \wedge d\bar{z}^n = 0 \\
&\Longleftrightarrow \frac{\partial \hat{f}_n}{\partial \bar{z}^h} = 0, \quad 1 \leq h \leq n-1 \\
&\Longleftrightarrow \hat{f}_n(z^1, \ldots, z^n) \text{ is holomorphic in } z^1, \ldots, z^{n-1}.
\end{aligned}
$$

Therefore, $\mathrm{T}_n \hat{f}_n(z^1, \ldots, z^n)$ is holomorphic in $(z^1, \ldots, z^{n-1})$ on Ω. Set

$$u = \mathrm{T}_n \hat{f}_n \in \mathscr{C}^\infty(\mathbf{C}^n).$$

It follows from Lemma 3.4.72 that on Ω

$$\bar{\partial} u = \frac{\partial \mathrm{T}_n \hat{f}_n}{\partial \bar{z}^n} d\bar{z}^n = f_n d\bar{z}^n = f.$$

We next assume $1 \le q \le n-1$ and that the case of $q+1$ holds. With a convex cylinder Ω' such that $\Omega \Subset \Omega' \Subset U$, we retake χ so that $\chi|_{\Omega'} = 1$. Recall

$$\hat{f} = \hat{f}_q d\bar{z}^q + \hat{f}_{q+1} d\bar{z}^{q+1} + \cdots + \hat{f}_n d\bar{z}^n. \tag{3.4.74}$$

Since $\bar{\partial}\hat{f} = \bar{\partial} f = 0$ on Ω', $\hat{f}_q$ is holomorphic in $(z^1, \ldots, z^{q-1})$ on Ω'. Set

$$g_q = \mathrm{T}_q \hat{f}_q \in \mathscr{C}^\infty(\mathbf{C}^n).$$

Then g_q is holomorphic in $(z^1, \ldots, z^{q-1})$ on Ω', and

$$\begin{aligned} \bar{\partial} g_q &= \frac{\partial \mathrm{T}_q \hat{f}_q}{\partial \bar{z}^q} d\bar{z}^q + \frac{\partial g_q}{\partial \bar{z}^{q+1}} d\bar{z}^{q+1} + \cdots + \frac{\partial g_q}{\partial \bar{z}^n} d\bar{z}^n \\ &= \hat{f}_q d\bar{z}^q + \frac{\partial g_q}{\partial \bar{z}^{q+1}} d\bar{z}^{q+1} + \cdots + \frac{\partial g_q}{\partial \bar{z}^n} d\bar{z}^n. \end{aligned}$$

Therefore, with

$$h := \hat{f} - \bar{\partial} g_q = h_{q+1} d\bar{z}^{q+1} + \cdots + h_n d\bar{z}^n,$$

h is $\bar{\partial}$-closed on Ω' by (3.4.70). It follows from the induction hypothesis that there is a function v on $\bar{\Omega}$ such that $\bar{\partial} v = h$. With $u := v + g_q$, $\bar{\partial} u = f$ on Ω. □

Theorem 3.4.75 (Oka, Dolbeault). *Let Ω be a holomorphically convex domain in $\mathbf{C}^n$. Then for any $\bar{\partial}$-closed differential form f (C^∞-class) of type $(0,1)$ on Ω, $\bar{\partial}$-equation*

$$\bar{\partial} u = f \tag{3.4.76}$$

has always a solution $u \in \mathscr{C}^\infty(\Omega)$.

Proof. For every point $a \in \Omega$ there is a polydisk neighborhood $U_a \Subset \Omega$ and by Lemma 3.4.73 there is a function $u_a \in \mathscr{C}^\infty(U_a)$ such that

$$\bar{\partial} u_a = f \quad (\text{on } U_a).$$

We take an open covering $\{U_\lambda\}$ of Ω by such U_a and those functions $u_\lambda \in \mathscr{C}^\infty(U_\lambda)$ satisfying $\bar{\partial} u_\lambda = f|_{U_\lambda}$. On $U_\lambda \cap U_\mu \neq \emptyset$ we have

$$\bar{\partial}(u_\lambda - u_\mu) = 0,$$

so that $u_\lambda - u_\mu \in \mathscr{O}(U_\lambda \cap U_\mu)$. That is, $\{(U_\lambda, u_\lambda)\}$ is continuous Cousin data on Ω: By Theorem 3.4.22 we have the (continuous) solution u. On each U_λ, $u - u_\lambda$ is holomorphic, so that u is of C^∞-class, and satisfies

$$\bar{\partial} u = \bar{\partial}(u - u_\lambda + u_\lambda) = \bar{\partial} u_\lambda = f.$$

Thus, u is a C^∞ solution of (3.4.76). □

In the above proof, we deduced the solvability of $\bar{\partial}$-equation (3.4.76) from that of the Continuous Cousin Problem. We show the converse:

Proposition 3.4.77. *Let Ω be a domain such that an arbitrary $\bar{\partial}$-equation satisfying the integrability condition is solvable. That is, we assume that for any $\bar{\partial}$-closed differential form f of C^∞-class and type $(0,1)$, (3.4.76) has a C^∞ solution u. Then an arbitrary Continuous Cousin Problem on Ω is solvable.*

Proof. Let $\mathscr{C} = \{(U_j, g_j)\}_{j\in\Lambda}$ be continuous Cousin data on Ω, where $\Lambda \subset \mathbf{N}$ and $\{U_j\}_{j\in\Lambda}$ is locally finite. In Proposition 3.4.16 there is a partition of unity χ_j $(j \in \Lambda)$ such that χ_j are of C^∞-class.[5] We consider $(g_k - g_j)\chi_j$, defined 0 on $U_k \setminus U_j$, an element of $\mathscr{C}^\infty(U_k)$, and set

$$u_k = \sum_{j\in\Lambda} (g_k - g_j)\chi_j \in \mathscr{C}^\infty(U_k).$$

On $U_k \cap U_l$ $(\neq \emptyset)$ we get

$$\begin{aligned} u_k - u_l &= g_k - g_l \in \mathscr{O}(U_k \cap U_l), \\ \bar{\partial} u_k &= \bar{\partial} u_l. \end{aligned} \tag{3.4.78}$$

Therefore, defining $f := \bar{\partial} u_k$ on each U_k, we have a C^∞ differential form f of type $(0,1)$ on Ω such that $\bar{\partial} f = 0$. It follows from the assumption on Ω that there is a solution $u \in \mathscr{C}^\infty(\Omega)$ of $\bar{\partial} u = f$. By (3.4.78)

$$u - u_k + g_k = u - u_l + g_l, \qquad \text{on } U_k \cap U_l.$$

With $G := u - u_k + g_k$ on each U_k we see that

$$G \in \mathscr{C}^0(\Omega), \qquad G - g_k = u - u_k \in \mathscr{C}^\infty(U_k),$$

and $\bar{\partial}(u - u_k) = f - f = 0$. We thus see that $G - g_k \in \mathscr{O}(U_k)$ and G is a solution of $\mathscr{C}$ on Ω. □

3.5 Analytic Interpolation Problem

Let $\Omega \subset \mathbf{C}^n$ be a domain.

Problem 3.5.1 (Analytic Interpolation). Let $T = \{a_\nu\}_{\nu\in\Gamma}$ $(a_\nu \in \Omega)$ be a closed discrete set of Ω, and let $\{A_\nu\}_{\nu\in\Gamma}$ be given arbitrarily. Then, find a holomorphic

[5] The proof for a domain in $\mathbf{C}^n$ is not difficult, and so left to the readers (cf. Lemma 4.3.49, or cf., e.g., Y. Matsushima (transl. by E. Kobayashi), Differentiable Manifolds, Chap. II §14, M. Dekker, New York, 1972; S. Murakami, Manifolds, Chap. 1 §1.6 (in Japanese), Kyoritsu Publ.).

function $F \in \mathscr{O}(\Omega)$ satisfying

$$F(a_\nu) = A_\nu, \qquad \nu \in \Gamma.$$

In the case when $n = 1$ and Γ is finite, the classical Lagrange polynomial interpolation is well-known: that is, setting

$$P(z) := \prod_{\nu \in \Gamma} (z - a_\nu), \tag{3.5.2}$$

$$Q(z) := \sum_{\nu \in \Gamma} \frac{A_\nu}{P'(a_\nu)} \cdot \frac{P(z)}{z - a_\nu},$$

we have $Q(a_\nu) = A_\nu$ ($\nu \in \Gamma$). Even if Γ is infinite and Ω is an arbitrary domain of $\mathbf{C}$, it is known and will be proved here that Interpolation Problem 3.5.1 is always solvable.

In the case of $n \geq 2$, Ω cannot be arbitrary for the solvability of Interpolation Problem 3.5.1. We first consider a necessary condition:

Proposition 3.5.3. *If Ω is a domain such that arbitrary Interpolation Problem 3.5.1 is solvable, then Ω is holomorphically convex.*

Proof. Let $K \Subset \Omega$ be a compact subset. Suppose that $\widehat{K}_\Omega \not\Subset \Omega$. Then there is a sequence of points $a_\nu \in \widehat{K}_\Omega$, $\nu = 1, 2, \ldots$ with no accumulation point in Ω. By the assumption there is a function $F \in \mathscr{O}(\Omega)$ such that

$$F(a_\nu) = \nu, \quad \nu = 1, 2, \ldots.$$

On the other hand,

$$|F(a_\nu)| \leq \|F\|_K < \infty, \quad \nu = 1, 2, \ldots.$$

This is absurd. □

A closed discrete subset T of Ω is a special case of complex submanifold (without connectedness) of dimension 0, and a map

$$f : a_\nu \in T \to A_\nu \in \mathbf{C}$$

is regarded as a holomorphic function on T. From this viewpoint we prove a generalized interpolation by Oka's Joku-Iko Principle:

Theorem 3.5.4 (Generalized Interpolation). *Let $\Omega \subset \mathbf{C}^n$ be a holomorphically convex domain (or equivalently a domain of holomorphy). Let $S \subset \Omega$ be a complex submanifold. Then the restriction map $\mathscr{O}(\Omega) \ni F \mapsto F|_S \in \mathscr{O}(S)$ is surjective; i.e., the following is exact:*

$$\mathscr{O}(\Omega) \ni F \longmapsto F|_S \in \mathscr{O}(S) \longrightarrow 0.$$

Proof. Take a function $f \in \mathscr{O}(S)$.

(1) Let $\mathrm{P}\ (\Subset \Omega)$ be an analytic polyhedral domain. We first find a function $F \in \mathscr{O}(\mathrm{P})$ with $F|_{S\cap \mathrm{P}} = f|_{S\cap \mathrm{P}}$. Consider the Oka map $\varphi_{\mathrm{P}} : \mathrm{P} \longrightarrow \mathrm{P}\Delta$. Then $\varphi_{\mathrm{P}}(S \cap \mathrm{P})$ is a complex submanifold of $\mathrm{P}\Delta$. By the Oka Extension Theorem 2.5.14 there is a function $G \in \mathscr{O}(\mathrm{P}\Delta)$ such that $G(\varphi_{\mathrm{P}}(z)) = f(z)$ $(z \in S \cap \mathrm{P})$. Then $F(z) := G(\varphi_{\mathrm{P}}(z)) \in \mathscr{O}(\mathrm{P})$, which is the required one.

(2) By Proposition 3.3.19 there is an increasing sequence of analytic polyhedral domains of Ω,

$$\mathrm{P}_1 \Subset \mathrm{P}_2 \Subset \cdots \Subset \mathrm{P}_\mu \Subset \cdots, \qquad \bigcup_{\mu=1}^{\infty} \mathrm{P}_\mu = \Omega.$$

For the Oka map of each P_μ we write $\varphi_\mu : \mathrm{P}_\mu \to \mathrm{P}\Delta_\mu$. We may assume that every φ_μ is defined in a slightly larger analytic polyhedral domain $\tilde{\mathrm{P}}_\mu \Supset \mathrm{P}_\mu$ with an Oka map

$$\varphi_\mu : \tilde{\mathrm{P}}_\mu \to \widetilde{\mathrm{P}\Delta}_\mu\ (\Supset \mathrm{P}\Delta_\mu).$$

From the result of (1) above we obtain

$$F'_\mu \in \mathscr{O}(\tilde{\mathrm{P}}_\mu), \quad F'_\mu|_{S\cap\tilde{\mathrm{P}}_\mu} = f|_{S\cap\tilde{\mathrm{P}}_\mu}, \quad \mu = 1, 2, \ldots. \tag{3.5.5}$$

Since the image $\varphi_\mu(S \cap \tilde{\mathrm{P}}_\mu)$ is a complex submanifold of $\widetilde{\mathrm{P}\Delta}_\mu$, it follows from the Geometric Syzygy Lemma 2.5.11 (i) that there is a finite generator system of $\mathscr{I}\langle \varphi_\mu(S \cap \tilde{\mathrm{P}}_\mu)\rangle$ on every $\overline{\mathrm{P}\Delta}_\mu$,

$$\sigma_{\mu h} \in \Gamma(\overline{\mathrm{P}\Delta}_\mu, \mathscr{I}\langle \varphi_\mu(S \cap \tilde{\mathrm{P}}_\mu)\rangle), \quad 1 \le h \le l_\mu.$$

The pull-backs $\sigma'_{\mu h} = \varphi^*_\mu \sigma_{\mu h}$ by φ_μ form a finite generator system of $\mathscr{I}\langle S\rangle$ on $\overline{\mathrm{P}}_\mu$:

$$\sigma'_{\mu h} \in \Gamma(\overline{\mathrm{P}}_\mu, \mathscr{I}\langle S\rangle), \quad 1 \le h \le l_\mu.$$

In order to avoid the complication of symbols we identify them with $\sigma'_{\mu h} \in \mathscr{O}(\overline{\mathrm{P}}_\mu)$.

We inductively define $F_\mu \in \mathscr{O}(\overline{\mathrm{P}}_\mu)$. First, set $F_1 = F'_1$. Suppose that $F_1, \ldots, F_\mu$ are determined so that

$$\begin{aligned} &F_\nu|_{\overline{\mathrm{P}}_\nu \cap S} = f|_{\overline{\mathrm{P}}_\nu \cap S}, \quad \nu = 2, \ldots, \mu, \\ &\|F_\nu - F_{\nu-1}\|_{\overline{\mathrm{P}}_{\nu-1}} < \frac{1}{2^{\nu-1}}, \quad \nu = 2, \ldots, \mu. \end{aligned} \tag{3.5.6}$$

In what follows we will implicitly use the arguments such that we identify $\overline{\mathrm{P}}_\mu$ with $\varphi(\overline{\mathrm{P}}_\mu) \subset \overline{\mathrm{P}\Delta}_\mu (\Subset \widetilde{\mathrm{P}\Delta}_\mu)$ through φ_μ, and by Extension Lemma 2.5.15 we extend functions on $\overline{\mathrm{P}}_\mu$ to those on $\overline{\mathrm{P}\Delta}_\mu$ and after some modifications restrict them to $\overline{\mathrm{P}}_\mu$.

By (3.5.5) and (3.5.6)

$$(F'_{\mu+1} - F_\mu)|_{\overline{\mathrm{P}}_\mu \cap S} = 0. \tag{3.5.7}$$

By the Geometric Syzygy Lemma 2.5.11 (ii) there are elements $a'_{\mu+1h} \in \mathscr{O}(\overline{\mathrm{P}}_\mu)$ $(1 \le h \le l_{\mu+1})$ such that

$$F'_{\mu+1}(z) - F_\mu(z) = \sum_{h=1}^{l_{\mu+1}} a'_{\mu+1h}(z)\sigma'_{\mu+1h}(z), \quad z \in \overline{\mathrm{P}}_\mu. \tag{3.5.8}$$

Note that it is an expression on $\overline{\mathrm{P}}_\mu$ by $\{\sigma'_{\mu+1h}\}$ defined over $\overline{\mathrm{P}}_{\mu+1}$. By the Oka–Weil Approximation Theorem 3.3.12 we approximate $a'_{\mu+1h}(z)$ sufficiently on $\overline{\mathrm{P}}_\mu$ by $a_{\mu+1h} \in \mathscr{O}(\Omega)$ $(1 \le h \le l_{\mu+1})$, and obtain

$$\left\| \sum_{h=1}^{l_{\mu+1}} a_{\mu+1h}\sigma'_{\mu+1h} - \sum_{h=1}^{l_{\mu+1}} a'_{\mu+1h}\sigma'_{\mu+1h} \right\|_{\overline{\mathrm{P}}_\mu} < \frac{1}{2^\mu}.$$

Setting

$$F_{\mu+1} = F'_{\mu+1} - \sum_{h=1}^{l_{\mu+1}} a_{\mu+1h}\sigma'_{\mu+1h} \in \mathscr{O}(\overline{\mathrm{P}}_{\mu+1}),$$

we have that $F_{\mu+1}|_{\overline{\mathrm{P}}_{\mu+1}\cap S} = f|_{\overline{\mathrm{P}}_{\mu+1}\cap S}$ and

$$\|F_{\mu+1} - F_\mu\|_{\overline{\mathrm{P}}_\mu} < \frac{1}{2^\mu}.$$

Thus (3.5.6) holds for $\nu = \mu + 1$.

Now the series

$$\begin{aligned} F(z) &= F_1(z) + \sum_{\mu=1}^{\infty} (F_{\mu+1}(z) - F_\mu(z)) \\ &= F_\nu(z) + \sum_{\mu=\nu}^{\infty} (F_{\mu+1}(z) - F_\mu(z)), \quad z \in \mathrm{P}_\nu, \ \forall \nu \in \mathbf{N} \end{aligned}$$

converges uniformly on compact subsets in Ω and satisfies

$$F \in \mathscr{O}(\Omega), \quad F|_S = f.$$ □

Remark 3.5.9. With the knowledge of Oka's Second Coherence Theorem (cf. §2.2.2) one sees immediately that Theorem 3.5.4 and the given proof remain valid for singular analytic subsets S. Furthermore, Theorem 3.5.4 together with the given proof extends to the case where Ω is a singular domain, i.e., a Stein space (for the notion of Steinness, see Definition 3.7.1; in general, cf., e.g., [39] §6.12). In fact, in Oka [48] IX (1953) he briefly wrote that such an extension from singular spaces for the Joku-Iko Principle had been the purpose to study the notion of "Coherence" around 1943, which did not exist then.

3.6 Unramified Domains over $\mathbf{C}^n$

(a) Unramified domain over $\mathbf{C}^n$. In the case of $n \geq 2$, there is a domain Ω of $\mathbf{C}^n$ such that all $f \in \mathcal{O}(\Omega)$ are analytically continued over a domain which is strictly larger than Ω (Hartogs' phenomenon). Furthermore, there is a case that they become infinitely multi-valued:

Example 3.6.1. With the coordinate $(z, w) \in \mathbf{C}^2$ we define Ω as follows. Fixing a branch of $\arg z$ with $\arg 1 = 0$, we set

$$\Omega: \frac{1}{2} < |z| < 1, \quad -\frac{\pi}{2} + \arg z < |w| < \frac{\pi}{2} + \arg z.$$

Let (z, w) be a point of the domain. As z turns around in the ring domain once in an anti-clockwise orientation with $\arg z$ increasing 2π, the domain of w is deformed, so that they do not overlap; Ω is a subdomain of $\mathbf{C}^2$. Then by making use of Laurent expansions, we see that every $f \in \mathcal{O}(\Omega)$ is analytically continued over a multivalent domain covering $\{1/2 < |z| < 1\} \times \mathbf{C}$ infinitely many times:

$$\tilde{\Omega}: \frac{1}{2} < |z| < 1, \quad |w| < \frac{\pi}{2} + \arg z$$

(cf. [39] Example 5.1.5). For example, $f_0(z, w) = \log z + w \in \mathcal{O}(\Omega)$ is one-valued on Ω, which is analytically continued to an infinitely many-valued function over $\tilde{\Omega}$.

If a point (z, w) of the above $\tilde{\Omega}$ is distinguished by the value $\arg z$, the function f_0 becomes a one-valued function on it. We formulate such a phenomenon in general:

Definition 3.6.2. Let $\mathfrak{D}$ be a Hausdorff topological space. We assume that $\mathfrak{D}$ is connected, unless otherwise mentioned, and that there is a continuous map $\pi : \mathfrak{D} \to \mathbf{C}^n$. If π is a local homeomorphism (i.e., every point $p \in \mathfrak{D}$ has a neighborhood U $(\subset \mathfrak{D})$ such that $V := \pi(U)$ is a neighborhood of $\pi(p)$ and the restriction $\pi|_U : U \to V$ is a homeomorphism), $\pi : \mathfrak{D} \to \mathbf{C}^n$ or simply $\mathfrak{D}$ is called an *unramified domain*, sometimes *unramified Riemann domain*. In the present book, we also simply say that $\mathfrak{D}$ is a *domain over* $\mathbf{C}^n$ and write $\mathfrak{D}/\mathbf{C}^n$.

Remark 3.6.3. (i) While we do not give the definition of complex manifolds, an unramified domain $\pi : \mathfrak{D} \to \mathbf{C}^n$ has the natural structure of a complex manifold such that π is a local biholomorphism (cf. [39] Definition 4.5.7).

(ii) If $\mathfrak{D}$ is a complex manifold or a complex space (cf. [39] §6.9), and the above $\pi|_U : U \to V$ is a finite map, (cf. [39] Definition 6.1.7), then $\pi : \mathfrak{D} \to \mathbf{C}^n$ is called in general a *domain over* $\mathbf{C}^n$, *possibly with ramification*. If $\mathfrak{D}$ is not an unramified domain, then $\mathfrak{D}$ is called a *ramified domain (over* $\mathbf{C}^n$*)*. In the present book, we mainly deal with unramified domains (over $\mathbf{C}^n$) and so domains (over $\mathbf{C}^n$) are assumed to be unramified, unless specially mentioned.

(iii) Let $\pi : \mathfrak{D} \to \mathbf{C}^n$ be a domain. For a point $p \in \mathfrak{D}$, $z = \pi(p)$ is called a *base point* of p, and $p \in \pi^{-1}z$ is said to be a *point over* z. If the point over every $z \in \mathbf{C}^n$ is at most one point, $\mathfrak{D}$ is called a *univalent* or *schlicht domain*; in this case, $\mathfrak{D}$ may be identified with $\pi(\mathfrak{D})$, and regarded as a subdomain of $\mathbf{C}^n$. In general, $\mathfrak{D}/\mathbf{C}^n$ is said to be *multivalent* or *multi-sheeted*. The number $N = \sup\{\mathrm{Card}\{\pi^{-1}\pi(p)\} : p \in \mathfrak{D}\} \le \infty$ is called the *sheet number* of $\pi : \mathfrak{D} \to \mathbf{C}^n$. If $N < \infty$, $\pi : \mathfrak{D} \to \mathbf{C}^n$ is said to be *finitely* (or N-) *sheeted*; otherwise, $\mathfrak{D}/\mathbf{C}^n$ is said to be *infinitely sheeted*.

Let $\pi : \mathfrak{D} \to \mathbf{C}^n$ be a domain. If there are a polydisk $\mathrm{P}\Delta(a;r) \subset \mathbf{C}^n$ (resp. an open ball $\mathrm{B}(a;R) \subset \mathbf{C}^n$) and an open set $W \subset \mathfrak{D}$ with homeomorphism $\pi|_W : W \to \mathrm{P}\Delta(a;r)$ (resp. $\mathrm{B}(a;R)$), then W is called a *polydisk* (resp. *open ball*) with center at $p = (\pi|_W)^{-1}(a)$ in $\mathfrak{D}$, also a *polydisk* (resp. *ball*) *neighborhood* of p, and we write $\mathrm{P}\Delta(p;r)\,(\cong \mathrm{P}\Delta(a;r))$ (resp. $\mathrm{B}(p,R)\,(\cong \mathrm{B}(a;R))$) for it.

While it is not possible to define directly the boundary $\partial\mathfrak{D}$ of $\mathfrak{D}$ (cf. §3.8), a sort of boundary distance can be defined from the inside: We fix a polydisk $\mathrm{P}\Delta$ of polyradius $r = (r_j)$ with center at 0 of $\mathbf{C}^n$ (resp. an open ball $\mathrm{B}(R)$ with radius R and center at 0). For $s > 0$ we consider a polydisk $\mathrm{P}\Delta(z;sr)$ of polyradius $sr = (sr_j)$ with center at $z \in \mathbf{C}^n$ (resp. an open ball $\mathrm{B}(z;sR)$).

Let $p \in \mathfrak{D}$. For small $s > 0$, there is a polydisk neighborhood $\mathrm{P}\Delta(p;sr) \subset \mathfrak{D}$ (resp. $\mathrm{B}(p;sR)) \subset \mathfrak{D}$) of p; as in (3.2.1) we define

$$\delta_{\mathrm{P}\Delta}(p,\partial\mathfrak{D}) = \sup\{s > 0 : \mathrm{P}\Delta(p;sr) \subset \mathfrak{D}\} > 0 \tag{3.6.4}$$

(resp. $\delta_{\mathrm{B}(R)}(p,\partial\mathfrak{D})$ is defined similarly). We call $\delta_{\mathrm{P}\Delta}(p,\partial\mathfrak{D})$ (resp. $\delta_{\mathrm{B}(R)}(p,\partial\mathfrak{D})$) the *boundary distance function* of $\mathfrak{D}$ with respect to $\mathrm{P}\Delta$ (resp. $\mathrm{B}(R)$). As in (3.2.2),

$$\begin{aligned} |\delta_{\mathrm{P}\Delta}(p',\partial\mathfrak{D}) - \delta_{\mathrm{P}\Delta}(p'',\partial\mathfrak{D})| &\le \|\pi(p') - \pi(p'')\|_{\mathrm{P}\Delta}, \\ & p',p'' \in \mathrm{P}\Delta(p;\delta_{\mathrm{P}\Delta}(p,\partial\mathfrak{D})). \end{aligned} \tag{3.6.5}$$

Therefore, $\delta_{\mathrm{P}\Delta}(p,\partial\mathfrak{D})$ is continuous; similarly, so is $\delta_{\mathrm{B}(R)}(p,\partial\mathfrak{D})$.

(b) Holomorphic functions in domains over $\mathbf{C}^n$. Let $\pi : \mathfrak{D} \to \mathbf{C}^n$ be a domain. For $p \in \mathfrak{D}$ the points q of $\mathrm{P}\Delta(p;\delta_{\mathrm{P}\Delta}(p,\partial\mathfrak{D}))$ are uniquely parametrized by $\pi(q) = z = (z_j)_{1 \le j \le n} \in \mathrm{P}\Delta(\pi(p);\delta_{\mathrm{P}\Delta}(p,\partial\mathfrak{D})r)$, which we use for a local complex coordinate system. We write $p(z)$ for a point $p \in \mathfrak{D}$ with the local complex coordinate system $\pi(p) = z = (z_1,\ldots,z_n) \in \mathbf{C}^n$.

Let $W \subset \mathfrak{D}$ be an open set. It makes sense to say that a function $f(p)$ defined in W is *holomorphic* if $f(p(z))$ is holomorphic locally in z. By definition π is a vector-valued holomorphic function on $\mathfrak{D}$. We denote by $\mathscr{O}(W)$ the set of all holomorphic functions in W.

We define functions of C^k-class $(0 \le k \le \infty)$ in W in the same way as above, and denote all of them by $\mathscr{C}^k(W)$.

All local properties of holomorphic functions in an open set of $\mathbf{C}^n$ also hold for holomorphic functions in W. Let $f \in \mathscr{O}(\mathrm{P}\Delta(p(a);r))$. Then $f(p(z))$ is expanded as a function of z to a power series with center at $z = a$:

$$f(p(z)) = \sum_{\alpha \in \mathbf{Z}_+^n} c_\alpha (z-a)^\alpha. \tag{3.6.6}$$

We denote the power series above by $\underline{f}_p$:

(3.6.7) $$\underline{f}_p = \sum_{\alpha \in \mathbf{Z}_+^n} c_\alpha (z-a)^\alpha.$$

We take a family $\mathscr{F}(\neq \emptyset) \subset \mathscr{O}(\mathfrak{D})$ and fix it.

Definition 3.6.8 (Holomorphic separation). A domain $\pi : \mathfrak{D} \to \mathbf{C}^n$ is said to be $\mathscr{F}$-*separable* if for arbitrary two distinct points $p, q \in \mathfrak{D}$ there is an element $f \in \mathscr{F}$ with $f(p) \neq f(q)$: In the case of $\mathscr{F} = \mathscr{O}(\mathfrak{D})$, $\mathfrak{D}$ is said to be *holomorphically separable.*

Since π is a vector-valued holomorphic function on $\mathfrak{D}$, the holomorphic separability condition makes sense for the case of $\pi(p) = \pi(q)$.

Proposition 3.6.9. *For a domain $\pi : \mathfrak{D} \to \mathbf{C}^n$ the following two conditions are equivalent.*

(i) *$\mathfrak{D}$ is holomorphically separable.*
(ii) *For arbitrary distinct points $p, q \in \mathfrak{D}$ over $a \in \mathbf{C}^n$, there is a function $f \in \mathscr{O}(\mathfrak{D})$ with $\underline{f}_p \neq \underline{f}_q$.*

Proof. (i) $\Rightarrow$ (ii): This is clear.

(ii) $\Rightarrow$ (i): As a holomorphic partial differential operator ∂^α is defined in (1.1.8), ∂^α is naturally defined for $f \in \mathscr{O}(\mathfrak{D})$ with respect to complex local coordinate z. It follows from the assumption that for distinct $p, q \in \mathfrak{D}$ with $\pi(p) = \pi(q)$ there is an element $f \in \mathscr{O}(\mathfrak{D})$ with $\underline{f}_p \neq \underline{f}_q \in \mathscr{O}_{\pi(p)}$; that is, there is a holomorphic partial derivation ∂^α such that $\partial^\alpha f(p) \neq \partial^\alpha f(q)$, and $\partial^\alpha f \in \mathscr{O}(\mathfrak{D})$. □

Let $\rho : \mathfrak{G} \to \mathbf{C}^m$ be another domain over $\mathbf{C}^m$. A continuous map $\psi : \mathfrak{D} \to \mathfrak{G}$ is called a *holomorphic map* if $\rho \circ \psi : \mathfrak{D} \to \mathbf{C}^m$ is a vector-valued holomorphic function. In that case, the pull-back by ψ

$$\psi^* : g \in \mathscr{O}(\mathfrak{G}) \to g \circ \psi \in \mathscr{O}(\mathfrak{D})$$

is defined, and ψ^* is a *(ring) homomorphism*: that is, for $g, h \in \mathscr{O}(\mathfrak{G})$,

$$\psi^*(g+h) = \psi^* g + \psi^* h, \quad \psi^*(g \cdot h) = \psi^* g \cdot \psi^* h.$$

In the case where ψ^* is bijective, ψ^* is called a *(ring) isomorphism.*

If $n = m$ and $\pi = \rho \circ \psi$, $\psi : \mathfrak{D} \to \mathfrak{G}$ is called a *relative map* over $\mathbf{C}^n$. A relative map over $\mathbf{C}^n$ is holomorphic. If a relative map $\psi : \mathfrak{D} \to \mathfrak{G}$ over $\mathbf{C}^n$ is bijective, the inverse ψ^{-1} is again a relative map over $\mathbf{C}^n$, and ψ is called a *relative isomorphism* over $\mathbf{C}^n$; also $\mathfrak{D}$ and $\mathfrak{G}$ are said to be *relatively isomorphic* over $\mathbf{C}^n$.

(c) Analytic continuation and envelope of holomorphy. Let $\pi : \mathfrak{D} \to \mathbf{C}^n$ be a domain. Let $\psi : \mathfrak{D} \to \mathfrak{G}$ be a relative map from $\mathfrak{D}$ into another domain $\mathfrak{G}$ over $\mathbf{C}^n$. If $f \in \mathscr{O}(\mathfrak{D})$ and $g \in \mathscr{O}(\mathfrak{G})$ satisfies $\psi^* g = f$, f is said to be *analytically continued* to g. If $\psi^* \mathscr{O}(\mathfrak{G}) = \mathscr{O}(\mathfrak{D})$, $\mathfrak{G}$ is called a *holomorphic extension* of $\mathfrak{D}$.

Let $\mathscr{F} \subset \mathcal{O}(\mathfrak{D})$ be a non-empty family. We would like to obtain a maximal domain, over which all functions of $\mathscr{F}$ are analytically continued simultaneously; that is, if $\mathscr{F} = \mathcal{O}(\mathfrak{D})$, it is a maximal domain among all the holomorphic extensions of $\mathfrak{D}$.

We fix a point $p_0 \in \mathfrak{D}$, and take a curve C in $\mathbf{C}^n$ with the initial point $a = \pi(p_0)$. For a function $f \in \mathscr{F}$ we consider the analytic function $f(p(z))$ of z about a, and denote by Γ the set of all curves C such that $f(p(z))$ is analytically continued along C. Let $C^b \in \Gamma$ be a curve with the terminal point $b \in \mathbf{C}^n$. Then an analytic function $f(p(z))$ $(f \in \mathscr{F})$ defined about a is analytically continued along C^b to an analytic function $f_{C^b}(z)$ about b. If C^b and $C'^b \in \Gamma$ are mutually homotopic within Γ, $\underline{f_{C^b}}_{\,b} = \underline{f_{C'^b}}_{\,b}$; in this sense, $\{C^b\}$ denotes the homotopy class of C. We write $\underline{f_{\{C^b\}}}_{\,b} := \underline{f_{C^b}}_{\,b}$.

Fix a polydisk $\mathrm{P}\Delta$ with center at the origin in $\mathbf{C}^n$. Let $s(\{C^b\}, f)$ be the supremum of $s > 0$ such that the power expansion of $\underline{f_{\{C^b\}}}_{\,b}(z)$ at $z = b$ converges in $b + s\mathrm{P}\Delta$ $(r > 0)$. We write $\Gamma^\dagger$ for the set of all $\{C^b\}$ with $\inf_{f \in \mathscr{F}} s(\{C^b\}, f) > 0$.

For two elements $\{C^b\}, \{C'^{b'}\}$ of $\Gamma^\dagger$ we define an equivalence relation $\{C^b\} \sim \{C'^{b'}\}$ by

$$b = b', \quad \underline{f_{\{C^b\}}}_{\,b} = \underline{f_{\{C'^{b'}\}}}_{\,b'}, \quad \forall f \in \mathscr{F}.$$

The equivalence class is denoted by $[\{C^b\}]$. The quotient set and the natural map are written as

$$\hat{\mathfrak{D}} = \Gamma^\dagger/\sim, \quad \hat{\pi} : [\{C^b\}] \in \hat{\mathfrak{D}} \to b \in \mathbf{C}^n.$$

By the construction, $\hat{\pi} : \hat{\mathfrak{D}} \to \mathbf{C}^n$ is a domain. Since $\mathfrak{D}$ is arc-wise connected, $\hat{\mathfrak{D}}$ is independent of the choice of $p_0 \in \mathfrak{D}$, and the relative map $\eta : \mathfrak{D} \to \hat{\mathfrak{D}}$ over $\mathbf{C}^n$ is naturally defined.

Definition 3.6.10. (i) The above-constructed domain $\hat{\pi} : \hat{\mathfrak{D}} \to \mathbf{C}^n$ is called the $\mathscr{F}$*-envelope* of $\mathfrak{D}$.

(ii) When $\mathscr{F} = \mathcal{O}(\mathfrak{D})$, the $\mathcal{O}(\mathfrak{D})$-envelope of $\mathfrak{D}$ is called the *envelope of holomorphy* of $\mathfrak{D}$.

(iii) When $\mathscr{F}$ consists of one function $f \in \mathcal{O}(\mathfrak{D})$, the $\{f\}$-envelope of $\mathfrak{D}$ is called the *domain of existence* of f.

(iv) A domain $\mathfrak{D}/\mathbf{C}^n$ is a *domain of holomorphy*, if $\mathfrak{D}$ is the envelope of holomorphy of itself.

Remark 3.6.11. (i) The domains defined in the above (i)–(iv) are all holomorphically separable.

(ii) The relative map $\eta : \mathfrak{D} \to \hat{\mathfrak{D}}$ ($\mathscr{F}$-envelope) over $\mathbf{C}^n$ is injective if and only if $\mathfrak{D}$ is $\mathscr{F}$-separable; in this case, $\mathfrak{D} \subset \hat{\mathfrak{D}}$.

Therefore we deduce:

Theorem 3.6.12. *Let $\pi : \mathfrak{D} \to \mathbf{C}^n$ be a holomorphically separable domain.*

(i) *$\mathfrak{D}$ carries the envelope of holomorphy $\hat{\pi} : \hat{\mathfrak{D}} \to \mathbf{C}^n$ containing $\mathfrak{D}$ as a subdomain with $\hat{\pi}|_{\mathfrak{D}} = \pi$. In particular, a univalent domain Ω $(\subset \mathbf{C}^n)$ has necessarily the envelope of holomorphy of Ω.*

(ii) $\mathcal{O}(\mathfrak{D}) = \{f|_{\mathfrak{D}} : f \in \mathcal{O}(\hat{\mathfrak{D}})\}$.

(iii) *Let* $\pi' : \mathfrak{D}' \to \mathbf{C}^n$ *be a holomorphically separable domain which contains* $\pi : \mathfrak{D} \to \mathbf{C}^n$ *and* $\pi'|_{\mathfrak{D}} = \pi$. *Assume that* $\mathcal{O}(\mathfrak{D}) = \{f|_{\mathfrak{D}} : f \in \mathcal{O}(\mathfrak{D}')\}$. *Then there is a relative injection* $\iota : \mathfrak{D}' \to \hat{\mathfrak{D}}$ $(\pi' = \hat{\pi} \circ \iota)$ *over* $\mathbf{C}^n$.

Proof. (i), (ii): These are immediate by the construction of $\hat{\mathfrak{D}}$.

(iii): It follows from the arc-wise connectedness of a domain over $\mathbf{C}^n$ and the construction of the envelope of holomorphy $\hat{\mathfrak{D}}$. □

Example 3.6.13. We give an example such that η in Remark 3.6.11 (ii) is not injective (cf. [39] Example 7.5.4).

In the unit polydisk $\mathrm{P}\Delta = \Delta(1)^2$ $(\subset \mathbf{C}^2)$ we set (describe the figure by oneself)

$$K = \left\{(z_1, z_2) \in \mathrm{P}\Delta : \tfrac{1}{4} \le |z_j| \le \tfrac{3}{4}, j = 1, 2\right\}, \quad \Omega = \mathrm{P}\Delta \setminus K,$$

$$\Omega_{\mathrm{H}} = \left(\Delta(1) \times \Delta\left(\tfrac{1}{4}\right)\right) \cup \left(\{\tfrac{3}{4} < |z_1| < 1\} \times \Delta(1)\right),$$

$$\omega = \Delta\left(\tfrac{1}{4}\right)^2, \quad U = \Delta(1) \times \Delta\left(\tfrac{1}{4}\right), \quad V = \Delta\left(\tfrac{1}{4}\right) \times \Delta(1).$$

Then U and V are subdomains of Ω, and $U \cap V = \omega$. We distinguish ω as a part of U and that as a part of V, so that we obtain a 2-sheeted domain $\pi : \mathfrak{D} \to \Omega \subset \mathbf{C}^2$. Note that Ω_{H} $(\subset \mathfrak{D})$ is a Hartogs domain, of which the envelope of holomorphy is $\mathrm{P}\Delta$. Therefore, the envelope of holomorphy of $\mathfrak{D}$ is $\hat{\mathfrak{D}} = \mathrm{P}\Delta$, and hence $\eta : \mathfrak{D} \to \mathrm{P}\Delta$ is not injective. There is no domain of holomorphy which contains $\mathfrak{D}$ as a subdomain.

Remark 3.6.14. As seen in Example 3.6.1, if $n \ge 2$, there is a univalent domain in $\mathbf{C}^n$ such that all holomorphic functions are analytically continued over an infinitely sheeted unramified domain over $\mathbf{C}^n$. Therefore it is theoretically necessary to deal with at least unramified domains over $\mathbf{C}^n$, multivalent in general. It is incomplete as a theory to deal only with univalent domains of $\mathbf{C}^n$: This is an important viewpoint of analytic function theory of several variables.

(d) σ-compact. Let $\pi : \mathfrak{D} \to \mathbf{C}^n$ be a domain. We take the euclidean metric on $\mathbf{C}^n$,

$$ds^2 = \sum_{j=1}^{n} \left(dx_j^2 + dy_j^2\right), \quad z_j = x_j + iy_j.$$

Through the local biholomorphism π we lift it to a Riemann metric on $\mathfrak{D}$, denoted by $\pi^* ds^2$, which is called the euclidean metric on $\mathfrak{D}$. For a curve $C(\phi : [t_0, t_1] \to \mathfrak{D})$ of piecewise C^1-class in $\mathfrak{D}$ we define the length with respect to $\pi^* ds^2$ by

$$L(C) = \int_C \pi^* ds = \int_{t_0}^{t_1} \sqrt{\sum_{j=1}^{n} \left((\phi'_{j1}(t))^2 + (\phi'_{j2}(t))^2\right)}\, dt,$$

$$\phi_j(t) = \phi_{j1}(t) + i\phi_{j2}(t) \quad \text{(local expression)}.$$

Fix a point $p_0 \in \mathfrak{D}$ and a polydisk $\mathrm{P}\Delta = \mathrm{P}\Delta(r)$ with polyradius r. With the boundary distance function $\delta_{\mathrm{P}\Delta}(p, \partial\mathfrak{D})$ of $\mathfrak{D}$ we define

$$\mathfrak{D}_\rho = \{p \in \mathfrak{D} : \delta_{\mathrm{P}\Delta}(p, \partial\mathfrak{D}) > \rho\}$$

for $\rho > 0$. We set

$$U_\sigma(p) = \mathrm{P}\Delta(p;\sigma r), \quad 0 < \sigma \le \delta_{\mathrm{P}\Delta}(p, \partial\mathfrak{D}).$$

Choose $\rho > 0$ so that $\mathfrak{D}_\rho \ni p_0$. It is noticed that every point p of the connected component of $\mathfrak{D}_\rho$ containing p_0 is connected by a piecewise C^1 curve $C(p)$ in $\mathfrak{D}$. Set

$$(3.6.15) \qquad d_\rho(p) = \inf_{C(p)\subset\mathfrak{D}_\rho} L(C(p)).$$

As easily checked, the function $d_\rho(p)$ satisfies the Lipschitz continuity condition:

$$(3.6.16) \qquad |d_\rho(p') - d_\rho(p'')| \le \|\pi(p') - \pi(p'')\| = \|p' - p''\|, \quad p', p'' \in U_{\delta_{\mathrm{P}\Delta}(p,\partial\mathfrak{D})}(p).$$

Here we identify points p', p'' contained in a univalent $\mathrm{P}\Delta(p;\delta_{\mathrm{P}\Delta}(p,\partial\mathfrak{D})r)$ with $\pi(p'), \pi(p'')$, respectively. In this way, unless confusion occurs, we write a point of a univalent domain of $\mathfrak{D}$ as a point of $\mathbf{C}^n$ for the sake of notational simplicity.

Lemma 3.6.17. *We have* $\{p \in \mathfrak{D}_\rho : d_\rho(p) < b\} \Subset \mathfrak{D}$ *for every* $b > 0$.

Proof. Without loss of generality we may assume $\mathrm{B} \subset \mathrm{P}\Delta$. For $b = \rho$ we have

$$\{p \in \mathfrak{D}_\rho : d_\rho(p) < \rho\} \Subset U_{\delta_{\mathrm{P}\Delta}(p_0,\partial\mathfrak{D})}(p_0).$$

Therefore, $\{p \in \mathfrak{D}_\rho : d_\rho(p) < \rho\} \Subset \mathfrak{D}$, and so the assertion holds for $b = \rho$.

Now, suppose that the assertion holds for a number $b \ge \rho$; i.e., $K := \{p \in \bar{\mathfrak{D}}_\rho : d_\rho(p) \le b\}$ is compact. For every $p \in K$, we get $\bar{U}_{\rho/2}(p) \Subset \mathfrak{D}$ and see that

$$K' = \bigcup_{p\in K} \bar{U}_{\rho/2}(p)$$

is compact. For, if we take a sequence of points $q_\nu \in K'$, $\nu \in \mathbf{N}$, then there are points $p_\nu \in K$ and $w_\nu \in \frac{\rho}{2}\overline{\mathrm{P}\Delta}$ such that

$$q_\nu = p_\nu + w_\nu, \quad \nu \in \mathbf{N}.$$

Since K is compact, after taking a subsequence we deduce that $\lim_{\nu\to\infty} p_\nu = x_0 \in K$ and $\lim_{\nu\to\infty} w_\nu = w_0$ with $w_0 \in \frac{\rho}{2}\overline{\mathrm{P}\Delta}$. It follows that

$$\lim_{\nu\to\infty} q_\nu = x_0 + w_0 \in K'.$$

Since $\{p \in \mathfrak{D}_\rho : d_\rho(p) < b + \rho/2\} \subset K'$, the assertion holds for $b + \rho/2$. Inductively, it holds for $\rho + \nu\rho/2$, $\nu = 1, 2, \ldots$; hence it holds for all $b > 0$. □

The following property is said to be *σ-compact*.

Proposition 3.6.18. *For a domain $\pi : \mathfrak{D} \to \mathbf{C}^n$, there is an open covering $\{\mathfrak{D}_\nu\}_{\nu=1}^{\infty}$ by an increasing sequence of subdomains $\mathfrak{D}_\nu$ of $\mathfrak{D}$ such that*

$$\mathfrak{D}_\nu \Subset \mathfrak{D}_{\nu+1}, \qquad \mathfrak{D} = \bigcup_{\nu=1}^{\infty} \mathfrak{D}_\nu.$$

Proof. We use the notation in the proof of Lemma 3.6.17. For a number $s > 0$ we denote by $\mathfrak{D}_{\rho.s}$ the set of all points $p \in \mathfrak{D}_\rho$ which is connected with p_0 by a curve C of piecewise C^1-class in $\mathfrak{D}_\rho$ with length $L(C) < s$. Then $\mathfrak{D}_{\rho,s}$ is naturally connected, and $\mathfrak{D}_{\rho,s} \Subset \mathfrak{D}$ by Lemma 3.6.17. Taking sequences $\rho_\nu \searrow 0$ and $s_\nu \nearrow \infty$ ($\nu = 1, 2, \ldots$), we obtain

$$\mathfrak{D}_\nu := \mathfrak{D}_{\rho_\nu, s_\nu} \Subset \mathfrak{D}_{\nu+1}, \qquad \bigcup_{\nu=1}^{\infty} \mathfrak{D}_\nu = \mathfrak{D}.$$ □

3.7 Stein Domains over $\mathbf{C}^n$

K. Oka dealt with the Approximation Problem and the Cousin Problem as well on unramified domains over $\mathbf{C}^n$ in his unpublished five papers of 1943 which historically first solved affirmatively the Pseudoconvexity Problem in general dimension,[6] and in the published paper Oka IX (1953) which was a revised version of the unpublished papers above, by making use of a part of the coherence theorems obtained in Oka VII and VIII.

In the present section we see how the main results on univalent domains of $\mathbf{C}^n$ extend to unramified domains over $\mathbf{C}^n$.

Let $\pi : \mathfrak{D} \to \mathbf{C}^n$ be a domain. The notions of a holomorphically convex hull and holomorphic convexity are defined in the same way as in the case of univalent domains (§3.1). For example, $\mathfrak{D}$ is *holomorphically convex* if for every compact subset $K \Subset \mathfrak{D}$,

$$\widehat{K}_{\mathfrak{D}} := \{p \in \mathfrak{D} : |f(p)| \le \|f\|_K, \ \forall f \in \mathscr{O}(\mathfrak{D})\} \Subset \mathfrak{D}.$$

Definition 3.7.1 (Stein domain). A domain $\mathfrak{D}/\mathbf{C}^n$ is called a *Stein domain* if:

(i) $\mathfrak{D}$ is holomorphically separable, and
(ii) $\mathfrak{D}$ is holomorphically convex.

An open set Ω of $\mathfrak{D}$ is said to be *Stein* if every connected component of Ω is a Stein domain.

The results shown in §3.2 are stated for a domain $\mathfrak{D}$ over $\mathbf{C}^n$ as follows: The proofs are the same. For example:

3.7.2. *The Cartan–Thullen Lemma 3.2.7 holds for $\mathfrak{D}$.*

[6] The case of univalent 2-dimensional domains had been affirmatively solved by Oka [47] (1941) and Oka VI (1942).

By the proof of Theorem 3.2.11 we have:

Theorem 3.7.3 (Cartan–Thullen). *A Stein domain is a domain of holomorphy.*

In the same way as we obtained Corollary 3.2.12 we have:

Corollary 3.7.4. *Let $\mathfrak{D}/\mathbf{C}^n$ be a domain. Let $\{\Omega_\gamma\}_{\gamma\in\Gamma}$ be a family of Stein open subsets of $\mathfrak{D}$. Then the interior of $\bigcap_{\gamma\in\Gamma}\Omega_\gamma$ is Stein.*

Remark 3.7.5. The converse in Theorem 3.7.3 holds in general, but we have to wait for the solution of the Pseudoconvexity Problem by Oka (see Theorem 5.3.2). The flow of the proof is as follows:

$$\mathfrak{D}/\mathbf{C}^n, \text{ domain of holomorphy} \Rightarrow \mathfrak{D}/\mathbf{C}^n, \text{ pseudoconvex} \Rightarrow \mathfrak{D}/\mathbf{C}^n, \text{ Stein.}$$

Here is an essential difference between the univalent case and the multivalent one.

An *analytic polyhedron* of $\mathfrak{D}$ is defined as in Definition 3.3.1 with $\Omega = \mathfrak{D}$. In the definition of Oka map (3.3.3), however, with an analytic polyhedron $\mathrm{P} \Subset \mathfrak{D}$ defined by $\varphi_j \in \mathscr{O}(\mathfrak{D})$ $(1 \le k \le l)$ we add a condition such that

$$\Phi_{\mathrm{P}} : p(z) \in \mathrm{P} \to \big(z, \varphi_1(p(z)), \ldots, \varphi_l(p(z))\big) \in \mathrm{P}\Delta, \tag{3.7.6}$$
$$\mathrm{P}\Delta := \mathrm{P}\Delta((r_j)) \times \mathrm{P}\Delta((\rho_j)) \ (\subset \mathbf{C}^n \times \mathbf{C}^l),$$

is *injective*. This is possible if $\mathfrak{D}$ is holomorphically separable (in particular, a domain of holomorphy: Cf. Remark 3.6.11). Then, $\Phi_{\mathrm{P}}(\mathrm{P})$ is a submanifold (not necessarily connected) of $\mathrm{P}\Delta$.

Theorem 3.7.7 (Oka–Weil Approximation). *Let $\mathfrak{D}/\mathbf{C}^n$ be a holomorphically separable domain. Let $K = \widehat{K}_{\mathfrak{D}} \Subset \mathfrak{D}$ be a holomorphically convex set. Then a holomorphic function in a neighborhood of K can be uniformly approximated on K by elements of $\mathscr{O}(\mathfrak{D})$.*

The proof is the same as that of Theorem 3.3.12.

The notion of a *Runge pair* $(\mathfrak{D}', \mathfrak{D})$ for a domain $\mathfrak{D}/\mathbf{C}^n$ and a subdomain $\mathfrak{D}' \subset \mathfrak{D}$ is defined as in Definition 3.3.16.

Theorem 3.7.8. *Let $\mathfrak{D}/\mathbf{C}^n$ be a domain. Let $\mathfrak{D}_\nu$ $(\nu \in \mathbf{N})$ be an increasing sequence of subdomains $\mathfrak{D}_\nu \subset \mathfrak{D}_{\nu+1}$ with $\mathfrak{D} = \bigcup_{\nu=1}^{\infty}\mathfrak{D}_\nu$. If for every $\nu \in \mathbf{N}$, $\mathfrak{D}_\nu$ is Stein and $(\mathfrak{D}_\nu, \mathfrak{D}_{\nu+1})$ is a Runge pair, then $\mathfrak{D}$ is Stein.*

The proof is left to Exercise 13 at the end of this chapter. The following main results hold for domains over $\mathbf{C}^n$:

Theorem 3.7.9 (Oka). *Let $\mathfrak{D}/\mathbf{C}^n$ be a Stein domain.*

(i) *The Continuous Cousin Problem on $\mathfrak{D}$ is always solvable* (Theorem 3.4.22).
(ii) *The Cousin I Problem on $\mathfrak{D}$ is always solvable* (Theorem 3.4.31).
(iii) *The Cousin II Problem on $\mathfrak{D}$ is analytically solvable, if and only if it is topologically solvable* (Theorem 3.4.51).

(iv) *The $\bar{\partial}$-equation, $\bar{\partial}u = f$, for a function u on $\mathfrak{D}$ with a C^∞ differential form f of type $(0,1)$ has a solution if f is $\bar{\partial}$-closed* (Oka–Dolbeault Theorem 3.4.75).
(v) *The general Interpolation Problem on $\mathfrak{D}$ is always solvable* (Theorem 3.5.4).

Similarly to Proposition 3.5.3 we have:

Proposition 3.7.10. *A domain $\mathfrak{D}/\mathbf{C}^n$ is Stein if and only if an arbitrary Interpolation Problem on $\mathfrak{D}$ is solvable.*

Remark 3.7.11. It is noted that up to the present argument, Theorem 3.7.9 or Proposition 3.7.10 is not yet proved for domains of holomorphy over $\mathbf{C}^n$ (cf. Remark 3.7.5). The notion of domains of holomorphy is natural from the viewpoint of analytic continuation, but difficult to deal with. On the other hand, the notion of holomorphic separability or holomorphic convexity is characterized only by the interior of the domain, so that it is easy for the abstraction. In fact the notion of Stein manifolds (cf. [39] Definition 4.5.7), consisting of holomorphic separability and holomorphic convexity, was introduced for abstract complex manifolds by K. Stein; the term "Stein domain" comes conversely from "Stein manifold". It is not the case that the term "Stein domain" is used in Oka's papers. Now, a "Stein neighborhood" or "Stein $**$" is a commonly used convenient wording, and so we follow it.

3.8 Supplement: Ideal Boundary

Let $\pi : \mathfrak{D} \to \mathbf{C}^n$ be a domain. As $\mathfrak{D}$ is defined an abstract Hausdorff topological space, $\mathfrak{D}$ is the total space by itself, and the boundary may not be considered. But, if we intuitively think of it as a space spreading over $\mathbf{C}^n$, we may consider its boundary. Here we will formulate it mathematically.

The content of this section is not absolutely necessary to read the present book, but is a part of the theoretical background. Also for those who have a concern about the boundary of $\mathfrak{D}$ it may be helpful for a better understanding.

We would like to define the "boundary" of $\mathfrak{D}$ by the information from the inside and π. A family $\tilde{\Upsilon} = \{\Upsilon_\nu\}_{\nu=1}^{\infty}$[7] of subsets of $\mathfrak{D}$ is a *filter base* if:

(i) every $\Upsilon_\nu \neq \emptyset$;
(ii) for every Υ_{ν_1} and Υ_{ν_2} of $\tilde{\Upsilon}$ there exists an element $\Upsilon_{\nu_3} \in \tilde{\Upsilon}$ such that $\Upsilon_{\nu_3} \subset \Upsilon_{\nu_1} \cap \Upsilon_{\nu_2}$.

Example 3.8.1. A countable fundamental neighborhood system of a point of a topological space is a filter base.

Two filter bases $\tilde{\Upsilon}$ and $\tilde{\Upsilon}'$ are defined to be mutually equivalent ($\tilde{\Upsilon} \sim \tilde{\Upsilon}'$) if for every element Υ_ν of $\tilde{\Upsilon}$ (resp. Υ'_ν of $\tilde{\Upsilon}'$) there is an element of $\tilde{\Upsilon}'$ (resp. $\tilde{\Upsilon}$) contained in Υ_ν (resp. Υ'_ν); this defines, in fact, an equivalence relation, and the equivalence class is denoted by $[\tilde{\Upsilon}]$.

[7] The index set may be uncountable in general, but here it is sufficient to assume it countable.

Definition 3.8.2. An equivalence class $[\tilde{\Upsilon}]$ of a filter base with $\tilde{\Upsilon} = \{\Upsilon_\nu\}_{\nu\in\mathbf{N}}$ is an *ideal boundary point* of $\mathfrak{D}$ (more precisely, the triple $(\mathfrak{D}, \pi, \mathbf{C}^n)$) if the following conditions are satisfied:

(i) All Υ_ν are connected.
(ii) $\tilde{\Upsilon} = \{\Upsilon_\nu\}_{\nu=1}^{\infty}$ has no accumulation point in $\mathfrak{D}$; that is, for any point $p \in \mathfrak{D}$ there is a neighborhood $U(p)$ such that $\Upsilon_\nu \cap U(p) = \emptyset$ except for finitely many Υ_ν.
(iii) (a) $\pi(\Upsilon_\nu)$ $(\nu = 1, 2, \ldots)$ has a unique accumulation point $z_0 \in \mathbf{C}^n$, or (b) $\pi(\Upsilon_\nu)$ $(\nu = 1, 2, \ldots)$ diverges to infinity; i.e., for every $R > 0$, $\pi(\Upsilon_\nu) \cap \mathrm{B}(R) = \emptyset$ except for finitely many Υ_ν.
(iv) In the case of (a) above, there is a fundamental neighborhood system $\{V_\nu\}_{\nu=1}^{\infty}$ of z_0 with connected V_ν such that Υ_ν is a connected component of $\pi^{-1}V_\nu$.

In the case of (iii) (a) above, the equivalence class $[\tilde{\Upsilon}]$ is called a *relative* or *ideal boundary point* of $\mathfrak{D}$ (or $\pi : \mathfrak{D} \to \mathbf{C}^n$) over z_0, and z_0 is called the base point of it. The set $\partial^*\mathfrak{D}$ of all relative (ideal) boundary points of $\mathfrak{D}$ is called the *relative* or *ideal boundary* of $\mathfrak{D}$ (over $\mathbf{C}^n$).

Remark 3.8.3. Even in the case of $\mathfrak{D} \subset \mathbf{C}^n$, as $\mathfrak{D}$ is regarded as a domain over $\mathbf{C}^n$ with the inclusion map $\iota : \mathfrak{D} \to \mathbf{C}^n$, the ideal boundary $\partial^*\mathfrak{D}$ does not coincide with the boundary $\partial\mathfrak{D}$ as a subset of $\mathbf{C}^n$ in general. For example, let $\mathfrak{D} = \mathbf{C} \setminus \{x \in \mathbf{R} : x \geq 0\}$. Then, $\partial\mathfrak{D} = \{x \in \mathbf{R} : x \geq 0\}$. But $\partial^*\mathfrak{D}$ has two ideal boundary points over $x \in \mathbf{R}$, $x > 0$; one is q_x^+ as a boundary point of the upper-half plane, and the other is q_x^- as a boundary point of the lower-half plane (cf. Fig. 3.6):

$$\partial^*\mathfrak{D} = \{0\} \cup \{q_x^+, q_x^- : x > 0\}.$$

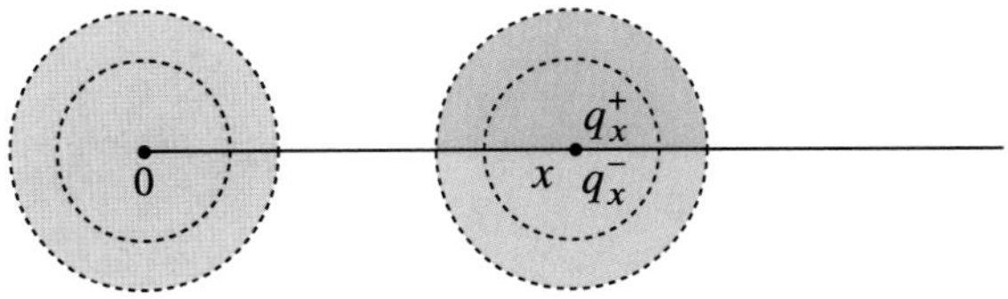

Fig. 3.6 Relative boundary.

***Note* 1.** The works of Cartan–Thullen in §3.2 were obtained in [11]; they are very important and provided the starting foundation of Oka's study.

Historically, Corollary 3.3.13 was announced in 1932 and proved in 1935 by A. Weil. Theorem 3.3.12 was proved by K. Oka in 1936/'37 (Oka I/II).

The Oka Principle (Theorem 3.4.51) says that the existence of an analytic solution is completely characterized by a topological condition; it was then accepted with a surprise, and thereafter such a characterization was termed as the *Oka Principle*. It has advanced so that there are now "Grauert's Oka Principle", "Gromov's Oka

Principle", etc.; it has served as a model case of developments of the theory of complex analysis and geometry (cf. F. Forstnerič [17]).

Dolbeault's Lemma 3.4.73 is a special case of the so-called $\bar{\partial}$-Poincaré Lemma for general (p, q)-forms, which is also referred to as Dolbeault's Lemma or the Dolbeault–Grothendieck Lemma.

By Theorem 3.4.75 and Proposition 3.4.77 we see that the solvability of the Continuous Cousin Problem is equivalent to that of $\bar{\partial}$-equation with integrable condition. In the present book (Oka theory) we will solve the Pseudoconvexity Problem by solving the Cousin I Problem, but in Hörmander [30] the approach is the opposite so that the solvability of $\bar{\partial}$-equations on a pseudoconvex domain is shown first, and then later the Cousin Problem is solved; the significant difference is that the order of the solutions is reversed. The intention is clearly written in the "Summary" of [30] Chap. 4:

> In this chapter we abandon the classical methods for solving the Cousin problems (that is, solving the Cauchy–Riemann equations), of which an example was given in section 2.7. Instead, we develop a technique for studying the Cauchy–Riemann equations where the main point is an L^2 estimate proved in sections .....

***Note* 2 (Weil condition).** Let $\Omega \subset \mathbf{C}^n$ be a domain of holomorphy, and let $\varphi \in \mathscr{O}(\Omega)$. As an immediate consequence of Exercise 10 at the end of the chapter, there are holomorphic functions $A_j(z, w) \in \mathscr{O}(\Omega \times \Omega)$ $(1 \leq j \leq n)$ satisfying

$$\varphi(z) - \varphi(w) = \sum_{j=1}^{n} A_j(z, w)(z_j - w_j). \tag{3.8.4}$$

There is an interesting history to this equation.

A. Weil (C.R. 1932) dealt with (3.8.4) for polynomial $\varphi(z)$, and obtained an integral formula (so-called Weil's integral formula) extending the Cauchy integral formula[8] and an approximation theorem of Runge type for polynomially convex subsets. He probably tried to prove it for holomorphic functions in domains of holomorphy, but (3.8.4) was left to be a condition (*Weil condition*) in his full paper, Math. Ann. 111 (1935).

Chronologically, H. Hefer took the problem and proved (3.8.4) in his Dissertation at Münster 1940; He died in the War in 1941, and the result was later published in Math. Ann. 122 (1950); his proof relied on the solution of the Cousin I Problem due to Oka I (1936) and II (1937). Without communications between Japan and Europe then, K. Oka V (1941) proved a weakened form

$$(\varphi(z) - \varphi(w))R(z, w) = \sum_{j=1}^{n} A_j(z, w)(z_j - w_j), \tag{3.8.5}$$

[8] In his essay "Souvenirs d'apprentissage, Birkhäuser 1991", A. Weil writes that he felt so much happiness by finding the formula to send a telegram on the finding from Aligarh to his friend in Calcutta, India.

where $R(z,w)$ is a holomorphic function in $\Omega \times \Omega$ with $R(z,z) = 1$, independent of $\varphi(z)$, and it is yet sufficient to deduce Weil's integral formula, which in $n = 2$ is stated roughly as

(3.8.6)
$$f(z) = \frac{1}{(2\pi i)^2} \sum_{j \neq k} \int_{\{|\varphi_j(w)|=1\} \cap \{|\varphi_k(w)|=1\}} \frac{(A_{j1}(w)A_{k2}(w) - A_{k1}(w)A_{j2}(w))f(w)}{(\varphi_j(w) - \varphi_j(z))(\varphi_k(w) - \varphi_k(z))} dw_1 dw_2,$$
$$z = (z_1, z_2) \in \mathrm{P},$$

where $\mathrm{P} \Subset \Omega$ is an analytic polyhedron defined by $|\varphi_j| < 1$ $(1 \le j \le l)$ with $\varphi_j \in \mathscr{O}(\Omega)$,

$$(\varphi_j(z) - \varphi_j(w))R(z.w) = A_{j1}(z_1 - w_1) + A_{j2}(z_2 - w_2), \quad 1 \le j \le l,$$

and $f(z)$ is a holomrophic function on $\bar{\mathrm{P}}$. Oka VI (1942) used the formula to solve the Pseudoconvexity Problem (Levi's Problem) for univalent domains in $n = 2$: At the end of the paper he wrote, as mentioned in the Preface,

> L'auteur pense que cette conclusion sera aussi indépendante des nombres de variables complexes.
>
> (The author thinks that this conclusion is true too, independently of the number of complex variables.)

S. Hitotsumatsu ([28], 1949) generalized Oka's result for univalent domains of general dimension $n \ge 2$ by making use of n variable versions of (3.8.5) and (3.8.6).

H.J. Bremermann ([8], 1954) extended Oka's result for univalent domains of general dimension $n \ge 2$ by making use of Weil's integral formula, referring to Hefer's result (3.8.4) with $R \equiv 1$, and independently F. Norguet ([45], 1954) for univalent domains of general dimension $n \ge 2$ by making use of Weil's integral formula with Weil condtion (3.8.4), referring to the coherence theorem mentioned at the beginning.

It is noted that Weil's integral formula is not yet established for multivalent domains over $\mathbf{C}^n$ (to the best of the author's knowledge). In the forthcoming Chap. 5 of the present book we will solve the Pseudoconvexity Problem for unramified multivalent domains over $\mathbf{C}^n$ by Oka's Joku-Iko Principle based on his First First Coherence Theorem 2.3.16, and the Cauchy integral (or Cousin integral) formula.

It is the same throughout these works to use the Fredholm integral equation of the second kind (type) formulated by Oka [47] (1941).

Exercises

1. Prove each item of Example 3.1.12.
2. Show that $\delta_{\mathrm{P}\Delta}(z)$ in (3.2.1) satisfies the axioms of a norm of the complex vector space $\mathbf{C}^n$: For $z, w \in \mathbf{C}^n$ and $\alpha \in \mathbf{C}$,

$$\|z+w\|_{\mathrm{P}\Delta} \le \|z\|_{\mathrm{P}\Delta} + \|w\|_{\mathrm{P}\Delta}, \quad \|\alpha z\|_{\mathrm{P}\Delta} = |\alpha| \cdot \|z\|_{\mathrm{P}\Delta};$$
$$\|z\|_{\mathrm{P}\Delta} \Longleftrightarrow z = 0.$$

3. Let $\Omega \subset \mathbf{C}^n$ be a domain, and let $K \Subset \Omega$ be a compact subset. Assume that there is no connected component of $\Omega \setminus K$ which is relatively compact in Ω. Prove that $\Omega \setminus K$ consists of finitely many connected components.
4. Prove that the relation defined by (3.4.3) is, in fact, an equivalence relation.
5. Let $\Omega \Subset \mathbf{C}$ be a bounded domain of which boundary consists of finitely many piecewise C^1 closed Jordan curves C_j, $(0 \le j \le l)$. We suppose that C_0 is the outer boundary, i.e., C_0 is the boundary of the unbounded connected component of $\mathbf{C} \setminus \bar{\Omega}$ and C_j $(1 \le j \le l)$ are the boundaries of the bounded connected components ω_j of $\mathbf{C} \setminus \bar{\Omega}$ $(\partial \omega_j = C_j$ $(1 \le j \le l)$, the inner boundaries). Take a point $p_j \in \omega_j$ for every $1 \le j \le l$.
   Show that a holomorphic function on $\bar{\Omega}$ can be approximated uniformly on compact subsets of Ω by rational functions with poles only at p_j, $1 \le j \le l$.
   Hint: Use the Cauchy integral formula $f(z) = \frac{1}{2\pi i} \sum_{j=0}^{l} \int \frac{f(\zeta)}{\zeta - z} d\zeta$ and the proof of Lemma 3.3.20. Note that the integral is Riemann's integral (cf. [38] §7.1).
6. Prove that (i) and (ii) of Definition 3.4.13 are equivalent.
7. Let $A \subset \mathbf{C}^n$ be a non-empty subset. Let $d(z, A)$ be the distance from a point $z \in \mathbf{C}^n$ to A defined by (3.1.3).
   Prove the following:
   a. $|d(z, A) - d(z', A)| \le \|z - z'\|$ $(z, z' \in \mathbf{C}^n)$. In particular, $d(z, A)$ is Lipschitz continuous.
   b. Let $\Omega \subset \mathbf{C}^n$ be an open subset, and let $E \subset \Omega$ be a closed subset. Prove that for $z \in \Omega$, $d(z, E) = 0$ if and only if $z \in E$.
   c. Let $E, F \subset \Omega$ be two closed subsets with $E \cap F = \emptyset$. Set

$$\rho(z) = \frac{d(z, F)}{d(z, E) + d(z, F)}, \quad z \in \Omega.$$

   Prove that $\rho \in \mathscr{C}^0(\Omega)$, $\rho(z) = 1$ on E, and $\rho(z) = 0$ on F.
8. Solve the interpolation problem of one variable by making use of the Weierstrass and Mittag-Leffler Theorems.
   Hint: Cf. (3.5.2).
9. Let $\Omega \subset \mathbf{C}^n$ be a holomorphically convex domain, and let $f_j \in \mathscr{O}(\Omega), 1 \le j \le q$, be finitely many holomorphic functions such that for every $z \in \Omega$ there is some $f_j(z) \ne 0$. Prove that there are holomorphic functions $a_j \in \mathscr{O}(\Omega), 1 \le j \le q$, satisfying

$$a_1(z)f_1(z)+\cdots+a_q(z)f_q(z)=1, \qquad z\in\Omega.$$

(This is due to H. Cartan [9] for compact cylinder domains.)
Hint: Use Exercise 6 and 7 of the previous chapter. By the Oka map $\varphi:\bar{\mathrm{P}}\to\overline{\mathrm{P}\Delta}$ we embed an analytic polyhedron $\mathrm{P}\Subset\Omega$ into a higher dimensional polydisk $\overline{\mathrm{P}\Delta}$, and extend f_j $(1\le j\le q)$ to $\tilde{f}_j\in\mathcal{O}(\overline{\mathrm{P}\Delta})$. Moreover we add finitely many $\tilde{f}_j\in\mathcal{O}(\overline{\mathrm{P}\Delta})$, $q+1\le j\le\tilde{q}$, which take the value 0 on $\varphi(\bar{\mathrm{P}})$ and have no common zero on the common zero set of $\tilde{f}_j$ $(1\le j\le q)$, and then obtain a finite system of holomorphic functions $\tilde{f}_j, 1\le j\le\tilde{q}$ without common zeros on $\overline{\mathrm{P}\Delta}$. Follow the proof of Theorem 3.5.4 for the rest.

10. Let $\Omega\subset\mathbf{C}^n$ be a domain of holomorphy with the natural coordinate system $(z_1,\ldots,z_n)$, and let $S=\Omega\cap\{z_1=z_2=\cdots=z_p=0\}$ with $1\le p\le n$.
Show that if $f\in\mathcal{O}(\Omega)$ satisfies $f|_S\equiv 0$, then there are holomorphic functions $f_j\in\mathcal{O}(\Omega), 1\le j\le p$, such that $f(z)=\sum_{j=1}^{p}f_j(z)\cdot z_j$ $(z\in\Omega)$.
11. Let $\Omega\subset\mathbf{C}^n$ be a domain and let $\mathscr{B}\mathcal{O}(\Omega)$ denote the set of all bounded holomorphic functions on Ω. For a compact subset $K\Subset\Omega$ the $\mathscr{B}\mathcal{O}(\Omega)$-hull is defined by

$$\widehat{K}_{\mathscr{B}\mathcal{O}(\Omega)}=\left\{z\in\Omega:|f(z)|\le\max_K|f|,\ \forall f\in\mathscr{B}\mathcal{O}(\Omega)\right\}.$$

Assume $K=\widehat{K}_{\mathscr{B}\mathcal{O}(\Omega)}$. Prove that a holomorphic function in a neighborhood of K is uniformly approximated on K by elements of $\mathscr{B}\mathcal{O}(\Omega)$.
12. Assume that a domain $\pi:\mathfrak{D}\to\mathbf{C}^n$ is finitely sheeted. Let $\mathrm{P}\Delta$ be a polydisk with center at the origin of $\mathbf{C}^n$. For $R,c>0$ we set

$$K=\{p\in\mathfrak{D}:\|\pi(p)\|\le R,\ \delta_{\mathrm{P}\Delta}(p,\partial\mathfrak{D})\ge c\},$$

where $\|\cdot\|$ is the euclidean norm.
Show that K is a compact subset of $\mathfrak{D}$.
Hint: Show that an arbitrary sequence of points of K has a convergent subsequence. Use $\bar{\mathrm{B}}:=\{\|z\|\le R\}\subset\bigcup_{z\in\bar{\mathrm{B}}}\left(z+\frac{c}{2}\cdot\mathrm{P}\Delta\right)$ (an open covering).
13. Prove Theorem 3.7.8.
14. Let $\mathfrak{D}/\mathbf{C}^n$ be a domain and let $\hat{\mathfrak{D}}/\mathbf{C}^n$ be the envelope of holomorphy of $\mathfrak{D}$.
    a. If $f\in\mathcal{O}(\mathfrak{D})$ does not take the value $\alpha\in\mathbf{C}$, then prove that the analytic continuation $\hat{f}\in\mathcal{O}(\hat{\mathfrak{D}})$ of f does not take the value α, either.
    b. With the same notation as above, if $|f|<M$ (or $|f|\le M$), show that $|\hat{f}|<M$ (or $|\hat{f}|\le M$).
15. Prove Proposition 3.4.15 for a general multivalent unramified domain $\mathfrak{D}/\mathbf{C}^n$.
16. Let $G:=\{x+iy:0<x<1,0<y<1\}\setminus\left\{\frac{1}{\nu}+iy:0<y<\frac{1}{2},\nu=1,2,\ldots\right\}\subset\mathbf{C}$. Describe the relative boundary ∂^*G with respect to the inclusion map $\iota:G\to\mathbf{C}$. Show that the set of base points of ∂^*G is *not* closed in $\mathbf{C}$.

Chapter 4
Pseudoconvex Domains I — Problem and Reduction

In the present and the next chapter we solve the last Pseudoconvexity Problem of the Three Big Problems on unramified domains over $\mathbf{C}^n$.[1] We fully employ the results obtained previously, but the path is yet long. Here, introducing the notion of plurisubharmonic (or pseudoconvex) functions, we formulate the Pseudoconvexity Problems and discuss their relations.

4.1 Plurisubharmonic Functions

4.1.1 Subharmonic Functions (One Variable)

With respect to the complex coordinate $z = x + iy \in \mathbf{C}$ the holomorphic partial differential operator $\frac{\partial}{\partial z}$ and the anti-holomorphic partial differential operator $\frac{\partial}{\partial \bar{z}}$ are defined by (1.1.6). Repeating them, we have

$$(4.1.1) \qquad \frac{\partial^2}{\partial z \partial \bar{z}} = \frac{\partial}{\partial z}\frac{\partial}{\partial \bar{z}} = \frac{1}{4}\left(\frac{\partial^2}{\partial x^2} + \frac{\partial^2}{\partial y^2}\right) = \frac{1}{4}\Delta.$$

Here, Δ is well-known as the Laplacian. With respect to the polar coordinate $z = re^{i\theta}$ $(r > 0)$ we have

$$(4.1.2) \qquad \Delta = \frac{\partial^2}{\partial r^2} + \frac{1}{r}\frac{\partial}{\partial r} + \frac{1}{r^2}\frac{\partial^2}{\partial \theta^2}$$

(cf. Exercise 2 at the end of this chapter).

Let $\varphi(z)$ be a function on $\Delta(a;R)$ $(R > 0)$. We define the mean integration of $\varphi(z)$ on the circle $\{|z-a| = r\}$ by

[1] For those studying the present subject for the first time, it is recommended to read this chapter and the next, assuming that the domains are univalent.

J. Noguchi, *Basic Oka Theory in Several Complex Variables*, Universitext,
https://doi.org/10.1007/978-981-97-2056-9_4

$$(4.1.3) \qquad M_\varphi(a;r)=\frac{1}{2\pi}\int_0^{2\pi}\varphi(a+re^{i\theta})d\theta, \qquad 0\le r<R,$$

provided that it exists.

Lemma 4.1.4 (Jensen's formula). *For $\varphi\in C^2(\Delta(a;R))$ we have*

$$M_\varphi(a;r)=\varphi(a)+\frac{1}{2\pi}\int_0^r\frac{dt}{t}\int_{\Delta(a;t)}\Delta\varphi\,d\lambda, \quad 0\le r<R.$$

Here, $d\lambda=dxdy$ with $z=x+iy$ denotes the standard surface measure on **C**.

Proof. By a parallel transform we may assume $a=0$. We set $M(r)=M_\varphi(0;r)$. It follows from (4.1.2) that

$$(4.1.5) \qquad \int_0^{2\pi}\Delta\varphi(re^{i\theta})d\theta=\int_0^{2\pi}\left(\frac{\partial^2}{\partial r^2}+\frac{1}{r}\frac{\partial}{\partial r}+\frac{1}{r^2}\frac{\partial^2}{\partial\theta^2}\right)\varphi(re^{i\theta})d\theta.$$

For each r, $\varphi(a+re^{i\theta})$ is a periodical function in θ with period 2π, and of C^2-class. The function $\frac{\partial^2}{\partial\theta^2}\varphi(a+re^{i\theta})$ has a primitive function $\frac{\partial}{\partial\theta}\varphi(a+re^{i\theta})$, which is periodic. Therefore we see that

$$\int_0^{2\pi}\frac{\partial^2}{\partial\theta^2}\varphi(re^{i\theta})d\theta=0,$$

$$\left(\frac{d^2}{dr^2}+\frac{1}{r}\frac{d}{dr}\right)M(r)=\frac{1}{2\pi}\int_0^{2\pi}\Delta\varphi(re^{i\theta})d\theta.$$

The left-hand side above is written as $\frac{1}{r}\frac{d}{dr}\left(r\frac{d}{dr}M(r)\right)$, and so

$$(4.1.6) \qquad \frac{d}{dr}\left(r\frac{d}{dr}M(r)\right)=\frac{r}{2\pi}\int_0^{2\pi}\Delta\varphi(re^{i\theta})d\theta.$$

On the other hand,

$$r\frac{d}{dr}M(r)=\frac{1}{2\pi}\int_0^{2\pi}\left(r\cos\theta\,\frac{\partial\varphi}{\partial x}+r\sin\theta\,\frac{\partial\varphi}{\partial y}\right)d\theta\to 0 \quad (r\to+0).$$

Thus, by integrating (4.1.6) in r, we get

$$r\frac{d}{dr}M(r)=\frac{1}{2\pi}\int_0^r sds\int_0^{2\pi}\Delta\varphi(se^{i\theta})d\theta=\frac{1}{2\pi}\int_{\Delta(r)}\Delta\varphi\,d\lambda;$$

$$\frac{d}{dr}M(r)=\frac{1}{2\pi r}\int_{\Delta(r)}\Delta\varphi\,d\lambda.$$

Since $M(0)=\varphi(0)$, we integrate the above equation again to deduce

$$M(r) - \varphi(0) = \frac{1}{2\pi} \int_0^r \frac{dt}{t} \int_{\Delta(t)} \Delta\varphi \, d\lambda. \qquad \square$$

Let $U \subset \mathbf{C}$ be an open set, and consider a function $\varphi : U \to [-\infty, \infty)$, which is allowed to take value $-\infty$.

Definition 4.1.7. The function φ is *subharmonic* if the following two conditions are satisfied:

(i) (Upper semi-continuity) φ is upper semi-continuous; that is, for every $c \in \mathbf{R}$, $\{z \in U : \varphi(z) < c\}$ is an open set; this is equivalent to

$$\overline{\lim_{z \to a}} \, \varphi(z) \le \varphi(a), \quad \forall a \in U.$$

(Cf. Exercise 3 at the end of this chapter.)

(ii) (Submean property) For every disk $\Delta(a;r) \Subset U$, $\varphi(a) \le M_\varphi(a;r)$.

By definition, $\varphi \equiv -\infty$ is a subharmonic function. If φ_j $(j = 1,2)$ are subharmonic and $c_j > 0$ $(j = 1,2)$ are constants, then $c_1\varphi_1 + c_2\varphi_2$ is subharmonic.

Theorem 4.1.8. *Let $\varphi \in C^2(U)$.*

(i) *φ is subharmonic if and only if $\Delta\varphi \ge 0$.*

(ii) *If φ is subharmonic and $\Delta(a;R) \subset U$, then $M_\varphi(a;r)$ is monotone increasing and continuous in $0 \le r < R$, and of C^1-class in $0 < r < R$.*

Proof. (i) Assume that φ is subharmonic. For a point $a \in U$, the Taylor expansion of $\varphi(a+\zeta)$ up to order 2 in sufficiently small $\zeta \in \mathbf{C}$ (i.e., small $|\zeta|$) is written as

$$\begin{aligned}(4.1.9) \qquad \varphi(a+\zeta) &= \varphi(a) + \frac{\partial\varphi}{\partial z}(a)\zeta + \frac{\partial\varphi}{\partial \bar z}(a)\bar\zeta \\ &\quad + \frac{1}{2}\frac{\partial^2\varphi}{\partial z^2}(a)\zeta^2 + \frac{1}{2}\frac{\partial^2\varphi}{\partial \bar z^2}(a)\bar\zeta^2 + \frac{\partial^2\varphi}{\partial z \partial \bar z}(a)|\zeta|^2 + o(|\zeta|^2).\end{aligned}$$

With $\zeta = re^{i\theta}$ we take the integration on $0 \le \theta \le 2\pi$:

$$0 \le \frac{1}{2\pi r^2} \int_0^{2\pi} (\varphi(a + re^{i\theta}) - \varphi(a)) d\theta = \frac{\partial^2\varphi}{\partial z \partial \bar z}(a) + o(1).$$

As $r \searrow 0$, we deduce

$$(4.1.10) \qquad \Delta\varphi(a) = 4\frac{\partial^2\varphi}{\partial z \partial \bar z}(a) \ge 0, \qquad a \in U.$$

Conversely, if $\Delta\varphi \ge 0$, Lemma 4.1.4 implies that $M_\varphi(a;r) \ge \varphi(0)$, and hence φ is subharmonic.

(ii) It immediately follows from (i) above and Lemma 4.1.4. $\square$

4.1.2 Plurisubharmonic Functions

Let $n \geq 2$ and let $U \subset \mathbf{C}^n$ be an open set. As in the previous subsection we consider a function $\varphi : U \to [-\infty, \infty)$.

Definition 4.1.11. The function φ is said to be *plurisubharmonic* or *pseudoconvex*[2] if the following conditions are satisfied:

(i) φ is upper semi-continuous.
(ii) For every point $z \in U$ and vector $v \in \mathbf{C}^n \setminus \{0\}$ the function

$$\zeta \in \mathbf{C} \to \varphi(z+\zeta v) \in [-\infty, \infty) \tag{4.1.12}$$

satisfies the submean property (subharmonic) in $\zeta \in \mathbf{C}$ where it is defined.

We denote by $\mathscr{P}(U)$ the set of all plurisubharmonic function on U, and set $\mathscr{P}^k(U) = \mathscr{P}(U) \cap \mathscr{C}^k(U)$ $(0 \leq k \leq \infty)$.

Suppose that $\varphi(z)$ $(z = (z_1, \ldots, z_n) \in U)$ is of C^2-class. In (ii) above, we have with $v = (v_1, \ldots, v_n)$

$$\left.\frac{\partial^2}{\partial \zeta \partial \bar{\zeta}}\right|_{\zeta=0} \varphi(z+\zeta v) = \sum_{j,k} \frac{\partial^2 \varphi}{\partial z_j \partial \bar{z}_k}(z) v_j \bar{v}_k,$$

$$L[\varphi](v) = L[\varphi](z; v) := \sum_{j,k} \frac{\partial^2 \varphi}{\partial z_j \partial \bar{z}_k}(z) v_j \bar{v}_k. \tag{4.1.13}$$

We call $L[\varphi](v)$ $(L[\varphi](z;v))$ the *Levi form* of φ. The Levi form $L[\varphi](v)$ is (resp. semi-) positive definite if

$$L[\varphi](v) > 0 \text{ (resp. } \geq 0), \qquad \forall v \in \mathbf{C}^n \setminus \{0\}.$$

We write $L[\varphi] > 0$ (resp. $L[\varphi] \geq 0$) for it.

It follows from Theorem 4.1.8 (i):

Theorem 4.1.14. *For $\varphi \in C^2(U)$, $\varphi \in \mathscr{P}(U)$ if and only if $L[\varphi] \geq 0$.*

Definition 4.1.15. We say that a function $\varphi \in C^2(U)$ is *strongly plurisubharmonic* or *strongly pseudoconvex* if $L[\varphi] > 0$.

By definition, $\varphi \equiv -\infty$ is plurisubharmonic. In general the following properties hold:

Theorem 4.1.16. (i) *If φ_j $(j = 1, 2)$ are plurisubharmonic and c_j $(j = 1, 2)$ are positive constants, then $c_1 \varphi_1 + c_2 \varphi_2$ is plurisubharmonic.*

---

[2] This notion of functions was first defined by K. Oka who termed it "fonction pseudoconvexe" (Oka VI, 1942). Around that time, P. Lelong also introduced the same notion, calling it "fonction plurisousharmonique"; nowadays, the latter is used more frequently.

(ii) *Let $\varphi \in \mathscr{P}(U)$ such that $\varphi(a) > -\infty$ at some point $a \in U$. Then φ is locally integrable in a connected component of U containing a with respect to the Lebesgue measure of $\mathbf{C}^n \cong \mathbf{R}^{2n}$.*

(iii) (**Maximum Principle**) *Let $\varphi \in \mathscr{P}(U)$. If φ attains the maximum at $a \in U$, then φ is a constant function in a connected component of U containing a.*

(iv) *Let $\varphi \in \mathscr{P}(U)$ and let ψ be a monotone increasing convex function defined on $[\inf \varphi, \sup \varphi)$. Then $\psi \circ \varphi \in \mathscr{P}(U)$; here, if $\inf \varphi = -\infty$, we set $\psi(-\infty) = \lim_{t \to -\infty} \psi(t)$.*

(v) *If $\varphi_\nu \in \mathscr{P}(U)$, $\nu = 1, 2, \ldots$, is a decreasing sequence of plurisubharmonic functions, then the limit $\varphi(z) = \lim_{\nu \to \infty} \varphi_\nu(z)$ is a plurisubharmonic function.*

(vi) *Let $\{\varphi_\lambda\}_{\lambda \in \Lambda}$ be a family of $\mathscr{P}(U)$. If $\varphi(z) := \sup_{\lambda \in \Lambda} \varphi_\lambda(z)$ is upper semi-continuous, then $\varphi(z) \in \mathscr{P}(U)$. In particular, for finite Λ, $\varphi(z)$ is upper semi-continuous, and so plurisubharmonic.*

Proof. (i) It is immediate from the definition.

(ii) Suppose that U is connected and $\varphi(a) > -\infty$ at some $a \in U$. Let $d\lambda$ denote the Lebesgue measure of $\mathbf{C}^n$. Take an open ball $\mathrm{B}(a;r) = a + \mathrm{B}(r) \Subset U$. Since φ is upper semi-continuous, φ has a upper bound $C \in \mathbf{R}$ on $\mathrm{B}(a;r)$. For $w \in \mathrm{B}(r)$ we get

$$-\infty < \varphi(a) \le \frac{1}{2\pi} \int_0^{2\pi} \varphi(a + e^{i\theta} w) d\theta \le C.$$

With $V(r) := \int_{\mathrm{B}(r)} d\lambda = \frac{\pi^n}{n!} r^{2n}$ we obtain

$$-\infty < \varphi(a) \le \frac{1}{2\pi V(r)} \int_0^{2\pi} \int_{\mathrm{B}(r)} d\lambda(w) \varphi(a + e^{i\theta} w) d\theta \le C. \tag{4.1.17}$$

It follows from the change of order of the integrations and the rotation invariance of the Lebesgue measure, $d\lambda(w) = d\lambda(e^{i\theta} w)$, that

$$-\infty < \varphi(a) \le \frac{1}{V(r)} \int_{\mathrm{B}(r)} \varphi(a + w) d\lambda(w) \le C. \tag{4.1.18}$$

Therefore φ is integrable over $\mathrm{B}(a;r) (\Subset U)$, so that $\varphi(b) > -\infty$ for almost all $b \in \mathrm{B}(a;r)$ with respect to $d\lambda$.

Let U' be the set of all $a \in U$ such that φ is integrable over every $\mathrm{B}(a;r) \Subset U$. From (4.1.18) we easily deduce that U' is non-empty, open and closed. Hence, $U' = U$, and φ is locally integrable on U.

(iii) Suppose that U is connected, and that $\varphi(a)$ is the maximum for a point $a \in U$. It follows from (4.1.18) that

$$\int_{\mathrm{B}(r)} (\varphi(a + w) - \varphi(a)) = 0.$$

If there were a point $w_0 \in \mathrm{B}(r)$ with $c := \varphi(a + w_0) - \varphi(a) < 0$, we deduce from the upper semi-continuity that $\{\varphi(a + w) - \varphi(a) < c/2\}$ is a non-empty open set, and since $\varphi(a+w)-\varphi(a)\le 0$ everywhere in $\mathrm{B}(r)$, we get

$$\int_{\mathrm{B}(r)}(\varphi(a+w)-\varphi(a))<0.$$

This contradicts (4.1.18). Therefore, $\varphi|_{\mathrm{B}(a;r)}\equiv\varphi(a)$. Set $V=\{b\in U:\varphi(b)=\varphi(a)\}$. It follows from the above arguments that V is a non-empty open subset. By the upper semi-continuity of $\varphi(z)$, V is closed; hence, $V=U$.

(iv) It is first noted that ψ is continuous by the condition. Therefore $\psi\circ\varphi$ is upper semi-continuous. Let $\mathrm{B}(a;R)\Subset U$ be any open ball, and let $v\in\mathbf{C}^n$ be a vector with $\|v\|=1$. It suffices to show the submean property of $\psi\circ\varphi(a+\zeta v)$ at $\zeta=0$. It follows from the convexity of ψ that

$$\frac{1}{2\pi}\int_0^{2\pi}\psi(\varphi(a+re^{i\theta}v))d\theta\ge\psi\left(\frac{1}{2\pi}\int_0^{2\pi}\varphi(a+re^{i\theta}v)d\theta\right),\quad 0<r\le R.$$

Because of the submean property of φ

$$\psi\left(\frac{1}{2\pi}\int_0^{2\pi}\varphi(a+re^{i\theta}v)d\theta\right)\ge\psi(\varphi(a)).$$

Therefore, the submean property of $\psi\circ\varphi(a+\zeta v)$ follows, and hence $\psi\circ\varphi$ is subharmonic.

(v) The limit function of a decreasing sequence of upper semi-continuous functions is upper semi-continuous. The submean property follows from Lebesgue's monotone convergence theorem.[3]

(vi) It is immediate from the definition. □

Example 4.1.19. (i) Let $f\in\mathscr{O}(U)$. By the computation of the Levi form we see $\log(c+|f|^2)\in\mathscr{P}(U)$ $(c>0)$. As $c\searrow 0$, $\log|f|\in\mathscr{P}(U)$. Therefore, $|f|^\rho\in\mathscr{P}^0(U)$ for all $\rho>0$.

(ii) It follows from the computations of the Levi forms that $\|z\|^2$ and $\log(c+\|z\|^2)$ with $c>$ are strongly plurisubharmonic in $z\in\mathbf{C}^n$. As $c\searrow 0$, $\log\|z\|\in\mathscr{P}(\mathbf{C}^n)$; hence, $\|z\|^\rho\in\mathscr{P}^0(\mathbf{C}^n)$ $(\rho>0)$.

(iii) $\varphi(z)=\sum_{j=1}^n(\Re z_j)^2$ (or $\sum_{j=1}^n(\Im z_j)^2$) is strongly plurisubharmonic.

4.1.3 Smoothing

We denote the Lebesgue measure of $\mathbf{C}^n\cong\mathbf{R}^{2n}(\ni(x_1,y_1,x_2,y_2,\dots,x_n,y_n))$ by

$$d\lambda=dx_1dy_1dx_2dy_2\cdots dx_ndy_n.$$

[3] Those who are not familiar with the Lebesgue integration theory may assume that the functions are continuous and the convergence is locally uniform; they do not cause trouble in understanding Oka theory itself.

For $\varepsilon > 0$ we put

$$U_\varepsilon = \{z \in U : d(z, \partial U) > \varepsilon\}$$

(cf. (3.1.3)). We take $\chi(z) = \chi(|z_1|, \ldots, |z_n|) \in C_0^\infty(\mathbf{C}^n)$ such that

$$\chi(z) \geq 0, \quad \operatorname{Supp} \chi \subset \mathrm{B}(= \mathrm{B}(1)), \quad \int \chi(z) d\lambda = 1,$$

and put

$$\chi_\varepsilon(z) = \chi(\varepsilon^{-1} z)\varepsilon^{-2n}, \quad \varepsilon > 0.$$

For a locally integrable function φ we define the *smoothing* by

$$\begin{aligned} \varphi_\varepsilon(z) = \varphi * \chi_\varepsilon(z) &= \int_{\mathbf{C}^n} \varphi(w)\chi_\varepsilon(w - z) d\lambda(w) \\ &= \int_{\mathbf{C}^n} \varphi(z + w)\chi_\varepsilon(w) d\lambda(w) \\ &= \int_{\mathrm{B}} \varphi(z + \varepsilon w)\chi(w) d\lambda(w), \quad z \in U_\varepsilon. \end{aligned} \tag{4.1.20}$$

Remark 4.1.21. While we used a unit open ball B, it is also possible to use a polydisk $\mathrm{P}\Delta$ with center at the origin 0; with $U_\varepsilon = \{z \in U : \delta_{\mathrm{P}\Delta}(z, \partial U) > \varepsilon\}$ we may define similarly χ_ε, and the arguments below work similarly.

Proposition 4.1.22. (i) *$\varphi_\varepsilon(z)$ is of C^∞-class in U_ε.*
(ii) *If φ is continuous, the convergence* $\lim_{\varepsilon \to +0} \varphi_\varepsilon(z) = \varphi(z)$ *is locally uniform.*

Proof. (i) It follows from the change of the integration and a partial differentiation in (4.1.20).
(ii) It is immediate from the last line of (4.1.20). □

Theorem 4.1.23. *Let $\varphi \in \mathscr{P}(U)$ such that $\varphi \not\equiv -\infty$ on each connected component of U.*
(i) *The smoothing $\varphi_\varepsilon(z)$ is a C^∞ plurisubharmonic function in U_ε.*
(ii) *As $\varepsilon \searrow 0$, $\varphi_\varepsilon(z)$ converges monotone decreasingly to $\varphi(z)$.*

Proof. (i) It has been already proved that φ_ε is of C^∞-class. For a vector $v \in \mathbf{C}^n \setminus \{0\}$ and $\zeta \in \mathbf{C}$, we show the submean property of $\varphi_\varepsilon(z + \zeta v)$ in ζ where it is defined. By a parallel transform it suffices to show it at $\zeta = 0$:

$$\varphi_\varepsilon(z) \leq \frac{1}{2\pi} \int_0^{2\pi} \varphi_\varepsilon(z + re^{i\theta} v) d\theta.$$

This follows from the interchange of the integrations.
(ii) With an arbitrary open set $V \Subset U$ we consider

$$z \in V, \quad 0 < \varepsilon < \varepsilon_0 := \frac{1}{2} \inf\{\|z' - z''\| : z' \in V, z'' \in \partial U\}.$$

By the rotational symmetries $d\lambda(w) = d\lambda(e^{i\theta}w)$, and $\chi(w) = \chi(e^{i\theta}w)$ $(0 \le \theta \le 2\pi)$ we have

$$\begin{aligned}\varphi_\varepsilon(z) &= \int_{\mathbf{C}^n} \varphi(z+\varepsilon w)\chi(w)d\lambda(w) \\ &= \int_{\mathbf{C}^n} d\lambda(w)\frac{1}{2\pi}\int_0^{2\pi} d\theta\, \varphi(z+\varepsilon e^{i\theta}w)\chi(w) \\ &\ge \varphi(z)\int_{\mathbf{C}^n} \chi(w)d\lambda(w) = \varphi(z).\end{aligned} \tag{4.1.24}$$

Note that φ has a upper bound $C_0 \in \mathbf{R}$ on $V + \mathrm{B}(\varepsilon_0)$ $(\Subset U)$. It follows that

$$\varphi(z) \le \varphi_\varepsilon(z) \le C_0, \qquad z \in V. \tag{4.1.25}$$

We first prove the point-wise convergence

$$\lim_{\varepsilon\to+0} \varphi_\varepsilon(z) = \varphi(z), \tag{4.1.26}$$

where the case of $\varphi(z) = -\infty$ is included. For, in the case of $\varphi(z) > -\infty$, the upper semi-continuity implies that $\{z' \in V : \varphi(z') < \varphi(z) + c\}$ is an open set containing z for $c > 0$. Therefore, for an arbitrary sufficiently small $\varepsilon > 0$,

$$\varphi(z) \le \varphi_\varepsilon(z) \le \varphi(z) + c\,; \quad \lim_{\varepsilon\to+0} \varphi_\varepsilon(z) = \varphi(z).$$

In the case of $\varphi(z) = -\infty$, the above c may be chosen arbitrarily small ($c < 0$, large $|c|$); hence

$$\varphi_\varepsilon(z) \le c\,; \quad \lim_{\varepsilon\to+0} \varphi_\varepsilon(z) = -\infty.$$

With a sufficiently small $\delta > 0$ we apply Theorem 4.1.8 (ii) for $\varphi_\delta(z+\zeta w)$ $(|\zeta| = r)$ to obtain

$$\begin{aligned}&\int_{\mathbf{C}^n} d\lambda(w)\frac{1}{2\pi}\int_0^{2\pi} d\theta \varphi_\delta(z+\varepsilon e^{i\theta}w) \\ &\le \int_{\mathbf{C}^n} d\lambda(w)\frac{1}{2\pi}\int_0^{2\pi} d\theta \varphi_\delta(z+\varepsilon' e^{i\theta}w), \quad 0 < \varepsilon < \varepsilon' < \varepsilon_0.\end{aligned} \tag{4.1.27}$$

Here, note that as $\delta \to +0$, $\varphi_\delta(z)$ $(z \in V)$ converges to $\varphi(z)$ point-wise, and (4.1.25) holds for φ_δ. By Lebesgue's convergence theorem[4] we deduce from (4.1.27) by letting $\delta \to 0$ that

$$\varphi_\varepsilon(z) \le \varphi_{\varepsilon'}(z), \qquad 0 < \varepsilon < \varepsilon' < \varepsilon_0. \tag{4.1.28}$$

This together with (4.1.26) implies the monotone convergence. □

[4] If φ is continuous, the convergence follows directly from 4.1.22 (ii). In fact, it is such a case when we use it later.

Theorem 4.1.29. (i) *The plurisubharmonicity is a local property.*

(ii) *Let $U \subset \mathbf{C}^m$ and $V \subset \mathbf{C}^n$ be open sets, and let $f : V \to U$ be a holomorphic map. For a plurisubharmonic function φ in U, the pull-back $f^*\varphi = \varphi \circ f$ is plurisubharmonic in V.*

Proof. By Theorem 4.1.23 it is sufficient to show both for C^∞ plurisubharmonic functions.

(i) Let $\varphi(z) \in C^\infty(U)$. The plurisubharmonicity of φ is characterized by the semi-positivity of the Levi form $L[\varphi](v)$, and so it is a local property.

(ii) Suppose $f : V \to U$ is given by $(z_j) = (f_j(\zeta_1, \ldots, \zeta_m))$. Then the Levi form is transformed to

$$L[f^*\varphi](v_1, \ldots, v_m) = L[\varphi]\left(\sum_{k=1}^{m} \frac{\partial f_1}{\partial \zeta_k} v_k, \ldots, \sum_{k=1}^{m} \frac{\partial f_n}{\partial \zeta_k} v_k\right).$$

Therefore, if φ is plurisubharmonic, $L[f^*\varphi]$ is semi-positive definite, so that $f^*\varphi$ is plurisubharmonic. □

Because of Theorem 4.1.29 we may define plurisubharmonic functions on a domain $\mathfrak{D}/\mathbf{C}^n$ as follows.

Definition 4.1.30. Let $\mathfrak{D}/\mathbf{C}^n$ be an unramified domain over $\mathbf{C}^n$. A function $\varphi : \mathfrak{D} \to [-\infty, \infty)$ is *plurisubharmonic* or *pseudoconvex* if for every point $p \in \mathfrak{D}$ there is a polydisk neighborhood $\mathrm{P}\Delta(p)$ of p such that, regarded as $\mathrm{P}\Delta(p) \subset \mathbf{C}^n$, $\varphi(z)$ is plurisubharmonic in $z \in \mathrm{P}\Delta(p)$. We denote the set of all plurisubharmonic functions on $\mathfrak{D}$ by $\mathscr{P}(\mathfrak{D})$, and set $\mathscr{P}^k(\mathfrak{D}) = \mathscr{P}(\mathfrak{D}) \cap \mathscr{C}^k(\mathfrak{D})$ $(0 \le k \le \infty)$.

4.2 Hartogs' Separate Analyticity

In this section we give a proof of the separate analyticity (cf. Definition 1.1.9 (ii)) due to F. Hartogs, proved in 1906 (see Theorem 4.2.9). The statement is very simple but the proof is not so simple. Those readers who know the Baire Category Theorem may skip the next subsection.

4.2.1 Baire Category Theorem

For a moment we deal with a general topological space, but in fact, we will soon assume it to be a metric space.

Definition 4.2.1. A topological space X is called a *Baire space*, if every countable union $\bigcup_{\nu \in \mathbf{N}} F_\nu$ of closed subsets $F_\nu (\subset X)$ $(\nu \in \mathbf{N})$ without interior points contains no interior point.

Remark 4.2.2. (i) As an example of non-Baire space, the set $\mathbf{Q}$ ($\subset \mathbf{R}$) endowed with the induced topology from the metric topology of $\mathbf{R}$ is *not* a Baire space. For a singleton $\{a\}$ ($\subset \mathbf{Q}$) is a closed set, and $\mathbf{Q}$ is a countable set; however, $\{a\}$ contains no interior point.

(ii) In Definition 4.2.1, if F_ν are only finitely many closed subsets, then the finite union $\bigcup F_\nu$ contains no interior point (see Exercise 1 at the end of the chapter).

Theorem 4.2.3 (Baire). (i) (Baire Category Theorem) *A complete metric space is Baire.*

(ii) *The euclidean n-space $\mathbf{R}^n$ with the standard metric is Baire.*

(iii) *Any closed subset of $\mathbf{R}^n$ is Baire.*

(iv) *Any open subset of $\mathbf{R}^n$ is Baire.*

Proof. (i) Let X be a space with a complete metric $d(x,y)$. Let $A_\nu \subset X$, $\nu = 1,2,\ldots$ be closed subsets without interior points, and set $A = \bigcup_{\nu=1}^{\infty} A_\nu$. For $a \in X$ and $r > 0$ we put

$$U(a;r) = \{x \in X : d(a,x) < r\}.$$

It suffices to show that $U(a;r) \not\subset A$ for every $U(a;r)$. Since $U(a;r) \setminus A_1 \neq \emptyset$, there is a point $x_1 \in U(a;r) \setminus A_1$. Because A_1 is closed, there is a number r_1 such that $0 < r_1 < 1/2$ and $\overline{U}(x_1;r_1) = \{x \in X : d(x,x_1) \le r_1\} \subset U(a;r) \setminus A_1$. Similarly, since $U(x_1;r_1) \setminus A_2 \neq \emptyset$, there are $x_2 \in U(x_1;r_1) \setminus A_2$ and $0 < r_2 < 1/2^2$ such that $\overline{U}(x_2;r_2) \subset U(x_1;r_1) \setminus A_2$. Inductively, we have

$$\overline{U}(x_\nu;r_\nu) \subset U(x_{\nu-1};r_{\nu-1}) \setminus \bigcup_{\mu=1}^{\nu} A_\mu, \quad 0 < r_\nu < \frac{1}{2^\nu}, \ \nu = 1,2,\ldots.$$

Since $\{x_\nu\}_{\nu=1}^{\infty}$ forms a Cauchy sequence, the limit $c = \lim_{\nu\to\infty} x_\nu \in X$ exists. It follows from the construction that for every $\nu \in \mathbf{N}$

$$c \in \overline{U}(x_\nu;r_\nu) \setminus \bigcup_{\mu=1}^{\nu} A_\mu.$$

Thus, $c \in U(a;r) \setminus \bigcup_{\mu=1}^{\infty} A_\mu$.

(ii) It is immediate from (i) above, because the standard metric of $\mathbf{R}^n$ is complete.

(iii) It is immediate from the definition.

(iv) Let $U \subset \mathbf{R}^n$ be an open subset. Let F_ν ($\nu \in \mathbf{N}$) be closed subsets of U without interior points. Suppose that there exists an interior point $a \in F := \bigcup_{\nu\in\mathbf{N}} F_\nu$. Then there is an open ball $\mathrm{B}(a;r) \subset F (\subset U)$. For any positive $r' < r$, the closed ball $\bar{\mathrm{B}}(a;r')$ is contained in F. It follows that

$$\bar{\mathrm{B}}(a;r') = \bigcup_{\nu\in\mathbf{N}} \left(F_\nu \cap \bar{\mathrm{B}}(a;r')\right).$$

Therefore a is an interior point of the union of the closed subsets $F_\nu \cap \bar{\mathrm{B}}(a;r')$ ($\nu \in \mathbf{N}$) of $\mathbf{R}^n$. But, the closed subsets $F_\nu \cap \bar{\mathrm{B}}(a;r')$ of $\mathbf{R}^n$ have no interior point by the supposition; this contradicts (iii) above. □

4.2.2 Separate Analyticity

We begin with a preparation on plurisubharmonic functions. Let $\mathrm{P}\Delta(r)$ be a polydisk of polyradius $r = (r_j)$ with center at 0 in $\mathbf{C}^n$, and let $\Gamma = \prod\{|z_j| = r_j\}$.

Lemma 4.2.4. *Let $u_\nu(z)$ ($\nu = 1,2,\ldots$) be a sequence of plurisubharmonic functions in a neighborhood of the closed polydisk $\overline{\mathrm{P}\Delta}(r)$. Let $\alpha > \beta > 0$, and assume that*

$$u_\nu(z) \le \alpha, \quad z \in \overline{\mathrm{P}\Delta}(r),\ \nu = 1,2,\ldots, \tag{4.2.5}$$

$$\overline{\lim_{\nu\to\infty}}\, u_\nu(z) \le \beta, \quad z \in \Gamma. \tag{4.2.6}$$

Then, for every compact subset $K \Subset \mathrm{P}\Delta(r)$ and an arbitrary $\varepsilon > 0$, there is a number $N \in \mathbf{N}$ such that

$$u_\nu(z) \le \beta + \varepsilon, \quad z \in K, \quad \nu \ge N.$$

Proof. For $\nu \in \mathbf{N}$ we set

$$F_\nu = \{z \in \Gamma : u_\mu(z) < \beta + \varepsilon,\ \mu \ge \nu\}.$$

Note that F_ν are Borel sets and

$$F_\nu \subset F_{\nu+1} \subset \cdots, \quad \bigcup_{\nu=1}^{\infty} F_\nu = \Gamma.$$

Let $m(*)$ denote the Lebesgue measure of the arguments $\theta_j = \arg z_j$ of $(z_j) \in \Gamma$. Then we have

$$\lim_{\nu\to\infty} m(\Gamma \setminus F_\nu) = 0. \tag{4.2.7}$$

With the Poisson kernel $P(r,\rho,\theta,\vartheta)$ of n variables we have

$$\begin{aligned} &u_\mu(\rho_1 e^{i\vartheta_1},\ldots,\rho_n e^{i\vartheta_n}) \\ &\quad\le \frac{1}{(2\pi)^n}\int_\Gamma P(r,\rho,\theta,\vartheta) u_\mu(r_1 e^{i\theta_1},\ldots,r_n e^{i\theta_n})\, d\theta_1\cdots d\theta_n \\ &\quad= \frac{1}{(2\pi)^n}\left(\int_{F_\nu} + \int_{\Gamma\setminus F_\nu}\right) P(r,\rho,\theta,\vartheta) u_\mu(r_1 e^{i\theta_1},\ldots,r_n e^{i\theta_n})\, d\theta_1\cdots d\theta_n, \end{aligned} \tag{4.2.8}$$

where $\rho = (\rho_j)$ with $0 < \rho_j < r_j$ ($1 \le j \le n$). We take a polydisk $\mathrm{P}\Delta(s)$ of polyradius $s = (s_j)$ such that $K \Subset \mathrm{P}\Delta(s) \Subset \mathrm{P}\Delta(r)$. It follows that for $\mu \ge \nu \in \mathbf{N}$ and $z \in K$

$$u_\mu(z) \le \beta + \varepsilon + \alpha m(\Gamma \setminus F_\nu) \prod_{j=1}^{n} \frac{r_j + s_j}{r_j - s_j}.$$

By (4.2.7) there is a number $N \in \mathbf{N}$ such that

$$\alpha m(\Gamma \setminus F_\nu) \prod_{j=1}^{n} \frac{r_j + s_j}{r_j - s_j} < \varepsilon, \quad \nu \ge N.$$

Therefore, $u_\mu(z) < \beta + 2\varepsilon$ for $z \in K$ and $\mu \ge N$; since ε is arbitrary, the proof is finished. □

Theorem 4.2.9 (Hartogs Separate Analyticity). *A separately holomorphic function $f(z)$ on an open set U of $\mathbf{C}^n$ is holomorphic in U.*

Proof. The problem is local; so it is sufficient to prove:

Claim 4.2.10. *For any point $a \in U$ there is a polydisk neighborhood $\mathrm{P}\Delta(a;r)$, where $f(z)$ is holomorphic.*

We may assume $a = 0$ by a parallel translation, and take a polydisk $\mathrm{P}\Delta(r) \Subset U$ of polyradius $r = (r_j)$. Then $f(z)$ is defined on the closed polydisk $\overline{\mathrm{P}\Delta}(r)$ such that:

4.2.11. for every j $(1 \le j \le n)$, the function $f(z_1, \ldots, z_j, \ldots, z_n)$ with arbitrarily fixed $z_k \in \bar{\Delta}(r_k)$ $(k \ne j)$ is a holomorphic function on the closed disk $\bar{\Delta}(r_j)$.

We are going to prove $f(z) \in \mathscr{O}(\mathrm{P}\Delta)$ by induction on n.

The case of $n = 1$ is trivial. We assume the case of $n-1$ $(n \ge 2)$ to hold. We write $z = (z_1, z') \in \bar{\Delta}(r_1) \times \overline{\mathrm{P}\Delta}(r')$ with $r' = (r_2, \ldots, r_n)$.

(a) (Bounded $f(z)$) We assume that $f(z)$ is bounded on $\overline{\mathrm{P}\Delta}(r)$; so, there is a constant $M > 0$ such that

(4.2.12) $$|f(z)| \le M, \quad z \in \overline{\mathrm{P}\Delta}(r).$$

With a fixed $z' \in \overline{\mathrm{P}\Delta}(r')$ we expand

(4.2.13) $$f(z_1, z') = \sum_{\nu=0}^{\infty} c_\nu(z') z_1^\nu, \quad z_1 \in \bar{\Delta}(r_1).$$

By the coefficient estimate we have

(4.2.14) $$|c_\nu(z')| \le \frac{M}{r_1^\nu}, \quad \nu = 0, 1, 2, \ldots.$$

From this estimate we easily infer that the convergence of (4.2.13) is uniform on every compact subset of $\Delta(r_1) \times \overline{\mathrm{P}\Delta}(r')$. Therefore it is sufficient to show the next claim in order to obtain $f(z) \in \mathscr{O}(\mathrm{P}\Delta(r))$:

Claim 4.2.15. $c_\nu(z') \in \mathscr{O}(\mathrm{P}\Delta(r'))$, $\nu = 0, 1, 2, \ldots$.

When $\nu = 0$, $c_0((z')) = f(0, z')$, which is holomorphic in $\mathrm{P}\Delta(r')$ by the induction hypothesis. Suppose that $c_\mu(z') \in \mathrm{P}\Delta(r')$ for $\mu = 0, 1, \ldots, \nu - 1$ ($\nu \geq 2$). We consider

$$(4.2.16) \qquad g_\nu(z_1, z') := \frac{f(z_1, z') - \sum_{\mu=0}^{\nu-1} c_\mu(z') z_1^\mu}{z_1^\nu} = c_\nu(z') + \sum_{\mu=1}^{\infty} c_{\nu+\mu}(z') z_1^\mu.$$

For each fixed $z_1 \neq 0$, $g_\nu(z_1, z')$ is holomorphic in $z' \in \mathrm{P}\Delta(r')$, and by (4.2.14)

$$\left| \sum_{\mu=1}^{\infty} c_{\nu+\mu}(z') z_1^\mu \right| \leq \sum_{\mu=1}^{\infty} \frac{M|z_1|^\mu}{r_1^{\nu+\mu}} = \frac{|z_1| M}{r_1^{\nu+1}(1 - |z_1|/r_1)} \to 0 \quad (|z_1| \to 0).$$

It follows from (4.2.16) that $c_\nu(z')$ is a uniform limit of holomorphic functions in $z' \in \mathrm{P}\Delta(r')$, and so holomorphic there.

(b) In the general case, we are going to prove first that there is a non-empty open subset ω in $\Delta(r_1)$ such that $f(z_1, z')$ is bounded on $\omega \times \overline{\mathrm{P}\Delta}(s')$ with any $n-1$ dimensional polydisk $\mathrm{P}\Delta(s') \Subset \mathrm{P}\Delta(r')$; then, $f(z_1, z') \in \mathcal{O}(\omega \times \mathrm{P}\Delta(s'))$ by (a) above, and hence $f(z_1, z') \in \mathcal{O}(\omega \times \mathrm{P}\Delta(r'))$.

Set

$$E_\nu = \bigcap_{z' \in \overline{\mathrm{P}\Delta}(s')} \{z_1 \in \bar{\Delta}(r_1) : |f(z_1, z')| \leq \nu\}, \quad \nu = 1, 2, \ldots.$$

Note that by the assumption (resp. the induction hypothesis), $f(z_1, z')$ is holomorphic in $z_1 \in \bar{\Delta}(r_1)$ (resp. $z' \in \mathrm{P}\Delta(r')$) with each fixed $z' \in \mathrm{P}\Delta(r')$ (resp. $z_1 \in \Delta(r_1)$). Therefore, E_ν are clearly closed sets, and

$$E_\nu \subset E_{\nu+1} \subset \cdots, \quad \bigcup_{\nu=1}^{\infty} E_\nu = \bar{\Delta}(r_1).$$

It follows from Baire Theorem 4.2.3 that some E_ν contains a non-empty open subset $\omega \subset \Delta(r_1)$.

Let $a_0 \in \omega$ and let $\phi : \bar{\Delta}(r_1) \to \bar{\Delta}(r_1)$ be a linear transform with $\phi(0) = a_0$. Note that $\phi(z_1)$ is holomorphic in a neighborhood of $\bar{\Delta}(r_1)$ and $\phi(\partial\Delta(r_1)) = \partial\Delta(r_1)$. We consider $f(\phi(z_1), z')$ in $(z_1, z') \in \bar{\Delta}(r_1) \times \overline{\mathrm{P}\Delta}(r')$. Then $f(\phi(z_1), z')$ satisfies the separate analyticity condition 4.2.11 on $\overline{\mathrm{P}\Delta}(r)$; moreover, there is a number $\delta_1 > 0$ such that $f(\phi(z_1), z')$ is holomorphic in $(z_1, z') \in \Delta(\delta_1) \times \mathrm{P}\Delta(r')$.

Thus, we may rewrite $f(z_1, z')$ for $f(\phi(z_1), z')$, and show $f(z_1, z') \in \mathcal{O}(\Delta(r_1) \times \mathrm{P}\Delta(r'))$. Taking smaller $\delta_1 > 0$ and $\mathrm{P}\Delta(r')$ if necessary, we may assume:

4.2.17. (i) $f(z_1, z')$ is separately holomorphic on $\bar{\Delta}(r_1) \times \overline{\mathrm{P}\Delta}(r')$,
(ii) $f(z_1, z')$ is holomorphic on $\bar{\Delta}(\delta_1) \times \overline{\mathrm{P}\Delta}(r')$.

Set $M = \|f(z_1, z')\|_{\bar{\Delta}(\delta_1) \times \overline{\mathrm{P}\Delta}(r')}$. As in (4.2.13) and (4.2.14) we have

$$(4.2.18) \qquad f(z_1, z') = \sum_{\nu=0}^{\infty} c_\nu(z') z_1^\nu, \quad |z_1| \leq r_1, \ z' \in \overline{\mathrm{P}\Delta}(r'),$$

(4.2.19) $$c_\nu(z') = \frac{1}{2\pi i}\int_{|z_1|=\delta_1} \frac{f(z_1,z')}{z_1^{\nu+1}}dz_1 \in \mathscr{O}(\overline{\mathrm{P}\Delta}(r')),\ \nu = 0,1,2,\ldots,$$

(4.2.20) $$|c_\nu(z')| \le \frac{M}{\delta_1^\nu},\quad z' \in \overline{\mathrm{P}\Delta}(r'),\ \nu = 0,1,2,\ldots.$$

Now we set

$$u_\nu(z') = \frac{1}{\nu}\log|c_\nu(z')|,\quad z' \in \overline{\mathrm{P}\Delta}(r'),\ \nu = 1,2,\ldots.$$

Because of the convergence radius of (4.2.18) we get

(4.2.21) $$\overline{\lim_{\nu\to\infty}}\, u_\nu(z') \le -\log r_1,\quad z' \in \overline{\mathrm{P}\Delta}(r').$$

By (4.2.20) we have a uniform bound

(4.2.22) $$u_\nu(z') \le -\log\delta_1 + \log M,\quad z' \in \overline{\mathrm{P}\Delta}(r'),\ \nu = 1,2,\ldots.$$

Let $\overline{\mathrm{P}\Delta}(s) \Subset \mathrm{P}\Delta(r)$ be given arbitrarily. By Lemma 4.2.4 with (4.2.21) and (4.2.22) there is a number $N \in \mathbf{N}$ such that

$$u_\nu(z') \le -\log s_1,\quad z' \in \overline{\mathrm{P}\Delta}(s'),\ \nu \ge N.$$

It follows that $|c_\nu(z')|s_1^\nu \le 1$ $(\nu \ge N)$; hence, the series (4.2.18) converges absolutely and uniformly on every compact subset of $\Delta(s_1)\times\overline{\mathrm{P}\Delta}(s')$. Since every term $c_\nu(z')z_1^\nu$ is holomorphic in $\mathrm{P}\Delta(s)$ by (4.2.19), so is the limit $f(z_1,z')$. Since $\overline{\mathrm{P}\Delta}(s) \Subset \mathrm{P}\Delta(r)$ is arbitrary, we see that $f(z)$ is holomorphic in $\mathrm{P}\Delta(r)$. □

Remark 4.2.23. (i) As seen in Remark 1.1.17 (ii), for real analytic functions, the separate analyticity even with boundedness condition does not imply the analyticity in all variables. Together with Hartogs' phenomenon, Hartogs' Theorem 4.2.9 represents a special feature of complex analytic functions of n (≥ 2) variables.

(ii) In the proof above, a little idea is to reduce $a_0 \in \omega$ to $a_0 = 0$ by making use of the linear transform $\phi(z_1)$, so that it is possible to claim the analyticity of f on the whole $\Delta(r_1)\times\mathrm{P}\Delta(r')$. Cf., e.g., the proofs of Hörmander [30], Chap. 2, and Nishino [37], Chap. 1.

4.3 Pseudoconvexity

4.3.1 Pseudoconvexity Problem

Let $\pi : \mathfrak{D} \to \mathbf{C}^n$ be an unramified domain over $\mathbf{C}^n$. We take and fix a polydisk $\mathrm{P}\Delta$ with center at the origin, and consider the boundary distance function $\delta_{\mathrm{P}\Delta}(p,\partial\mathfrak{D})$.

As a fundamental basis we prove *Oka's Theorem of Boundary Distance Functions*, which serves as a key towards the solution of the Pseudoconvexity Problem:

Theorem 4.3.1 (Oka). *If $\mathfrak{D}$ is a domain of holomorphy, then*

$$-\log \delta_{\mathrm{P}\Delta}(p, \partial\mathfrak{D}) \in \mathscr{P}^0(\mathfrak{D}).$$

Proof. It follows from (3.6.5) that $-\log \delta_{\mathrm{P}\Delta}(p, \partial\mathfrak{D})$ is continuous.

Since the plurisubharmonicity is a local property, it suffices to deal with it in a neighborhood U of $a \in \mathfrak{D}$. Since $\pi : \mathfrak{D} \to \mathbf{C}^n$ is locally biholomorphic, we may regard $U \subset \mathbf{C}^n$. Take a complex line $L \subset \mathbf{C}^n$ ($L \cong \mathbf{C}$) passing through a. We then prove that the restriction $-\log \delta_{\mathrm{P}\Delta}(p, \partial\mathfrak{D})|_{L\cap U}$ satisfies the submean property on a circle, of which the center may be assumed to be a by a translation.

We take a closed disk and the circle in $L \cap \mathfrak{D}$ with center at a:

$$E = \{a + \zeta v : |\zeta| \le R\} \subset L \cap \mathfrak{D}, \quad K = \{a + \zeta v : |\zeta| = R\},$$

where $v \in \mathbf{C}^n \setminus \{0\}$ is a directional vector of L. Noting that $-\log \delta_{\mathrm{P}\Delta}(a + \zeta v, \partial\mathfrak{D})$ is a continuous function on $|\zeta| \le R$, we take the Poisson integral $u(\zeta)$ of $-\log \delta_{\mathrm{P}\Delta}(p, \partial\mathfrak{D})$ on the closed disk $|\zeta| \le R$ (cf. [38] Chap. 3 §6). Then $u(\zeta)$ is continuous on $|\zeta| \le R$, harmonic ($\Delta u = 0$) in the interior $|\zeta| < R$, and is equal to $-\log \delta_{\mathrm{P}\Delta}(a + \zeta v, \partial\mathfrak{D})$ on the circle $|\zeta| = R$. We get

$$(4.3.2) \qquad u(0) = \frac{1}{2\pi} \int_0^{2\pi} -\log \delta_{\mathrm{P}\Delta}(a + Re^{i\theta} v, \partial\mathfrak{D}) d\theta.$$

For an arbitrarily small $\varepsilon > 0$,

$$-\log \delta_{\mathrm{P}\Delta}(a + Re^{i\theta} v, \partial\mathfrak{D}) < u(re^{i\theta}) + \varepsilon,$$

where $r < R$ is sufficiently close to R. With the adjoint harmonic function $u^*(\zeta)$ of $u(\zeta)$ in $|\zeta| < R$, we obtain a holomorphic function $g(\zeta) = u(\zeta) + iu^*(\zeta)$. Let

$$(4.3.3) \qquad g(\zeta) = \sum_{\nu=0}^{\infty} c_\nu \zeta^\nu, \quad |\zeta| < R$$

be the power series expansion. Since it converges uniformly on $|\zeta| = r$, the partial sum $P(\zeta) = \sum_{\nu=0}^{N} c_\nu \zeta^\nu$ with sufficiently large $N \in \mathbf{N}$ satisfies $|g(\zeta) - P(\zeta)| < \varepsilon$ ($|\zeta| = r$). Therefore,

$$(4.3.4) \qquad -\log \delta_{\mathrm{P}\Delta}(a + \zeta v, \partial\mathfrak{D}) < \Re P\left(\frac{r}{R}\zeta\right) + 2\varepsilon, \quad |\zeta| = R.$$

Since L is an affine linear subspace of $\mathbf{C}^n$, there is a polynomial $\hat{P}(z)$ on $\mathbf{C}^n$ with $\hat{P}(a + \zeta v) = P\left(\frac{r}{R}\zeta\right)$. It follows from (4.3.4) that

$$\delta_{\mathrm{P}\Delta}(z, \partial\mathfrak{D}) > \left|e^{-\hat{P}(z) - 2\varepsilon}\right|, \quad z \in K.$$

The Maximum Principle implies $\widehat{K}_{\mathfrak{D}} \supset E$, and then by Lemma 3.2.7

$$\delta_{P\Delta}(z, \partial\mathfrak{D}) \geq \left|e^{-\hat{P}(z)-2\varepsilon}\right|, \quad z \in E.$$

In particular, at $z = a$ we get

$$\delta_{P\Delta}(a, \partial\mathfrak{D}) \geq \left|e^{-\hat{P}(a)-2\varepsilon}\right| = e^{-u(0)-2\varepsilon}.$$

This together with (4.3.2) implies

$$-\log \delta_{P\Delta}(a, \partial\mathfrak{D}) \leq \frac{1}{2\pi}\int_0^{2\pi} -\log \delta_{P\Delta}(a + Re^{i\theta}v, \partial\mathfrak{D})d\theta + 2\varepsilon.$$

Now as $\varepsilon > 0$ is arbitrary,

$$-\log \delta_{P\Delta}(a, \partial\mathfrak{D}) \leq \frac{1}{2\pi}\int_0^{2\pi} -\log \delta_{P\Delta}(a + Re^{i\theta}v, \partial\mathfrak{D})d\theta.$$

Thus $-\log \delta_{P\Delta}(a + \zeta v, \partial\mathfrak{D})$ satisfies the submean property. □

Together with Theorem 3.7.3 we immediately have:

Corollary 4.3.5. *If $\mathfrak{D}$ is holomorphically convex, then* $-\log \delta_{P\Delta}(p, \partial\mathfrak{D}) \in \mathscr{P}^0(\mathfrak{D})$.

4.3.6. (**Pseudoconvexity Problem I**) Does the converse of Theorem 4.3.1 hold? Furthermore, is $\mathfrak{D}$ Stein?

Definition 4.3.7. (i) A real-valued function $\psi : \mathfrak{D} \to [-\infty, \infty)$ with $-\infty$ allowed as a value is an *exhaustion function* if $\mathfrak{D}_c := \{p \in \mathfrak{D} : \psi(p) < c\} \Subset \mathfrak{D}$ for every $c \in \mathbf{R}$. The set $\mathfrak{D}_c$ is called a *sublevel set*.
(ii) A domain $\mathfrak{D}/\mathbf{C}^n$ is *pseudoconvex* if there exists a pseudoconvex exhaustion function ψ on $\mathfrak{D}$.[5] In the case when ψ is chosen to be $\mathbf{R}$-valued and of C^k-class ($0 \leq k \leq \infty$), $\mathfrak{D}$ is said to be C^k*-pseudoconvex*.

Remark 4.3.8. (i) In many references, the definition of an exhaustion function includes the continuity. Also "C^0-pseudoconvex" or "C^∞-pseudoconvex" is simply termed "pseudoconvex". The reader should confer other books.
(ii) Since a pseudoconvex exhaustion function $\psi : \mathfrak{D} \to [-\infty, \infty)$ is only upper semi-continuous, we do not have necessarily $\mathfrak{D}_c \Subset \mathfrak{D}_{c'}$ for real numbers $c < c'$. If c' is chosen sufficiently large, naturally $\mathfrak{D}_c \Subset \mathfrak{D}_{c'}$. Therefore $1/(c - \psi(p))$ does not serve as an exhaustion function on the sublevel set $\mathfrak{D}_c$; it is not known immediately if $\mathfrak{D}_c$ is pseudoconvex or not. (With the continuity assumption it is immediate.)

Remark 4.3.9. The notions defined for a domain $\mathfrak{D}/\mathbf{C}^n$ are listed as follows:

[5] Note that ψ is not assumed to be continuous and that the holomorphic separation is not assumed for $\mathfrak{D}$.

(i) $\mathfrak{D}$ is a domain of holomorphy (with the holomorphic separation).
(ii) $\mathfrak{D}$ is holomorphically convex.
(iii) $\mathfrak{D}$ is Stein (holomorphically convex + holomorphically separable).
(iv) $-\log \delta_{\mathrm{P}\Delta}(p, \partial \mathfrak{D}) \in \mathscr{P}^0(\mathfrak{D})$.
(v) $\mathfrak{D}$ is C^0-pseudoconvex.
(vi) $\mathfrak{D}$ is pseudoconvex.

(i) is the most natural notion from the viewpoint of the domain of existence of analytic functions, and also historically it is the oldest. But it is defined by the analytic continuation which requires the "outside" of the given domain; the existence of the "outside" is implicitly assumed (extrinsic). In comparison, (ii) is completely characterized only by the interior information of the domain (intrinsic). In the case of univalent domains, the two are equivalent (Theorem 3.2.11), but it is not yet proved here for multivalent domains in general at this moment (cf. Remark 3.7.5 and Theorem 5.3.2). The holomorphic separation is also a fundamental property of domains; added with it, one gets (iii). (v)⇒(vi) is clear.

It is "Pseudoconvexity Problem" to ask if (i)—(vi) are mutually equivalent: In particular, (vi)⇒(iii) is crucial.

4.3.10. **(Pseudoconvexity Problem II)** Is a pseudoconvex domain over $\mathbf{C}^n$ Stein?

Pseudoconvexity Problems I 4.3.6, and II 4.3.10 together with their relations were affirmatively solved finally by K. Oka; the aim of this chapter and the next is to give the proofs.

Before going into the proofs, we investigate the mutual relations among the notions listed in Remark 4.3.9.

Theorem 4.3.11 (Oka). *If $\mathfrak{D}/\mathbf{C}^n$ is a pseudoconvex domain, then*

$$-\log \delta_{\mathrm{P}\Delta}(p, \partial \mathfrak{D}) \in \mathscr{P}^0(\mathfrak{D}).$$

Proof. We follow the proof of Theorem 4.3.1. Take a point $p_0 \in \mathfrak{D}$ and its polydisk neighborhood $\mathrm{P}\Delta \Subset \mathfrak{D}$. We regard $\mathrm{P}\Delta \subset \mathbf{C}^n$. With a vector $v \in \mathbf{C}^n \setminus \{0\}$ we consider a closed disk

$$\bar{\Delta} = \{p_0 + \zeta v : |\zeta| \le R\} \subset \mathrm{P}\Delta \subset \mathfrak{D}.$$

Let $u(\zeta)$ be the Poisson integral of $-\log \delta(p_0 + \zeta v, \partial \mathfrak{D})$ on the circle $|\zeta| = R$. Take the holomorphic function $g(\zeta)$ in (4.3.3) so that

$$\Im g(0) = 0, \quad g(0) = u(0). \tag{4.3.12}$$

Note that $\Re g = u$. For every small $\varepsilon > 0$ we take $0 < t < 1$, sufficiently close to 1, so that $g(t\zeta)$ is holomorphic in a neighborhood of $|\zeta| \le R$ and

$$|\Re g(t\zeta) - u(\zeta)| < \varepsilon, \quad |\zeta| = R.$$

Thus we have

(4.3.13) $$\delta_{P\Delta}(p_0+\zeta v, \partial\mathfrak{D}) \geq |e^{-g(t\zeta)-\varepsilon}|, \quad |\zeta| = R.$$

We want to show that this inequality holds also at $\zeta = 0$. With an arbitrarily chosen vector $w \in P\Delta$ we consider a holomorphic map

$$\Psi : (\zeta, \xi) \mapsto p_0 + \zeta v + \xi w e^{-g(t\zeta)-\varepsilon}$$

for $\xi \in \mathbf{C}$ with $|\xi| \leq 1$. If $\xi = 0$, then $\Psi(\{|\zeta| \leq R\}, 0) = \bar{\Delta} \subset \mathfrak{D}$ by definition. If $|\zeta| = R$, then it follows from (4.3.13) that

$$\Psi(\{|\zeta| = R\} \times \{|\xi| \leq 1\}) \subset \mathfrak{D}.$$

Set

(4.3.14) $$K = (\{|\zeta| \leq R\} \times \{0\}) \cup (\{|\zeta| = R\} \times \{|\xi| \leq 1\}) \ (\subset \mathbf{C}^2).$$

Note that $\Psi(K) \Subset \mathfrak{D}$.

We take an exhaustion function $\phi \in \mathscr{P}(\mathfrak{D})$, and a sublevel set $\mathfrak{D}_c := \{\phi < c\} \Supset \Psi(K)$. Since the map $\Psi : \Omega_{\mathrm{H}} \to \mathfrak{D}_c$ is holomorphic in a neighborhood Ω_{H} (a Hartogs domain) of $K \subset \mathbf{C}^2$, we see that $\phi \circ \Psi \in \mathscr{P}(\Omega_{\mathrm{H}})$. Let S denote the set of all s with $0 \leq s \leq 1$, satisfying the condition

$$\Psi(\{|\zeta| \leq R\} \times \{|\xi| \leq s\}) \subset \mathfrak{D}.$$

Since $S \ni 0$, $S \neq \emptyset$. By the condition, S is an open set. Also, if the condition is satisfied, the Maximal Principle implies that $\Psi(\{|\zeta| \leq R\} \times \{|\xi| \leq s\}) \subset \mathfrak{D}_c$. For an accumulation point s' of $s \in S$

$$\Psi(\{|\zeta| \leq R\} \times \{|\xi| \leq s'\}) \subset \bar{\mathfrak{D}}_c.$$

Therefore, S is closed, and so $S = [0, 1]$. In particular, with $\zeta = 0$ and $\xi = 1$ one obtains

$$p_0 + w e^{-g(0)-\varepsilon} \in \mathfrak{D}.$$

Since $w \in P\Delta$ is arbitrary, it follows together with (4.3.12) that

$$\begin{aligned}
&p_0 + e^{-u(0)-\varepsilon} \cdot P\Delta \subset \mathfrak{D};\\
&\quad \delta_{P\Delta}(p_0, \partial\mathfrak{D}) \geq e^{-u(0)-\varepsilon};\\
-\log\delta_{P\Delta}(p_0, \partial\mathfrak{D}) &\leq \frac{1}{2\pi}\int_0^{2\pi} -\log\delta_{P\Delta}(p_0 + Re^{i\theta}v)d\theta + \varepsilon;\\
-\log\delta_{P\Delta}(p_0, \partial\mathfrak{D}) &\leq \frac{1}{2\pi}\int_0^{2\pi} -\log\delta_{P\Delta}(p_0 + Re^{i\theta}v)d\theta.
\end{aligned}$$

Thus, $-\log\delta_{P\Delta}(p, \partial\mathfrak{D}) \in \mathscr{P}^0(\mathfrak{D})$. □

We prove the converse of the above theorem.

Theorem 4.3.15 (Oka). *Let $\pi : \mathfrak{D} \to \mathbf{C}^n$ be a domain. If $-\log \delta_{\mathrm{P}\Delta}(p, \partial\mathfrak{D})$ is plurisubharmonic, then $\mathfrak{D}$ is C^0-pseudoconvex.*

Proof. We construct a continuous pseudoconvex exhaustion function on $\mathfrak{D}$.

(1) If $\pi : \mathfrak{D} \to \mathbf{C}^n$ is finitely sheeted, then the function defined by

$$\lambda(p) = \max\{-\log \delta_{\mathrm{P}\Delta}(p, \partial\mathfrak{D}), \|\pi(p)\|\}$$

is a continuous pseudoconvex exhaustion function on $\mathfrak{D}$.

(2) The proof in the case when $\mathfrak{D}/\mathbf{C}^n$ is infinitely sheeted is elementary, but a little bit long, so that we divide it into several steps.

(a) Take a point $p_0 \in \mathfrak{D}$ and fix it. In what follows we consider a number $\rho > 0$ with $\rho < \delta_{\mathrm{P}\Delta}(p_0, \partial\mathfrak{D})$. Set

$$\mathfrak{D}_\rho = \text{the connected component of } \{p \in \mathfrak{D} : \delta_{\mathrm{P}\Delta}(p, \partial\mathfrak{D}) > \rho\} \text{ containing } p_0.$$

For $0 < \rho' < \rho$, $\mathfrak{D}_\rho \subset \mathfrak{D}_{\rho'}$, and $\bigcup_{\rho>0} \mathfrak{D}_\rho = \mathfrak{D}$.

As stated in Remark 4.1.21, $d\lambda$ denotes the Lebesgue measure of $\mathbf{C}^n$. We choose a C^∞-function $\chi(z) \geq 0$ such that

$$\text{Supp}\, \chi \subset \mathrm{P}\Delta, \quad \int_{w \in \mathbf{C}^n} \chi(w)\, d\lambda(w) = 1. \tag{4.3.16}$$

For $\varepsilon > 0$ we put $\chi_\varepsilon(w) = \chi(w/\varepsilon)\varepsilon^{-2n}$. Then

$$\text{Supp}\, \chi_\varepsilon \subset \varepsilon \mathrm{P}\Delta, \quad \int_{\mathbf{C}^n} \chi_\varepsilon(w)\, d\lambda(w) = 1.$$

Let $d_\rho(p)$ $(p \in \mathfrak{D}_\rho)$ be the continuous function defined by (3.6.15). As far as a point in a univalent subdomain of $\mathfrak{D}$ is concerned, we write it as a point of $\mathbf{C}^n$, unless confusion occurs. For $0 < \varepsilon \leq \rho$ we obtain the smoothing of d_ρ,

$$(d_\rho)_\varepsilon(p) = (d_\rho) * \chi_\varepsilon(p) = \int_{w \in \varepsilon \mathrm{P}\Delta} d_\rho(p+w)\chi_\varepsilon(w)\, d\lambda(w), \quad p \in \mathfrak{D}_\rho.$$

Let (r_{0j}) be the polyradius of $\mathrm{P}\Delta$, and set $C_0 = \sqrt{\sum_j r_{0j}^2}$. It follows from the definition and (3.6.16) that

$$|(d_\rho)_\varepsilon(p) - d_\rho(p)| \leq \varepsilon C_0, \quad p \in \mathfrak{D}_\rho.$$

By Lemma 3.6.17 we see that

$$\{p \in \mathfrak{D}_\rho : (d_\rho)_\varepsilon(p) < b\} \Subset \mathfrak{D}, \quad \forall b > 0. \tag{4.3.17}$$

(b) For $p \in \mathfrak{D}$ we write the local complex coordinate system $\pi(p) = (z_j) = (x_j + i y_j)$. Let ξ $(\|\xi\| = 1)$ be one of the directional vectors in the space $\mathbf{C}^n \cong \mathbf{R}^{2n}$ of $x_j, y_j, 1 \leq j \leq n$, and write $\frac{\partial}{\partial \xi}$ for the directional derivative. With $h \in \mathbf{R}$ we have

$$\lim_{h\to 0}\frac{(d_\rho)_\varepsilon(p+h\xi)-(d_\rho)_\varepsilon(p)}{h}=\frac{\partial (d_\rho)_\varepsilon}{\partial \xi}(p).$$

On the other hand, we deduce from (3.6.16) the following estimate:

$$\begin{aligned}
&\left|\frac{(d_\rho)_\varepsilon(p+h\xi)-(d_\rho)_\varepsilon(p)}{h}\right|\\
&=\left|\frac{1}{h}\int_w \{d_\rho(p+h\xi+w)-d_\rho(p+w)\}\chi_\varepsilon(w)d\lambda(w)\right|\\
&\le \frac{1}{|h|}\int_w |d_\rho(p+h\xi+w)-d_\rho(p+w)|\,\chi_\varepsilon(w)d\lambda(w)\\
&\le \frac{1}{|h|}|h|\cdot\|\xi\|=1.
\end{aligned}$$

Therefore we have

$$\left|\frac{\partial (d_\rho)_\varepsilon}{\partial \xi}(p)\right|\le 1,\quad p\in \mathfrak{D}_\rho,\ 0<\varepsilon\le\rho. \tag{4.3.18}$$

With $0<2\varepsilon\le\rho$ we consider

$$\tilde{d}_{\rho,\varepsilon}(p)=\left((d_\rho)_\varepsilon\right)_\varepsilon(p),\quad p\in\mathfrak{D}_\rho.$$

It follows that

$$\begin{aligned}
\frac{\partial \tilde{d}_{\rho,\varepsilon}}{\partial \xi}(p)&=\int_w \frac{\partial (d_\rho)_\varepsilon}{\partial \xi}(p+w)\chi_\varepsilon(w)d\lambda(w)\\
&=\int_w \frac{\partial (d_\rho)_\varepsilon}{\partial \xi}(w)\chi\left(\frac{w-p}{\varepsilon}\right)\frac{1}{\varepsilon^{2n}}d\lambda(w).
\end{aligned}$$

Similarly to $\frac{\partial}{\partial\xi}$, we take a directional derivative $\frac{\partial}{\partial\eta}$ with respect to one of $x_j, y_j, 1\le j\le n$, and then have

$$\frac{\partial^2 \tilde{d}_{\rho,\varepsilon}}{\partial\eta\partial \xi}(p)=\int_w \frac{\partial (d_\rho)_\varepsilon}{\partial \xi}(w)\frac{\partial\chi}{\partial\eta}\left(\frac{w-p}{\varepsilon}\right)\frac{-1}{\varepsilon^{2n+1}}d\lambda(w).$$

Together with (4.3.18) we see that

$$\begin{aligned}
\left|\frac{\partial^2 \tilde{d}_{\rho,\varepsilon}}{\partial\eta\partial \xi}(p)\right|&\le \int_w \left|\frac{\partial (d_\rho)_\varepsilon}{\partial \xi}(w)\right|\cdot\left|\frac{\partial\chi}{\partial\eta}\left(\frac{w-p}{\varepsilon}\right)\right|\frac{1}{\varepsilon^{2n+1}}d\lambda(w)\\
&\le\frac{1}{\varepsilon}\int_w\left|\frac{\partial\chi}{\partial\eta}(w)\right|d\lambda(w)=\frac{C_1}{\varepsilon}.
\end{aligned}\tag{4.3.19}$$

Here, C_1 is a positive constant independent from ε and ρ. We set

$$\hat{d}_\rho(p) = \tilde{d}_{\rho,\frac{\rho}{2}}(p), \quad p \in \mathfrak{D}_\rho.$$

By (4.3.17) we have for $\hat{d}_\rho(p)$ that

(4.3.20) $$\{p \in \mathfrak{D}_\rho : \hat{d}_\rho(p) < b\} \Subset \mathfrak{D}, \quad \forall b > 0.$$

With $C_2 \gg \frac{2C_1}{\rho}$ we set

(4.3.21) $$\varphi_\rho(p) = \hat{d}_\rho(p) + C_2 \|\pi(p)\|^2.$$

Then from (4.3.19) we obtain

$$\sum_{j,k} \frac{\partial^2 \varphi_\rho}{\partial z_j \partial \bar{z}_k} \xi_j \bar{\xi}_k \geq \|(\xi_j)\|^2.$$

Summarizing the above, we get:

Lemma 4.3.22. *There is a strongly plurisubharmonic function $\varphi_\rho(p) > 0$ of C^∞-class on $\mathfrak{D}_\rho$ such that*

$$\{p \in \mathfrak{D}_\rho : \varphi_\rho(p) < b\} \Subset \mathfrak{D}, \quad \forall b > 0.$$

(c) Here we use the property, "$-\log \delta_{\mathrm{P}\Delta}(p, \partial\mathfrak{D}) \in \mathscr{P}^0(\mathfrak{D})$". Choose $a_1 > 0$ so that $\delta_{\mathrm{P}\Delta}(p_0) > e^{-a_1}$, and take a divergent monotone increasing sequence, $a_1 < a_2 < \cdots < a_j \nearrow \infty$. We set $\mathfrak{D}_j = \mathfrak{D}_{e^{-a_j}}$ $(j = 1, 2, \ldots)$, which satisfy $\mathfrak{D}_j \subset \mathfrak{D}_{j+1}$ and $\mathfrak{D} = \bigcup_{j=1}^\infty \mathfrak{D}_j$.

Applying Lemma 4.3.22 for each $\mathfrak{D}_j$ $(\rho = e^{-a_j})$, we obtain a strongly plurisubharmonic function $\varphi_j(p)$ of C^∞-class such that for every $b > 0$

$$\{p \in \mathfrak{D}_j : \varphi_j(p) < b\} \Subset \mathfrak{D}_{j+1}.$$

We choose a monotone increasing sequence, $b_1 < b_2 < \cdots \nearrow \infty$ as follows. With an arbitrarily chosen $b_1 > 0$ we set

$$\Delta_1 = \{p \in \mathfrak{D}_1 : \varphi_4(p) < b_1\}.$$

It follows that

(4.3.23) $$\partial \Delta_1 \subset \{-\log \delta_{\mathrm{P}\Delta}(p) = a_1\} \cup \{\varphi_4(p) = b_1\}.$$

Since $\Delta_1 \Subset \mathfrak{D}_2$, there is $b_2 \gg \max\{2, b_1\}$ such that

$$\Delta_2 = \{p \in \mathfrak{D}_2 : \varphi_5(p) < b_2\} \Supset \Delta_1.$$

Inductively, we choose $b_j > \max\{j, b_{j-1}\}$ so that

(4.3.24) $$\Delta_j = \{p \in \mathfrak{D}_j : \ \varphi_{j+3}(p) < b_j\} \Supset \Delta_{j-1}.$$

It follows that

$$\mathfrak{D} = \bigcup_{j=1}^{\infty} \Delta_j. \tag{4.3.25}$$

We set $\Phi_1(p) = \varphi_4(p) + 1\,(> 1)$ for $p \in \Delta_4$. For $j \geq 1$, we assume that $\Phi_h(p)$, $1 \leq h \leq j$, are defined so that the following conditions are satisfied:

4.3.26. (i) $\Phi_h(p) \in \mathscr{P}^0(\Delta_{h+3})$.
(ii) $\Phi_h(p) > h$, $\forall p \in \Delta_{h+2} \setminus \Delta_{h+1}$, $1 \leq h \leq j$.
(iii) $\Phi_h(p) = \Phi_{h-1}(p)$, $\forall p \in \Delta_h$, $2 \leq h \leq j$.

We define a continuous plurisubharmonic function on Δ_{j+4} by

$$\psi_{j+1}(p) = \max\{-\log \delta_{\mathrm{P}\Delta}(p, \partial\mathfrak{D}) - a_{j+1}, \varphi_{j+4}(p) - b_{j+1}\}, \quad p \in \Delta_{j+4}.$$

Then

$$\begin{gathered} \psi_{j+1}(p) \leq 0, \quad p \in \bar{\Delta}_{j+1}, \\ \min_{\bar{\Delta}_{j+3} \setminus \Delta_{j+2}} \psi_{j+1}(p) > 0. \end{gathered} \tag{4.3.27}$$

From this it follows that for a sufficiently large $k_{j+1} > 0$

$$\min_{\bar{\Delta}_{j+3} \setminus \Delta_{j+2}} k_{j+1}\psi_{j+1}(p) > \max\left\{ j+1, \max_{\bar{\Delta}_{j+2}} \Phi_j(p) \right\}. \tag{4.3.28}$$

We define

$$\Phi_{j+1}(p) = \begin{cases} \max\{\Phi_j(p), k_{j+1}\psi_{j+1}(p)\}, & p \in \Delta_{j+2}, \\ k_{j+1}\psi_{j+1}(p), & p \in \Delta_{j+4} \setminus \Delta_{j+2} \end{cases}$$

(cf. Fig. 4.1). We see by (4.3.28) that in a neighborhood of $\partial\Delta_{j+2}$

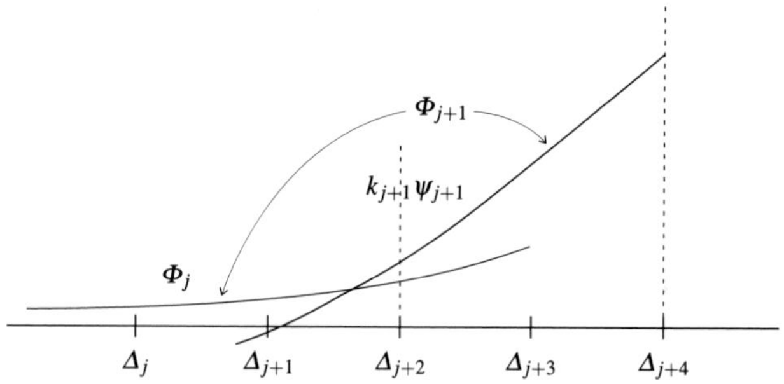

Fig. 4.1 The defining graph of Φ_{j+1}.

$$\Phi_{j+1}(p) = k_{j+1}\psi_{j+1}(p),$$

and so that $\Phi_{j+1}(p)$ is a continuous plurisubharmonic function on Δ_{j+4}. It follows from (4.3.27) and (4.3.28) that

$$\Phi_{j+1}(p) = \Phi_j(p), \quad p \in \Delta_{j+1},$$
$$\Phi_{j+1}(p) > j+1, \quad p \in \Delta_{j+4} \setminus \Delta_{j+2}.$$

Therefore, inductively we obtain $\Phi_j(p)$, $j = 1,2,\ldots$, satisfying 4.3.26. Now we define

$$\Phi(p) = \lim_{j\to\infty} \Phi_j(p), \quad p \in \mathfrak{D}.$$

Then, $\Phi(p)$ is a continuous plurisubharmonic function on $\mathfrak{D}$. By 4.3.26 (ii)

$$\Phi(p) > j, \quad p \in \mathfrak{D} \setminus \Delta_{j+1}, \quad j = 1,2,\ldots,$$

so that $\Phi(p)$ is an exhaustion function. □

Corollary 4.3.29. *If a domain $\mathfrak{D}(/\mathbf{C}^n)$ is pseudoconvex, then $\mathfrak{D}$ is C^0-pseudoconvex.*

Proof. This readily follows from Theorems 4.3.11 and 4.3.15. □

Remark 4.3.30. (i) By virtue of Oka's Theorem 4.3.15, we see that Pseudoconvexity Problem I 4.3.6 follows from II 4.3.10.

(ii) In terms of the items stated in Remark 4.3.9, we have shown the following implications:

- (iii) ⇒ (i) (Theorem 3.7.3);
- (iii) ⇒ (ii) (Obvious);
- (i) ⇒ (iv) (Theorem 4.3.1);
- (ii) ⇒ (iv) (Corollary 4.3.5);
- (vi) ⇒ (iv) (Theorem 4.3.11);
- (iv) ⇒ (v) (Theorem 4.3.15);
- (v) ⇒ (vi) (Trivial).

The final difficulty is the implication of (vi) ⇒ (iii) (Pseudoconvexity Problem II 4.3.10), which we will solve in the next chapter (cf. Afterword for historical comments).

Theorem 4.3.31. *Let $\mathfrak{D}(/\mathbf{C}^n)$ be a domain, and let Ω_α $(\alpha \in \Gamma)$ be a family of subdomains of $\mathfrak{D}$. If all Ω_α are pseudoconvex, then every connected component of the interior $(\bigcap_{\alpha\in\Gamma} \Omega_\alpha)^\circ$ of the intersection is pseudoconvex.*

Proof. It follows from Theorem 4.3.11 that $-\log \delta_{\mathrm{P}\Delta}(z, \partial\Omega_\alpha)$ is plurisubharmonic in Ω_α. For any connected component ω of $(\bigcap_{\alpha\in\Gamma} \Omega_\alpha)^\circ$,

$$-\log \delta_{\mathrm{P}\Delta}(z, \partial\omega) = \sup_{\alpha\in\Gamma} -\log \delta_{\mathrm{P}\Delta}(z, \partial\Omega_\alpha), \quad z \in \omega,$$

and $-\log \delta_{\mathrm{P}\Delta}(z, \partial\omega)$ is continuous. By Theorem 4.1.16 (vi), we see that $-\log \delta_{\mathrm{P}\Delta}(z, \partial\omega)$ $(z \in \omega)$ is plurisubharmonic. Then Theorem 4.3.15 implies the (C^0-) pseudoconvexity of ω. □

4.3.2 Bochner's Tube Theorem

We have introduced several notions of "convexity". There is a class of domains for which they coincide. We would like to discuss it here. The topic is a little a way from the Three Big Problems, but has an interest of its own and as an application of Oka's Theorem 4.3.1 of Boundary Distance Functions.

Let $\varphi : R \to \mathbf{R}^n$ be a real unramified domain; i.e., R is a Hausdorff topological space, and φ is a local homeomorphism. We consider a domain over $\mathbf{C}^n$ of the following type:

(4.3.32) $$\pi : T_R = R \times \mathbf{R}^n \cong R + i\mathbf{R}^n \ni x + iy \to \varphi(x) + iy \in \mathbf{C}^n,$$

which is called a *tube domain*, or simply a *tube*. We call R the *base* of the tube domain T_R. In many references, the case of univalent $R \hookrightarrow \mathbf{R}^n$ is dealt with, but here we consider the generalized case as a special unramified domain over $\mathbf{C}^n$. The univalent domains of this type are of fundamental importance in the theories of bounded symmetric domains and Laplace integral transforms in partial differential equations.

As a model case of tube domains, we consider the convergent domain $\Omega(f)$ of a power series in §1.1.4; using the same notation as there, we set

$$T := \{(z_1, \ldots, z_n) \in \mathbf{C}^n : (e^{z_1}, \ldots, e^{z_n}) \in \Omega(f)\} = \log \Omega^*(f) + i\mathbf{R}^n.$$

We see by Theorem 1.1.25 that T is a convex tube domain (univalent) with real convex domain $\log \Omega^*(f)$ as the base. The function $f(e^{z_1}, \ldots, e^{z_n})$ is a holomorphic function with period $2\pi i\mathbf{Z}$. An analytic function f given in any neighborhood of an arbitrary shape of the origin has the convergent domain which is necessarily analytically continued over the convex hull of the given domain in the expression of $f(e^{z_1}, \ldots, e^{z_n})$.

In the univalent case, T_R is convex if and only if $R = \mathrm{ch}(R)$, where $\mathrm{ch}(R)$ stands for the convex hull of R (see §1.1.1).

In general, let $\pi : \Omega \to \mathbf{C}^n$ be a domain. For distinct two points $p, q \in \Omega$, we take a line segment $L[\pi(p), \pi(q)]$ $(\subset \mathbf{C}^n)$ connecting $\pi(p)$ and $\pi(q)$. Let L_p be the connected component of $\pi^{-1}L[\pi(p), \pi(q)]$ containing p. In the case $q \in L_p$, we write $L_p = L[p, q]$ for L_p (note that it does not necessarily exist). If $L[p, q]$ exists, it carries the natural parameterization,

(4.3.33) $$t \in [0, 1] \to \pi|_{L[p,q]}^{-1}((1-t)\pi(p) + t\pi(q)) \in L[p, q].$$

Lemma 4.3.34. *Ω is univalent and convex if and only if for arbitrary distinct two points $p, q \in \Omega$, there exists the line segment $L[p,q]$ connecting p and q.*

Proof. The necessity is obvious. We show the sufficiency. For $L[p,q] \subset \Omega$, the restriction $\pi|_{L[p,q]} : L[p,q] \to L[\pi(p), \pi(q)]$ is a bijection, so that $\pi(p) \neq \pi(q)$; this means the injectivity of π. The convexity is the definition itself. □

Definition 4.3.35. A real-valued function $\psi : \Omega \to \mathbf{R}$ on Ω is said to be *convex* if for an arbitrary $L[p,q] \subset \Omega$ $(p \neq q)$ above, the restricted function $\psi|_{L[p,q]} : L[p,q] \to \mathbf{R}$ is convex with respect to the parameter t of (4.3.33).

We consider the boundary distance $\delta_{\mathrm{P}\Delta}(z, \partial T_R)$ of a tube domain T_R over $\mathbf{C}^n$. It follows that

$$\delta_{\mathrm{P}\Delta}(p+iy, \partial T_R) = \delta_{\mathrm{P}\Delta}(p, \partial T_R), \qquad \forall y \in \mathbf{R}^n. \tag{4.3.36}$$

That is, with $p = x + iy$ $(x \in R, y \in \mathbf{R}^n)$, $\delta_{\mathrm{P}\Delta}(p, \partial T_R)$ is a function only of the real part x.

Lemma 4.3.37. *If $-\log \delta_{\mathrm{P}\Delta}(p, \partial T_R)$ is plurisubharmonic, then it is convex.*

Proof. We first assume that $\delta_{\mathrm{P}\Delta}(p, \partial T_R)$ is of C^2-class. Set $\varphi(p) = -\log \delta_{\mathrm{P}\Delta}(p, \partial T)$. With $\pi(p) = (z_j) = (x_j + iy_j)$, the plurisubharmonicity with respect to the local coordinates (z_j) leads to

$$L[\varphi](p; v) = \sum_{j,k} \frac{\partial^2 \varphi}{\partial z_j \partial \bar{z}_k}(p) v_j \bar{v}_k \geq 0, \qquad \forall (v_j) \in \mathbf{C}^n. \tag{4.3.38}$$

By (4.3.36) we have

$$L[\varphi](p; v) = \frac{1}{4} \sum_{j,k} \frac{\partial^2 \varphi}{\partial x_j \partial x_k}(p) v_j \bar{v}_k \geq 0.$$

Therefore, $\frac{d^2 \varphi}{dt^2}(t) \geq 0$ with respect to t of (4.3.33), so that $\varphi(t)$ is convex.

In general, as φ is not necessarily of C^2-class, we first take $L[p,q]$ arbitrarily. Since $L[p,q] \Subset T_R$, the smoothing $\varphi_\varepsilon(p)$ with sufficiently small $\varepsilon > 0$ is defined in a neighborhood of $L[p,q]$, and is C^∞ plurisubharmonic there. Since $\varphi_\varepsilon(p)$ satisfies (4.3.36) too, $\varphi_\varepsilon|_{L[p,q]}$ is a convex function as shown above. The convergence $\varphi_\varepsilon \searrow \varphi$ (as $\varepsilon \searrow 0$) is uniform on $L[p,q]$, and hence $\varphi|_{L[p,q]}$ is convex. □

Theorem 4.3.39. *For a tube domain T_R $(\varphi : R \to \mathbf{R}^n)$ over $\mathbf{C}^n$, the following conditions are equivalent:*

(i) *T_R is a domain of holomorphy.*
(ii) *T_R is holomorphically convex.*
(iii) *T_R is pseudoconvex.*
(iv) *$-\log \delta_{\mathrm{P}\Delta}(z, \partial T_R)$ is (continuous) plurisubharmonic.*
(v) *T_R is univalent and convex; that is, the base R is a convex domain of $\mathbf{R}^n$.*

Proof. The implications, "(i), (ii), (iii) ⇒ (iv)", and "(v) ⇒ (i), (ii), (iii)", are the special cases of what have been proved.

(iv) ⇒ (v): Assume that there is a line segment $L[p,q] \subset R$ connecting the two points $p,q \in R$. In the case $p = q$, we consider $L[p,q] = \{p\}$ to be a degenerate line segment. It follows from Lemma 4.3.37 that $-\log \delta_{\mathrm{P}\Delta}(p, \partial T_R)$ is a continuous convex function, so that

$$\max_{z \in L[p,q]} -\log \delta_{\mathrm{P}\Delta}(z, \partial T_R) = \max_{p,q} -\log \delta_{\mathrm{P}\Delta}(z, \partial T_R),$$

$$\min_{z \in L[p,q]} \delta_{\mathrm{P}\Delta}(z, \partial T_R) = \min_{p,q} \delta_{\mathrm{P}\Delta}(z, \partial T_R). \tag{4.3.40}$$

Set

$$S = \{(p,q) \in R^2 : \exists L[p,q] \subset R\} \subset R^2.$$

It suffices to to prove $S = R^2$; then, by Lemma 4.3.34 the proof is finished.

Obviously, S is a non-empty open set. We are going to show that S is closed in R^2. Let $(p,q) \in R^2$ be an accumulation point of S. Then there is a sequence of points $(p_\nu, q_\nu) \in S$ $(\nu = 1, 2, \ldots)$ such that

$$\lim_{\nu \to \infty} p_\nu = p, \quad \lim_{\nu \to \infty} q_\nu = q, \quad L[p_\nu, q_\nu] \subset R. \tag{4.3.41}$$

Let $0 < \rho < \min_{p,q} \delta_{\mathrm{P}\Delta}(z, \partial T_R)$. Then we deduce from (4.3.41) and (4.3.40) that

$$U_\nu := \bigcup_{z \in L[p_\nu, q_\nu]} (z + \rho \mathrm{P}\Delta) \subset R. \tag{4.3.42}$$

With a sufficiently large $\nu_0 \in \mathbf{N}$, $U_\nu \supset L[p,q]$ for $\nu \geq \nu_0$ (see Fig. 4.2). Therefore, $L[p,q] \subset R$, and hence $(p,q) \in S$. □

The next is known as *Bochner's Tube Theorem.*

Theorem 4.3.43 (S. Bochner, K. Stein ($n = 2$)). *The envelope of holomorphy of a univalent tube T_R $(\subset \mathbf{C}^n)$ is the convex hull $T_{\mathrm{ch}(R)}$.*

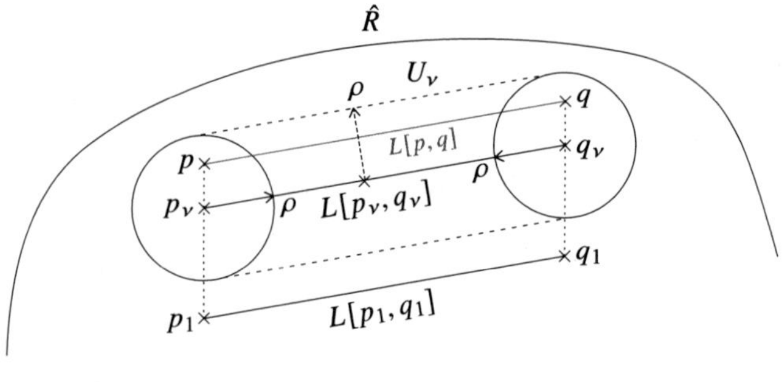

Fig. 4.2 Line segment $L[p_\nu, q_\nu]$.

Proof. By Theorem 3.6.12, there exists the envelope of holomorphy $\pi : \hat{T} \to \mathbf{C}^n$ of T_R, which is possibly multivalent, and contains T_R as a subdomain.

By the definitions of tube and envelope of holomorphy, we have that $\hat{T} = \hat{T} + iy$ ($\forall y \in \mathbf{R}^n$), and hence $\hat{R} = \pi^{-1}\mathbf{R}^n$ (here, $\mathbf{R}^n$ is the real part of $\mathbf{C}^n = \mathbf{R}^n + i\mathbf{R}^n$). Setting $\varpi = \pi|_{\hat{R}}$, we have a real unramified domain

$$\varpi : \hat{R} \longrightarrow \mathrm{ch}(R) \subset \mathbf{R}^n$$

such that $\hat{T} = \hat{R} + i\mathbf{R}^n$: $\hat{T}$ is a tube domain over $\mathbf{C}^n$. By Theorem 4.3.39, $\hat{T}$ has to be univalent and convex. Therefore it follows that $\hat{T} = \mathrm{ch}(R) + i\mathbf{R}^n$. □

4.3.3 Pseudoconvex Boundary

In order to consider the boundary of a domain $\mathfrak{D}/\mathbf{C}^n$, we assume the following condition through this subsection, unless otherwise mentioned:

Condition 4.3.44. $\mathfrak{D}$ is a univalent domain, or a relatively compact subdomain of a larger domain $\mathfrak{G}/\mathbf{C}^n$; in the latter case, $\mathfrak{D}$ is called a *bounded domain* of $\mathfrak{G}$.

Under this condition, $\partial\mathfrak{D}$ is defined as a closed set. Some properties which we describe hold for more general domains, but the discussions are left to the readers.

Theorem 4.3.45. *For a domain $\mathfrak{D}$, the following are equivalent:*

(i) *$\mathfrak{D}$ is pseudoconvex.*
(ii) *For every boundary point $a \in \partial\mathfrak{D}$, there is a neighborhood U of a such that $U \cap \mathfrak{D}$ is pseudoconvex.*

Proof. Assume that $\mathfrak{D}$ is pseudoconvex. Let $\varphi \in \mathscr{P}(\mathfrak{D})$ be a pseudoconvex exhaustion function on $\mathfrak{D}$. For a boundary point $a \in \partial\mathfrak{D}$ we take a Stein neighborhood $U \ni a$. Then, $-\log\delta_{\mathrm{P}\Delta}(p, \partial U) \in \mathscr{P}^0(U)$. Set $\psi(p) := \max\{\varphi(p), -\log\delta_{\mathrm{P}\Delta}(p, \partial U)\}$ ($p \in U \cap \mathfrak{D}$). It follows that $\psi \in \mathscr{P}(U \cap \mathfrak{D})$ and ψ is an exhaustion function on $U \cap \mathfrak{D}$.

We show the converse. Assume that every point $a \in \partial\mathfrak{D}$ has a neighborhood U with pseudoconvex $U \cap \mathfrak{D}$. By Theorem 4.3.11, $-\log\delta_{\mathrm{P}\Delta}(p, \partial(U \cap \mathfrak{D})) \in \mathscr{P}^0(U \cap \mathfrak{D})$. Taking a sufficiently small neighborhood $V \Subset U$ of a, we have

$$-\log\delta_{\mathrm{P}\Delta}(p, \partial(U \cap \mathfrak{D})) = -\log\delta_{\mathrm{P}\Delta}(p, \partial\mathfrak{D}), \quad p \in V \cap \mathfrak{D}.$$

Therefore we see that $-\log\delta_{\mathrm{P}\Delta}(p, \partial\mathfrak{D}) \in \mathscr{P}^0(V \cap \mathfrak{D})$. Covering $\partial\mathfrak{D}$ by such V, we get a neighborhood W of $\partial\mathfrak{D}$ such that $-\log\delta_{\mathrm{P}\Delta}(p, \partial\mathfrak{D}) \in \mathscr{P}^0(W \cap \mathfrak{D})$.

When $\mathfrak{D}$ is a bounded domain of $\mathfrak{G}$, there is a small $c > 0$ such that

$$\{p \in \mathfrak{D} : \delta_{\mathrm{P}\Delta}(p, \partial\mathfrak{D}) < c\} \subset W \cap \mathfrak{D}.$$

With $\varphi(p) := \max\{-\log\delta_{\mathrm{P}\Delta}(p, \partial\mathfrak{D}), -\log c\}$, $\varphi \in \mathscr{P}^0(\mathfrak{D})$ and φ is an exhaustion function.

In the case of $\mathfrak{D} \subset \mathbf{C}^n$, we set $F = \mathfrak{D} \setminus W$, which is closed in $\mathfrak{D}$. Let $\chi(t)$ $(t \geq 0)$ be a monotone increasing convex function in t such that

$$-\log \delta_{\mathrm{P}\Delta}(z, \partial\mathfrak{D}) < \chi(\|z\|), \quad z \in F. \tag{4.3.46}$$

For example, we set $F(r) = F \cap \bar{\mathrm{B}}(r)$ and take

$$c_1 > \max_{F(1)} -\log \delta_{\mathrm{P}\Delta}(z, \partial\mathfrak{D}).$$

Set $\chi(t) = c_1 + k_2(t-1)^+$ $(0 \leq t \leq 2,\ k_2 > 1)$ with $(t-1)^+ = \max\{t-1, 0\}$. Taking a large $k_2 > 1$, we have

$$-\log \delta_{\mathrm{P}\Delta}(z, \partial\mathfrak{D}) < \chi(\|z\|), \quad z \in F(2).$$

With setting $\chi(t) := c_1 + k_2(t-1)^+ + k_3(t-2)^+$ $(0 \leq t \leq 3)$ for a suitable $k_3 > k_2$, we have

$$-\log \delta_{\mathrm{P}\Delta}(z, \partial\mathfrak{D}) < \chi(\|z\|), \quad z \in F(3).$$

Repeating this procedure, we get $\chi(t)$ with $\chi(\|z\|) \in \mathscr{P}^0(\mathbf{C}^n)$.

It follows from (4.3.46) that

$$\psi(z) := \max\{-\log \delta_{\mathrm{P}\Delta}(z, \partial\mathfrak{D}), \chi(\|z\|)\}, \quad z \in \mathfrak{D}$$

is a continuous pseudoconvex exhaustion function on $\mathfrak{D}$. □

N.B. The property of Theorem 4.3.45 (ii) is called the *local pseudoconvexity* of $\partial\mathfrak{D}$. We saw the equivalence of the (global) pseudoconvexity of $\mathfrak{D}$ and the local pseudoconvexity of $\partial\mathfrak{D}$. After all, the Pseudoconvexity Problem is to ask if the statement remains valid after replacing "pseudoconvex" with "Stein".

Now we assume that a domain $\mathfrak{D}/\mathbf{C}^n$, multivalent in general, is C^0-pseudoconvex, and that φ is a C^0-pseudoconvex exhaustion function on $\mathfrak{D}$. Then the sublevel set $\mathfrak{D}_c = \{\varphi < c\}$ with $1/(c - \varphi(p))$ is C^0-pseudoconvex (in the sense that every connected component of it is C^0-pseudoconvex). Here, $\mathfrak{D}_c$ satisfies a better pseudoconvexity than that of $\mathfrak{D}$ because the defining function φ is defined over the outside of $\partial\mathfrak{D}_c$ as a plurisubharmonic function. In fact, in such a case as $a \in \partial\mathfrak{D}_c$, Theorem 4.3.74 proved later implies that $U \cap \mathfrak{D}_c$ is Stein for a Stein neighborhood U $(\subset \mathfrak{D})$ of a. By making use of this local Steinness on the boundary $\partial\mathfrak{D}_c$ we deduce that $\mathfrak{D}_c$ is Stein, and then as the limit of the increasing Stein domains we conclude the Steinness of $\mathfrak{D}$: This is the flow of the argument to solve the problem.

Definition 4.3.47. Let $\mathfrak{G}/\mathbf{C}^n$ be a domain, possibly infinitely sheeted. Let $\mathfrak{D}/\mathbf{C}^n$ be a subdomain of $\mathfrak{G}$, not necessarily bounded but with non-empty boundary $\partial\mathfrak{D}$ in $\mathfrak{G}$.

(i) A point $a \in \partial\mathfrak{D}$ is called a *pseudoconvex boundary point* if there are a neighborhood $U(\subset \mathfrak{G})$ of a and $\varphi \in \mathscr{P}^0(U)$ with $U \cap \mathfrak{D} = \{\varphi < 0\}$. If every point $a \in \partial\mathfrak{D}$ is a pseudoconvex boundary point, $\partial\mathfrak{D}$ is said to be a *pseudoconvex boundary*.

(ii) In (i) above, if φ is chosen to be strongly plurisubharmonic, we call a a *strongly pseudoconvex boundary point*. If every $a \in \partial\mathfrak{D}$ is a strongly pseudoconvex boundary point, $\partial\mathfrak{D}$ is called a *strongly pseudoconvex boundary*. (Up to here, the boundedness of $\mathfrak{D}$ is not required.)

(iii) If $\mathfrak{D}$ is bounded and has a strongly pseudoconvex boundary $\partial\mathfrak{D}$, $\mathfrak{D}$ is called a *strongly pseudoconvex domain*.[6]

The function φ used above is called the *defining function of* $\mathfrak{D}$ about $a \in \partial\mathfrak{D}$.

Let $\mathfrak{D}$ be C^0-pseudoconvex, and let $\varphi : \mathfrak{D} \to [-\infty, \infty)$ be a C^0-pseudoconvex exhaustion function on $\mathfrak{D}$. Then the boundary $\partial\mathfrak{D}'_c$ of any connected component of $\mathfrak{D}_c$ is a pseudoconvex boundary.

Proposition 4.3.48. *Let* $\mathfrak{D}$ *be a domain. If* $\partial\mathfrak{D}$ *is a pseudoconvex boundary, then* $\mathfrak{D}$ *is pseudoconvex. In particular, a strongly pseudoconvex domain is pseudoconvex.*

Proof. By Theorem 4.3.45, it suffices to show that for every point $a \in \partial\mathfrak{D}$, $\mathrm{B}(a;r) \cap \mathfrak{D}$ with an open ball neighborhood $\mathrm{B}(a;r)$ is pseudoconvex. (Here we identify a neighborhood of a with that of the base point of a in $\mathbf{C}^n$.) Taking $\mathrm{B}(a;r)$ smaller if necessary, we take the defining function $\varphi \in \mathscr{P}^0(\mathrm{B}(a;r))$ of $\mathrm{B}(a;r) \cap \partial\mathfrak{D}$. Set

$$\psi(z) = \max\{1/(r - \|z - a\|), -1/\varphi(z)\}, \qquad z \in \mathrm{B}(a;r) \cap \mathfrak{D}.$$

Then, $\psi(z)$ is a pseudoconvex exhaustion function on $\mathrm{B}(a;r) \cap \mathfrak{D}$. □

The next lemma on differentiable functions is frequently used.

Lemma 4.3.49. *For a pair of open sets* $V \Subset U \subset \mathfrak{D}$, *there is a function* $\rho \in C^\infty(\mathfrak{D})$ *satisfying*

$$\rho \geq 0, \quad \mathrm{Supp}\,\rho \Subset U, \quad \rho(p) = 1 \ (p \in V). \tag{4.3.50}$$

Proof. Let W be an open subset such that $V \Subset W \Subset U$, and define the set function of W by

$$\tau(p) = \begin{cases} 1, & p \in W, \\ 0, & p \in \mathfrak{D} \setminus W. \end{cases}$$

There is a local biholomorphic map $\pi : \mathfrak{D} \to \mathbf{C}^n$, and $\overline{W}$ is compact. Therefore, for a sufficiently small $\varepsilon > 0$, the smoothing $\tau_\varepsilon(p)$ of $\tau(p)$ is defined at least for $p \in \overline{W}$. With sufficiently small ε, $\mathrm{Supp}\,\tau_\varepsilon \Subset U$; the other properties follow immediately. □

In what follows, $\langle (v_j), (w_j) \rangle = \sum_j v_j \bar{w}_j$ denotes the standard hermitian inner product of $\mathbf{C}^n$.

Proposition 4.3.51. *For a bounded domain* $\mathfrak{D} \Subset \mathfrak{G}$ *the following are equivalent:*

(i) ($\partial\mathfrak{D}$ local) $\mathfrak{D}$ *is strongly pseudoconvex.*

---

6 There are references where domains with the strong Levi pseudoconvex boundary condition (see Definition 4.3.58) are called strongly pseudoconvex domains; however, the equivalence will be shown.

(ii) ($\partial\mathfrak{D}$ global) *There are a neighborhood* U *of* $\partial\mathfrak{D}$ *and a function* $\psi \in \mathscr{P}^0(U \cup \mathfrak{D})$ *such that* $\mathfrak{D} = \{\psi < 0\}$ *and* ψ *is strongly plurisubharmonic in* U.

Proof. (ii)⇒(i) is trivial.

(i)⇒(ii). By the assumption, there are a finite open covering $\{U_\nu\}_{\nu=1}^l$ of $\partial\mathfrak{D}$, and strongly plurisubharmonic functions φ_ν on U_ν such that

$$U_\nu \cap \mathfrak{D} = \{\varphi_\nu < 0\}, \qquad 1 \le \nu \le l.$$

We take an open covering $\{V_\nu\}_{\nu=1}^l$ of $\partial\mathfrak{D}$ with $V_\nu \Subset U_\nu$, and real-valued functions $\chi_\nu \in C^\infty(\mathfrak{G})$ such that

$$\chi \ge 0, \quad \chi_\nu|_{V_\nu} \equiv 1, \quad \operatorname{Supp}\chi_\nu \Subset U_\nu.$$

With $c > 0$ we set

$$\varphi(z) = \sum_{\nu=1}^{l} \left(\chi_\nu^2(z)\varphi_\nu(z) + c\chi_\nu(z)\varphi_\nu^2(z)\right).$$

About an arbitrary point $a \in \partial\mathfrak{D}$ we compute the Levi from $L[\varphi](v) = L[\varphi](a;v)$ ($v \in \mathbf{C}^n \setminus \{0\}$). Assume $\|v\| = 1$. We identify a small neighborhood of a with that of $\pi(a)$ in $\mathbf{C}^n$, and use the complex coordinates $(z_1,\dots,z_n)$. For notation we write:

$$\partial\varphi_\nu = \partial\varphi_\nu(a) = \left(\frac{\partial\varphi_\nu}{\partial z_1}(a),\dots,\frac{\partial\varphi_\nu}{\partial z_n}(a)\right),$$

$$\bar\partial\varphi_\nu = \bar\partial\varphi_\nu(a) = \left(\frac{\partial\varphi_\nu}{\partial \bar z_1}(a),\dots,\frac{\partial\varphi_\nu}{\partial \bar z_n}(a)\right),$$

$$\langle v, \bar\partial\varphi_\nu\rangle = \langle v, \bar\partial\varphi_\nu(a)\rangle = \sum_{j=1}^n v_j \frac{\partial\varphi_\nu}{\partial z_j}(a), \quad v = (v_1,\dots.v_n).$$

We have

$$\partial\varphi(p) = \sum_{\nu=1}^l \chi_\nu^2(p)\partial\varphi_\nu(p), \quad p \in \partial\mathfrak{D}. \tag{4.3.52}$$

By computations we further get:

$$\begin{aligned}
L[\varphi](v) &= \sum_{\nu=1}^l \left(2c\chi_\nu|\langle v,\bar\partial\varphi_\nu\rangle|^2 + 4\chi_\nu\Re\left(\langle v,\bar\partial\varphi_\nu\rangle\cdot\overline{\langle v,\bar\partial\chi_\nu\rangle}\right) + \chi_\nu^2 L[\varphi_\nu](v)\right)\\
&\ge \sum_{\nu=1}^l \left(2c\chi_\nu|\langle v,\bar\partial\varphi_\nu\rangle|^2 - 4\chi_\nu|\langle v,\bar\partial\varphi_\nu\rangle|\cdot|\langle v,\bar\partial\chi_\nu\rangle| + \chi_\nu^2 L[\varphi_\nu](v)\right)\\
&= \sum_{\nu=1}^l 2c\chi_\nu|\langle v,\bar\partial\varphi_\nu\rangle|^2 - \sum_{\nu=1}^l 4\chi_\nu|\langle v,\bar\partial\varphi_\nu\rangle|\cdot|\langle v,\bar\partial\chi_\nu\rangle| + \sum_{\nu=1}^l \chi_\nu^2 L[\varphi_\nu](v).
\end{aligned}$$

Since the last term is positive, with a sufficiently small $\eta > 0$, the terms with $|\langle v, \bar{\partial}\varphi_\nu\rangle| \le \eta$ satisfies

$$-\sum_{\nu:|\langle v,\bar{\partial}\varphi_\nu\rangle|\le\eta} 4\chi_\nu|\langle v,\bar{\partial}\varphi_\nu\rangle|\cdot|\langle v,\bar{\partial}\chi_\nu\rangle| + \sum_{\nu=1}^{l}\chi_\nu^2 L[\varphi_\nu](v) > 0.$$

On the other hand, for terms $|\langle v, \bar{\partial}\varphi_\nu\rangle| > \eta$, with a sufficiently large $c > 0$ we obtain

$$\begin{aligned}
&\sum_{\nu:|\langle v,\bar{\partial}\varphi_\nu\rangle|>\eta}\left(2c\chi_\nu|\langle v,\bar{\partial}\varphi_\nu\rangle|^2 - 4\chi_\nu|\langle v,\bar{\partial}\varphi_\nu\rangle|\cdot|\langle v,\bar{\partial}\chi_\nu\rangle|\right)\\
&= \sum_{\nu:|\langle v,\bar{\partial}\varphi_\nu\rangle|>\eta} 2\chi_\nu|\langle v,\bar{\partial}\varphi_\nu\rangle|\left(c|\langle v,\bar{\partial}\varphi_\nu\rangle| - 2|\langle v,\bar{\partial}\chi_\nu\rangle|\right)\\
&\ge \sum_{\nu:|\langle v,\bar{\partial}\varphi_\nu\rangle|>\eta} 2\chi_\nu\eta(c\eta - 2|\langle v,\bar{\partial}\chi_\nu\rangle|) \ge 0.
\end{aligned}$$

Thus, $L[\varphi]$ is positive definite at $a \in \partial\mathfrak{D}$, and so $L[\varphi]$ is positive definite in a neighborhood of a. Since $\partial\mathfrak{D}$ is compact, with sufficiently large $c > 0$, $L[\varphi]$ is positive definite in a neighborhood W of $\partial\mathfrak{D}$.

Let $\delta > 0$ be sufficiently small such that $U := \{p \in W : |\varphi(p)| < \delta\} \Subset W$. Now we set

$$\psi(z) = \begin{cases} -\delta, & z \in \mathfrak{D} \cup \{p \in W : \varphi(p) \le -\delta\},\\ \varphi(z), & z \in \{p \in W : \varphi(p) > -\delta\}.\end{cases}$$

Then U and ψ satisfy the required properties. □

The above proof is due to H. Grauert [23]. The notion of a strongly pseudoconvex boundary (point) plays an important role in a number of places.

Proposition 4.3.53. *Let $\mathfrak{D}/\mathbf{C}^n$ be a pseudoconvex domain. Then there is an increasing sequence of strongly pseudoconvex domains $\mathfrak{D}_j \Subset \mathfrak{D}_{j+1} \Subset \mathfrak{D}$ ($j = 1, 2, \ldots$) with $\mathfrak{D} = \bigcup_{j=1}^\infty \mathfrak{D}_j$.*

Proof. By Theorem 4.3.11, $\mathfrak{D}$ satisfies the assumption of Theorem 4.3.15. Let $\mathfrak{D} = \bigcup_{j=1}^\infty \Delta_j$ be the covering obtained by (4.3.25). By (4.3.24) every point of $\partial\Delta_j$ is a strongly pseudoconvex boundary point. We fix a point $p_0 \in \Delta_1$ and denote by $\mathfrak{D}_j$ the connected component of Δ_j containing p_0 for $j \ge 1$. For every point $p \in \mathfrak{D}$ there is a curve C connecting p and p_0. Since C is compact, there is a Δ_j with $\Delta_j \supset C$; thus, $\mathfrak{D}_j \ni p$. We have $\mathfrak{D} = \bigcup_j \mathfrak{D}_j$ with strongly pseudoconvex domains $\mathfrak{D}_j \Subset \mathfrak{D}_{j+1}$. □

Because of Proposition 4.3.53 the following provides an important step in the solution of the Pseudoconvexity Problem.

4.3.54. (**Pseudoconvexity Problem III**) Is a strongly pseudoconvex domain Stein?

4.3.4 Levi Pseudoconvexity

We keep Condition 4.3.44. The content of this subsection is not absolutely necessary for the development of Oka theory. It is, however, useful to understand the notion of a (strongly) pseudoconvex boundary of Definition 4.3.47 and the historical advances.

Definition 4.3.47 of a (strongly) pseudoconvex boundary point is dependent on the choice of $\chi \in \mathscr{P}^0(U)$, and not determined only by $\partial\mathfrak{D}$. Assuming the regularity of $\partial\mathfrak{D}$ (i.e., no singularities), we study here the dependence (or the independence) from the viewpoint of the shape (geometry) of $\partial\mathfrak{D}$ itself.

For a moment we assume $\chi \in C^1(U)$ ($U \ni a$) to satisfy the following about $a \in \partial\mathfrak{D}$:

$$
\begin{aligned}
&U \cap \mathfrak{D} = \{\chi < 0\}, && U \cap \partial\mathfrak{D} = \{\chi = 0\},\\
&\operatorname{grad}\chi(a') \neq 0, && \forall a' \in U \cap \partial\mathfrak{D},\\
&\partial\chi(a') = \left(\frac{\partial\chi}{\partial z_1}(a'), \ldots, \frac{\partial\chi}{\partial z_n}(a')\right) \neq 0, && \forall a' \in U \cap \partial\mathfrak{D}.
\end{aligned}
\tag{4.3.55}
$$

The conditions of the second and the third lines are equivalent. By the implicit function Theorem 1.4.6, $U \cap \partial\mathfrak{D}$ is a real $(2n-1)$-dimensional submanifold (real hypersurface) of C^1-class.

For $a \in U \cap \partial\mathfrak{D}$ we set

$$
\mathbf{T}_a(\partial\mathfrak{D}) = \left\{ (v_j) \in \mathbf{C}^n : \sum_{j=1}^{n} v_j \frac{\partial\chi}{\partial z_j}(a) = 0 \right\},
\tag{4.3.56}
$$

which is called the *holomorphic tangent space* of $\partial\mathfrak{D}$ at a. We consider the restriction $L[\chi]|_{\mathbf{T}_a(\partial\mathfrak{D})}$ of the Levi form $L[\chi](v)$ of χ to the holomorphic tangent space $\mathbf{T}_a(\partial\mathfrak{D})$.

Proposition 4.3.57. *The holomorphic tangent space $\mathbf{T}_a(\partial\mathfrak{D})$ is independent from the choice of $\chi \in \mathscr{C}^1(U)$.*

Proof. Let $\chi_1 \in \mathscr{C}^1(U)$ be a function satisfying (4.3.55). By the implicit function Theorem 1.4.6, $\tau(z) := \chi(z)/\chi_1(z) \in \mathscr{C}^0(U)$, and

$$
\chi(z) = \tau(z)\chi_1(z), \quad \tau(z) > 0, \qquad z \in U.
$$

Since $\chi(a) = \chi_1(a) = 0$, with $h \in \mathbf{R}$ we have

$$
\begin{aligned}
\frac{\partial\chi}{\partial x_j}(a) &= \lim_{h\to 0} \frac{\chi(a_1, \ldots, a_j + h, \ldots, a_n)}{h}\\
&= \lim_{h\to 0} \tau(a_1, \ldots, a_j + h, \ldots, a_n) \frac{\chi_1(a_1, \ldots, a_j + h, \ldots a_n)}{h}\\
&= \tau(a) \frac{\partial\chi_1}{\partial x_j}(a).
\end{aligned}
$$

Similarly, $\frac{\partial \chi}{\partial y_j}(a) = \tau(a)\frac{\partial \chi_1}{\partial y_j}(a)$. Therefore,

$$\frac{\partial \chi}{\partial z_j}(a) = \tau(a)\frac{\partial \chi_1}{\partial z_j}(a). \quad \partial\chi(a) = \tau(a)\partial\chi_1(a).$$

Thus, $\mathbf{T}_a(\partial\mathfrak{D})$ coincides with that defined by χ_1. □

Definition 4.3.58. Suppose that $\chi \in \mathscr{C}^2(U)$ satisfies (4.3.55). Then a point $a \in U \cap \partial\mathfrak{D}$ is called a *Levi pseudoconvex point* if

$$L[\chi](a; v) \geq 0, \qquad v \in \mathbf{T}_a(\partial\mathfrak{D}) \setminus \{0\}; \tag{4.3.59}$$

if it is positive definite, a is called a *strongly Levi pseudoconvex point*. If every $b \in U \cap \partial\mathfrak{D}$ is a (resp. strongly) Levi pseudoconvex point, $U \cap \partial\mathfrak{D}$ is called (resp. *strongly*) *Levi pseudoconvex*.

Condition (4.3.59) is called the *Levi condition* or the *Levi–Krzoska condition*; if it is positive definite, it is called the *strong Levi condition* or the *strong Levi–Krzoska condition*.

Lemma 4.3.60. *The property of $a \in \partial\mathfrak{D}$ being a (strongly) Levi convex point does not depend on the choice of the defining function χ of $\partial\mathfrak{D}$ about a.*

Proof. Let $\chi, \chi_1 \in \mathscr{C}^2(U)$ be the defining functions of $\mathfrak{D}$ about $a \in \partial\mathfrak{D}$, satisfying (4.3.55). Assume that a is a strongly pseudoconvex point with respect to χ. Set

$$\chi_1(z) = \tau(z)\chi(z), \quad \tau(z) > 0, \quad z \in U.$$

By the implicit function Theorem 1.4.6 and a computation, we have that $\tau \in C^1(U)$. We compute the Levi form

$$L[\chi_1](a; v) = \left.\frac{\partial^2}{\partial t \partial \bar{t}}\right|_{t=0} \chi_1(a + tv)$$

for a vector $v \in \mathbf{T}_a(\partial\mathfrak{D}) \setminus \{0\}$. We have

$$\frac{\partial}{\partial \bar{t}}\chi_1(a+tv) = \tau(a+tv)\sum_{j=1}^n \frac{\partial \chi}{\partial \bar{z}_j}(a+tv)\bar{v}_j + \chi(a+tv)\sum_{j=1}^n \frac{\partial \tau}{\partial \bar{z}_j}(a+tv)\bar{v}_j.$$

Next we compute

$$\left.\frac{\partial^2}{\partial t \partial \bar{t}}\right|_{t=0} \chi_1(a+tv) = \lim_{t \to 0}\frac{1}{t}\left(\frac{\partial \chi_1}{\partial \bar{t}}(a+tv) - \frac{\partial \chi_1}{\partial \bar{t}}(a)\right).$$

Noting that $\chi(a) = \sum_{j=1}^n \frac{\partial \chi}{\partial \bar{z}_j}(a)\bar{v}_j = \sum_{j=1}^n \frac{\partial \chi}{\partial z_j}(a)v_j = 0$, we obtain

$$L[\chi_1](a; v) = \tau(a)L[\chi](a; v), \quad v \in \mathbf{T}_a(\partial\mathfrak{D}) \setminus \{0\}.$$

Then, the assertion follows immediately. □

Lemma 4.3.61. *For a point* $a \in \partial\mathfrak{D}$ *with* (4.3.55) *the following conditions are equivalent:*

(i) *a is a strongly pseudoconvex point.*
(ii) *a is a strongly Levi pseudoconvex point.*

Proof. (i)⇒(ii) is trivial.

Assume (ii). Let $\chi \in \mathscr{C}^2(U)$ be a defining function of $\mathfrak{D}$ about $a \in \partial\mathfrak{D}$, and let $c > 0$ be a constant determined later. We set

$$\varphi(z) = e^{c\chi(z)} - 1, \quad z \in U.$$

Then, φ is a defining function of $\mathfrak{D}$ about $a \in \partial\mathfrak{D}$. For $v \in \mathbf{C}^n \setminus \{0\}$, we have

$$L[\varphi](a;v) = ce^{c\chi}\Big(L[\chi](a;v) + c|\langle v, \bar{\partial}\chi(a)\rangle|^2\Big).$$

Since $\mathbf{T}_a(\partial\mathfrak{D})$ is the orthogonal complement space of $\bar{\partial}\chi(a)$, we decompose v into the orthogonal factors:

$$v = w \oplus \zeta\bar{\partial}\chi(a), \quad w \in \mathbf{T}_a(\partial\mathfrak{D}),\ \zeta \in \mathbf{C}.$$

It follows that

$$\begin{aligned} (4.3.62) \qquad L[\varphi](a;v) = c\Bigg(& L[\chi](a;w) + 2\Re\zeta\left\langle \bar{\partial}\chi(a), \left(\sum_k \frac{\partial^2\chi}{\partial z_k \partial \bar{z}_j}(a) w_k\right)_j \right\rangle \\ & + |\zeta|^2 L[\chi](a;\bar{\partial}\chi(a)) + c|\zeta|^2 \|\bar{\partial}\chi(a)\|^4 \Bigg). \end{aligned}$$

By the assumption of $L[\varphi]|_{\mathbf{T}(\partial\mathfrak{D})} \gg 0$, there are positive constants C_1 and C_2 such that

$$\begin{aligned} & L[\chi](a;w) \geq C_1\|w\|^2, \\ & \left|\left\langle \bar{\partial}\chi(a), \left(\sum_k \frac{\partial^2\chi}{\partial z_k \partial \bar{z}_j}(a) w_k\right)_j \right\rangle\right| \leq C_2\|w\|. \end{aligned}$$

Then, it follows from (4.3.62) that

$$\begin{aligned} L[\varphi](a;v) &\geq c\big(C_1\|w\|^2 - 2C_2\|w\||\zeta| + (c\|\bar{\partial}\chi(a)\|^4 - |L[\chi](a;\bar{\partial}\chi(a))|)|\zeta|^2\big) \\ &= c\big(C_1\|w\|^2 - 2C_2\|w\||\zeta| + C_3|\zeta|^2\big), \end{aligned}$$

where $C_3 := c\|\bar{\partial}\chi(a)\|^4 - |L[\chi](a;\bar{\partial}\chi(a))| > 0$ with a large $c > 0$. Note that C_3 may be chosen arbitrarily large. Now, we have

$$L[\varphi](a;v) \geq \frac{c}{2}(C_1\|w\|^2 + C_3|\zeta|^2) + \frac{c}{2}(C_1\|w\|^2 - 4C_2\|w\||\zeta| + C_3|\zeta|^2).$$

With a large C_3 such that $C_1C_3 \geq 4C_2^2$, the second quadric term of the right-hand side above is non-negative. It follows that

$$L[\varphi](a;v) \geq \frac{c}{2}(C_1\|w\|^2 + C_3|\zeta|^2) > 0.$$

Hence we deduce that φ is strongly pseudoconvex in a neighborhood about a. □

Remark 4.3.63. It is known that the above lemma remains valid under the semi-positive definite condition, but the proof is more involved, not simply computations, but needs arguments similar to the proof of Theorem 4.3.45 (cf. [30] Theorem 2.6.12). Here we omit it because it is not particularly necessary.

Proposition 4.3.64. *Let $\mathfrak{D}$ be a bounded domain of $\mathfrak{G}/\mathbf{C}^n$. If every point $a \in \partial\mathfrak{D}$ is a strongly Levi pseudoconvex boundary point, then there is a strongly pseudoconvex function φ in a neighborhood U ($\subset \mathfrak{G}$) of $\partial\mathfrak{D}$ such that $\partial\varphi \neq 0$ on $\partial\mathfrak{D}$, and $U \cap \mathfrak{D} = \{\varphi < 0\}$. In particular, φ satisfies a strong Levi condition at every point of $\partial\mathfrak{D}$.*

Proof. It follows from Lemma 4.3.61 and Proposition 4.3.51. Here the condition, $\partial\varphi \neq 0$ on $\partial\mathfrak{D}$, follows from (4.3.52). □

Historically, the Pseudoconvexity Problem was posed in the following form:

4.3.65 (**Levi's Problem**, [4] Kap. IV). Assume that the boundary of a bounded domain $\mathfrak{D}$ ($\Subset \mathfrak{G}$) is strongly Levi pseudoconvex. Then, is $\mathfrak{D}$ a domain of holomorphy?

Note. Thereafter a number of notions of pseudoconvexity were proposed, and their relations were investigated; it might be evidence that the notion of "pseudo-convex" was ambiguous (this does not mean each definition of "pseudoconvex" was ambiguous). Oka went back to Hartogs' phenomenon, and then obtained the notion of pseudoconvex functions (equivalently, plurisubharmonic functions), with which he solved the Pseudoconvexity Problem. Because of the procedure, he termed the problem *Hartogs' Inverse Problem*, formulating Pseudoconvexity Problems I 4.3.6, II 4.3.10, and III 4.3.54. As observed already, Levi's Problem above is implied by Pseudoconvexity Problem III 4.3.54.

In the next chapter we will give the solution of Levi's Problem, where we will not use Levi's condition; but, we solve Pseudoconvexity Problem III 4.3.54 stated in terms of strongly pseudoconvex (plurisubharmonic) functions. This circumstance is the same in the other proofs by Grauert's method and also by the L^2-$\bar{\partial}$ method of Hörmander.

4.3.5 Strongly Pseudoconvex Boundary Points and Stein Domains

Let $\pi : \mathfrak{G} \to \mathbf{C}^n$ be a domain. We consider a bounded domain $\mathfrak{D} \Subset \mathfrak{G}$ and its boundary $\partial\mathfrak{D}$. Suppose that there is a real continuous function λ defined in a neighborhood U of $\partial\mathfrak{D}$, and that

$$\mathfrak{D} = \{\lambda < 0\}, \qquad \partial\mathfrak{D} = \{\lambda = 0\}.$$

Now we assume that λ is strongly plurisubharmonic in a neighborhood V of a boundary point $p_0 \in \partial\mathfrak{D}$.

For a while we consider only in a neighborhood of p_0, so that we regard $V \subset \mathbf{C}^n$, and by translation, we assume $p_0 = 0 \in \mathbf{C}^n$. Then $\lambda(z)$ is developed to a Taylor expansion with center at $z = 0$:

$$\text{(4.3.66)} \qquad \lambda(z) = \Re\left\{2\sum_{j=1}^{n} \frac{\partial\lambda}{\partial z_j}(0) z_j + \sum_{j,k} \frac{\partial^2\lambda}{\partial z_j \partial z_k}(0) z_j z_k\right\} + \sum_{j,k} \frac{\partial^2\lambda}{\partial z_j \partial \bar{z}_k}(0) z_j \bar{z}_k + o(\|z\|^2).$$

With a small $r_0 > 0$,

$$\sum_{j,k} \frac{\partial^2\lambda}{\partial z_j \partial \bar{z}_k}(0) z_j \bar{z}_k + o(\|z\|^2) \geq \frac{1}{2}\sum_{j,k} \frac{\partial^2\lambda}{\partial z_j \partial \bar{z}_k}(0) z_j \bar{z}_k > 0, \quad 0 < \|z\| < r_0.$$

Introducing a new parameter $t \in \mathbf{C}$, we define holomorphic functions in z and (t,z) by

$$\text{(4.3.67)} \qquad P(z) = 2\sum_{j=1}^{n} \frac{\partial\lambda}{\partial z_j}(0) z_j + \sum_{j,k} \frac{\partial^2\lambda}{\partial z_j \partial z_k}(0) z_j z_k,$$
$$\sigma(t,z) = t - P(z).$$

Note that $P(z)$ is a *non-zero* quadric. The function $P(z)$ (resp. $\sigma(t,z)$) is caled the *Levi* (resp. *Levi–Oka) polynomial* of λ (or the boundary $\partial\mathfrak{D}$) at $p_0 = 0$.[7]

We consider a family of complex hypersurfaces defined by

$$\Sigma_t = \{z \in \mathbf{C}^n : \sigma(t,z) = 0\} \subset \mathbf{C}^n, \quad t \in \mathbf{C}$$

which are called the *Oka hypersurfaces* of λ (or the boundary $\partial\mathfrak{D}$) at $p_0 = 0$ (cf. Fig. 4.3).[8]

---

[7] For $P(z)$ the term is due to Range [53] Chap. II, §2.8, and Lieb [33] §12; the introduction of $\sigma(t,z)$ and Σ_t is inspired by Oka IX of 1943 (unpublished) in [50], and Oka IX (1953) in [48], [49]; cf. Remark 4.3.73 and Note at the end of this chapter.

[8] The term is due to Oka VI (1942), IX (1953) in [48], [49] and IX of 1943 (unpublished) in [50]; cf. ibid.

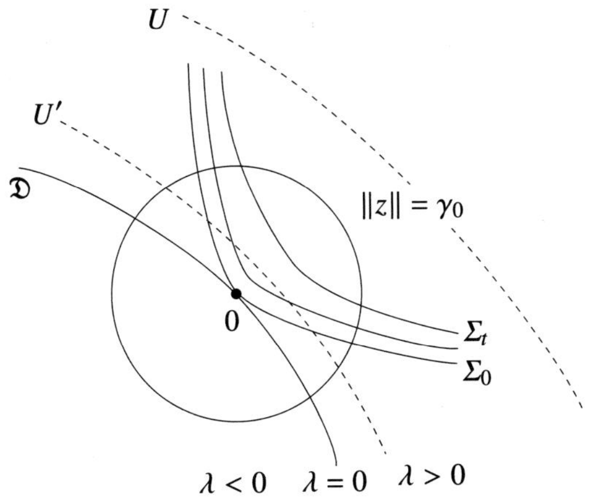

Fig. 4.3 Oka hypersurface Σ_t.

In a neighborhood of $(0,0)$, we restrict t to the real axis; if $t > 0$, $\sigma(t,z) = 0$ implies $\lambda(z) > t > 0$, so that

$$\Sigma_t \cap \bar{\mathfrak{D}} \cap \{\|z\| < r_0\} = \emptyset,$$

and if $t = 0$, then

$$\bar{\mathfrak{D}} \cap \{\sigma(0,z) = 0\} \cap \{\|z\| < r_0\} = \{0\}.$$

For small $\delta_0 > 0$ we have

$$\left(\bigcup_{0 \le t \le \delta_0} \Sigma_t\right) \cap \{\|z\| = r_0\} \cap \bar{\mathfrak{D}} = \emptyset. \tag{4.3.68}$$

Taking a neighborhood $U' \subset U$ of $\bar{\mathfrak{D}}$ sufficiently close to $\bar{\mathfrak{D}}$ (cf. Fig. 4.3), and a sufficiently small neighborhood V of the closed internal $[0,\delta_0]$ in complex plane $\mathbf{C}$, we have

$$\{(t,z) \in V \times U' : \sigma(t,z) = 0\} \Subset V \times \{\|z\| < r_0\}.$$

Set $W_0 = V \times (U' \cap \{\|z\| < r_0\})$. Then, the complex hypersurface $\{\sigma(t,z) = 0\} \cap W$ of $W := V \times U'$ is that of W_0. With $W_1 := (V \times U') \setminus \{\sigma(t,z) = 0\}$, we have an open covering $W = W_0 \cup W_1$. Set

$$g_0 = \frac{1}{\sigma(t,z)} \quad \big((t,z) \in W_0\big); \qquad g_1 = 0 \ \ (\text{on } W_1). \tag{4.3.69}$$

Then $\{(W_j, g_j)\}_{j=0,1}$ is Cousin I data on W.

Summarizing the above, we obtain:

Lemma 4.3.70. *Let the notation be as above. Assume that $\lambda(p)$ is strongly plurisubharmonic in a neighborhood of a boundary point $p_0 \in \partial\mathfrak{D}$ ($\pi(p_0) = z_0$).*

Then we have:

(i) *There are a small neighborhood U_0 of p_0 and a small $\delta_0 > 0$ with the identification of $p \in U_0$ and $\pi(p) = z$ such that the Oka hypersurfaces Σ_t of λ at $p_0 = z_0$ satisfy the following:*

$$(4.3.71)\qquad \begin{aligned} &a.\ z_0 \in \Sigma_0.\\ &b.\ \Sigma_t \cap \Sigma_{t'} = \emptyset, \quad t \neq t' \in [0, \delta_0].\\ &c.\ \left(\bigcup_{0\le t\le \delta_0} \Sigma_t\right) \cap \bar{\mathfrak{D}} \cap U_0 = \{z_0\}.\\ &d.\ \left(\bigcup_{0\le t\le \delta_0} \Sigma_t\right) \cap \partial U_0 \cap \bar{\mathfrak{D}} = \emptyset. \end{aligned}$$

Here we may take $U_0 = \{z : \|z - z_0\| < r_0\}$ with an arbitrarily small $r_0 > 0$.

(ii) *There are neighborhoods U' of $\bar{\mathfrak{D}}$ and ω ($\subset \mathbf{C}$) of $[0, \delta_0]$ such that on $W := \omega \times U'$ there is Cousin I data with polar set $\{\sigma(t,z) = 0\} \cap W$ defined by (4.3.69), where $\sigma(t,z)$ is the Levi–Oka polynomial of λ at $p_0 = z_0$.*

(iii) *If $\bar{\mathfrak{D}}$ has a Stein neighborhood, then with Stein U' above, the above Cousin I data has a solution $G(t,p) \in \mathscr{M}(W)$ on $W = \omega \times U'$. In particular, we have:*
   a. *$G(0,p)$ has poles on the Oka hypersurface Σ_0 touching $\partial\mathfrak{D}$ only at p_0 from the outside,*
   b. *$G(0,p)$ is holomorphic on $\bar{\mathfrak{D}} \setminus \{p_0\}$,*
   c. *$\underline{\lim}_{p\to p_0, z\in\bar{\mathfrak{D}}\setminus\{p_0\}} |G(0,p)| = \infty$.*

Example 4.3.72. We consider the unit open ball $\mathrm{B}(1) = \{\|z\| < 1\}$ ($\Subset \mathbf{C}^n$), a typical example of strongly pseudoconvex domains. Let $z = (z_1, \ldots, z_n) \in \mathbf{C}^n$ be the coordinate system and let $p_0 = (p_{0j}) = (1, 0, \ldots, 0) \in \partial\mathrm{B}(1)$. We compute the Oka hypersurfaces Σ_t in Lemma 4.3.70 at p_0. Note that $\lambda(z) := \|z\|^2 - 1$ is strongly plurisubharmonic and $\mathrm{B}(1) = \{\lambda(z) < 0\}$, and that Σ_t is given by $\sigma(t,z)$ of (4.3.67). Since $\frac{\partial\lambda}{\partial z_j} = \bar{z}_j$, the Levi–Oka polynomial and the Oka hypersurfaces of $\partial\mathrm{B}(1)$ at p_0 are written as

$$\sigma(t,z) = t - 2(z_1 - 1), \quad \Sigma_t = \left\{z_1 = 1 + \frac{t}{2}\right\}.$$

In particular, $\Sigma_0 = \{z_1 = 1\}$ and

$$\left(\bigcup_{t\ge 0} \Sigma_t\right) \cap \overline{\mathrm{B}(1)} = \{p_0\}.$$

Remark 4.3.73. The method of constructing the Oka hypersurfaces Σ_t above will be used in a number of places henceforward. The idea to consider not only Σ_0 but Σ_t with a parameter t and to apply the Cousin I Problem in the total space of z and t is due to the last one of Oka's unpubished five papers of 1943 (cf. [44], Part II, §8).

The purpose is to obtain a holomorphic function on $\mathfrak{D}$ which diverges to infinity at p_0, and then deform it so that it is holomorphic at p_0 but has finite values with large moduli. Here we showed that it is possible to construct Cousin I data on $\mathfrak{D}$ with such a divergence property at a strongly pseudoconvex boundary point $p_0 \in \partial\mathfrak{D}$. If $n = 1$, it suffices to take a rational function $1/(z - p_0)$ for any boundary point $p_0 \in \partial\mathfrak{D}$. In the case of $n \geq 2$, the function $1/P(z)$ with the Levi polynomial $P(z)$ which is constructed from the strongly plurisubharmonic function $\lambda(p)$ about p_0, plays the role of $1/(z - p_0)$. For this purpose Oka invented in VI (1942) the notion of (strongly) pseudoconvex (plurisubharmonic) functions. However, while $1/(z - p_0)$ is already defined on $\mathbf{C}$, $1/P(z)$ is defined only locally about p_0; the gap is considerably large.

Theorem 4.3.74.[9] *Let $\pi : \mathfrak{D} \to \mathbf{C}^n$ be a Stein domain, and let $\lambda : \mathfrak{D} \to [-\infty, \infty)$ be a plurisubharmonic function. Then, $\mathfrak{D}_c = \{p \in \mathfrak{D} : \lambda(p) < c\}$ is $\mathscr{O}(\mathfrak{D})$-convex, and Stein.*

Proof. **(a)** We first assume that λ is strongly plurisubharmonic. Take any compact subset $K \Subset \mathfrak{D}_c$. Since $\mathfrak{D}$ is Stein, $\widehat{K} := \widehat{K}_{\mathfrak{D}} \Subset \mathfrak{D}$. We are going to show $\widehat{K} \subset \mathfrak{D}_c$. Set

$$c_0 = \max_{\widehat{K}} \lambda = \lambda(z_0), \quad z_0 \in \widehat{K}.$$

it suffices to prove $c_0 < c$. Suppose $c_0 \geq c$. It is noted that $z_0 \notin K$. We take an $\mathscr{O}(\mathfrak{D})$-analytic polyhedron $\mathrm{P} \Supset \widehat{K}$. Applying Lemma 4.3.70 to $\mathfrak{D} = \mathfrak{D}_{c_0}$ and $p_0 = z_0$, we obtain the Levi–Oka polynomial $\sigma(t, z)$ of $\partial\mathfrak{D}_{c_0}$ at z_0, and the Oka hypersurfaces $\Sigma_t = \{z : \sigma(t, z) = 0\}$ satisfying (4.3.71). For U' of Lemma 4.3.70 (ii) we can choose a sufficiently small P such that $U' \supset \mathrm{P}$. With a connected neighborhood $\omega \supset [0, \delta_0]$ (closed interval) in $\mathbf{C}$, the open set $W := \omega \times \mathrm{P}$ is Stein. We consider the Cousin I data $\{(W_j, g_j)\}_{j=0,1}$ on W defined by (4.3.69). By Theorem 3.4.31 there is a meromorphic function $G(t, p)$ on W such that

$$G(t, z) - \frac{1}{\sigma(t, z)} \in \mathscr{O}(W_0). \tag{4.3.75}$$

Since $G(t, p)$ is holomorphic on $[0, \delta_0] \times K$,

$$M := \max_{[0, \delta_0] \times K} |G(t, p)| < \infty.$$

It follows from (4.3.75) that there is a point $t_0 \in (0, \delta_0)$ with $|G(t_0, z_0)| > M + 1$ and $G(t_0, z) \in \mathscr{O}(\widehat{K})$. The construction implies that

$$\max_{p \in K} |G(t_0, p)| \leq M < M + 1 < |G(t_0, z_0)|,$$
$$K \cup \{z_0\} \subset \widehat{K}.$$

---

[9] This is a well-known fact, but it is new in the sense that the proof is given before the solution of the Pseudoconvexity Problem, or Levi's Problem.

By the Oka–Weil Approximation Theorem 3.7.7 we approximate $G(t_0,p)$ by elements of $\mathscr{O}(\mathfrak{D})$, uniformly on $\widehat{K}$. Then we see that $z_0 \notin \widehat{K}$; this is a contradiction.

(b) Let λ be a plurisubharmonic function in general. Taking a compact set $K \Subset \mathfrak{D}_c$, we set $\widehat{K} := \widehat{K}_{\mathfrak{D}} \Subset \mathfrak{D}$. Take an $\mathscr{O}(\mathfrak{D})$-analytic polyhedron P with $\widehat{K} \Subset \mathrm{P} \Subset \mathfrak{D}$. Since P is $\mathscr{O}(\mathfrak{D})$-convex, $\widehat{K}_{\mathrm{P}} = \widehat{K}$.

Take $c' \in \mathbf{R}$ such that

$$\max_K \lambda < c' < c.$$

With $\varepsilon_0 := \min\left\{\delta_{\mathrm{B}}(z,\partial\mathfrak{D}) : z \in \bar{\mathrm{P}}\right\}$, we consider the smoothing λ_ε of λ, where $0 < \varepsilon < \varepsilon_0$. We would like to show:

Claim 4.3.76. *For a sufficiently small* $\varepsilon > 0$, $K \subset \{\lambda_\varepsilon(p) < c'\} \cap \mathrm{P}$.

$\because$) Take any point $p \in K$, and c'' with $\lambda(p) < c'' < c'$. Since λ is upper semi-continuous, $\lambda < c''$ in a neighborhood U of p. For an arbitrarily small $\varepsilon > 0$ we have

$$\lambda_\varepsilon(p) \le c'' < c'.$$

Therefore there is a neighborhood $U'(\subset U)$ of p such that $\lambda_\varepsilon(p') < c'$ $(\forall p' \in U')$. Because of the compactness of K, there are finitely many such U' covering K. Hence, for sufficiently small $\varepsilon > 0$ we obtain that $\{\lambda_\varepsilon < c'\} \cap \mathrm{P} \supset K$. △

In a neighborhood of $\bar{\mathrm{P}}$ we set

$$\tilde{\lambda}_\varepsilon^{\varepsilon'}(p) = \lambda_\varepsilon(p) + \varepsilon'\|\pi(p)\|^2, \qquad \varepsilon' > 0.$$

Since $\bar{\mathrm{P}}$ is compact,

$$M := \max_{\bar{\mathrm{P}}} \|\pi(p)\|^2 < \infty.$$

The function $\tilde{\lambda}_\varepsilon^{\varepsilon'}(p)$ is strongly plurisubharmonic in a neighborhood of $\bar{\mathrm{P}}$, and satisfies

$$\lambda_\varepsilon(p) \le \tilde{\lambda}_\varepsilon^{\varepsilon'}(p) \le \lambda_\varepsilon(p) + \varepsilon' M, \qquad p \in \bar{\mathrm{P}}.$$

With $\varepsilon' > 0$ such that $c' + \varepsilon' M < c$ the following implications hold:

$$\begin{aligned} K \Subset \{\lambda_\varepsilon < c'\} \cap \mathrm{P} &\subset \left\{\tilde{\lambda}_\varepsilon^{\varepsilon'} < c' + \varepsilon' M\right\} \cap \mathrm{P} \subset \{\lambda_\varepsilon < c' + \varepsilon' M\} \cap \mathrm{P} \\ &\subset \{\lambda < c\} \cap \mathrm{P}. \end{aligned}$$

Since P is Stein, it follows from (a) above that

$$\widehat{K}_{\mathrm{P}} \Subset \left\{\tilde{\lambda}_\varepsilon^{\varepsilon'} < c' + \varepsilon' M\right\} \cap \mathrm{P} \subset \{\lambda < c\} \cap \mathrm{P}.$$

Therefore, $\widehat{K} = \widehat{K}_{\mathrm{P}} \Subset \mathfrak{D}_c$. □

Corollary 4.3.77. *Let* $\lambda : \mathbf{C}^n \to [-\infty, \infty)$ *be a plurisubharmonic function. Then for every* $c \in \mathbf{R}$*, each connected component of* $\{z \in \mathbf{C}^n : \lambda(z) < c\}$ *is a domain of holomorphy.*

That is, in the case of a univalent domain, the Pseudoconvexity Problem is solved by Corollary 4.3.77, if the pseudoconvex function defining the pseudoconvex domain is defined on $\mathbf{C}^n$.

Let $\mathfrak{D}/\mathbf{C}^n$ be domain and let $K \subset \mathfrak{D}$ be a subset. As the holomorphically convex hull $\widehat{K}_{\mathfrak{D}}$ is defined in (3.1.10), we define the *pseudoconvex hull* of K in $\mathfrak{D}$ by

$$(4.3.78) \qquad \widehat{K}_{\mathscr{P}(\mathfrak{D})} = \left\{ p \in \mathfrak{D} : |\varphi(p)| \le \sup_K \varphi,\ \forall \varphi \in \mathscr{P}(\mathfrak{D}) \right\}.$$

Theorem 4.3.79. *Let $\mathfrak{D}/\mathbf{C}^n$ be a Stein domain. If $K \Subset \mathfrak{D}$ is a compact subset, then, $\widehat{K}_{\mathscr{P}(\mathfrak{D})} = \widehat{K}_{\mathfrak{D}}$.*

Proof. By definition, $\widehat{K}_{\mathscr{P}(\mathfrak{D})} \subset \widehat{K}_{\mathfrak{D}}$. Suppose that the equality does not hold. There is a point $a \in \widehat{K}_{\mathfrak{D}}$ such that there is a function $\varphi \in \mathscr{P}(\mathfrak{D})$ with $\varphi(a) > \sup_K \varphi$. Take a number $c \in \mathbf{R}$ so that $\varphi(a) > c > \sup_K \varphi$. Then $K \Subset \mathfrak{D}_c = \{\varphi < c\}$. It follows from Theorem 4.3.74 that $\mathfrak{D}_c$ is $\mathscr{O}(\mathfrak{D})$-convex, and then Theorem 3.3.18, which holds for Stein domains over $\mathbf{C}^n$, implies $\widehat{K}_{\mathfrak{D}} = \widehat{K}_{\mathfrak{D}_c} \not\ni a$; this is absurd. □

We consider a univalent domain $\Omega \subset \mathbf{C}^n$. For a compact subset $K \Subset \Omega$ we have the following relations for its hulls in general:

$$(4.3.80) \qquad \begin{gathered} K \subset K^{\star}_{\Omega} \subset \widehat{K}_{\Omega} \subset \widehat{K}_{\mathscr{P}(\Omega)} \\ d(K, \partial\Omega) = d(K^{\star}_{\Omega}, \partial\Omega) \ge d(\widehat{K}_{\Omega}, \partial\Omega) \ge d(\widehat{K}_{\mathscr{P}(\Omega)}, \partial\Omega). \end{gathered}$$

Cf. Proposition 3.1.7 for the equality above. From Theorem 4.3.79 we immediately obtain:

Corollary 4.3.81. *Let $K \Subset \Omega$ $(\subset \mathbf{C}^n)$ be as above. Assume that Ω is a domain of holomorphy. Then,*

$$(4.3.82) \qquad \begin{gathered} K \subset K^{\star}_{\Omega} \subset \widehat{K}_{\Omega} = \widehat{K}_{\mathscr{P}(\Omega)} \\ d(K, \partial\Omega) = d(K^{\star}_{\Omega}, \partial\Omega) = d(\widehat{K}_{\Omega}, \partial\Omega) = d(\widehat{K}_{\mathscr{P}(\Omega)}, \partial\Omega). \end{gathered}$$

Remark 4.3.83. Theorem 4.3.74 suggests a close relationship between holomorphic functions and plurisubharmonic functions. It is interesting to notice that Theorems 4.3.74 and 4.3.79 imply the solution of the Pseudoconvexity Problem if the domain is univalent and the given plurisubharmonic function is defined on a domain of holomorphy, or equivalently, a holomorphically convex domain; i.e., they are proved before the solution of the Psudoconvexity Problem. In Hörmander [30] §4.3 the statements similar to those above are proved for a pseudoconvex domain of $\mathbf{C}^n$ by making use of the solution of the Pseudoconvexity Problem by means of the $\bar{\partial}$-L^2 method.

Note. The pseudoconvexity properties (iv)–(vi) stated in Remark 4.3.9 are described only in terms of real functions, where only local complex coordinates are used in the definition of plurisubharmonicity. Moreover, as described after §4.3.3, the pseudoconvexity of $\mathfrak{D}$ is a local property of the boundary $\partial\mathfrak{D}$. Thus, the properties appear to be far from the existence of the global holomorphic functions. These difficulties might be the reason why the Pseudoconvexity Problem was first believed to be unsolvable (R. Remmert, [49] Vorwort).

We have summarized the Pseudoconvexity Problem as in the present chapter, but it should be reminded that such a notion as "plurisubharmonic functions" or "pseudoconvex functions" was not yet invented when the problem was proposed. K. Oka introduced the new notion of "pseudoconvex functions" in Oka VI (1942) in order to solve the problem; this was the critical difference in the notions of the pseudoconvexity of Levi and Oka (see also IX (unpublished, 1943) in [50], and IX §13 (published, 1953) in [48], [49]): Thanks to the notion of Oka's pseudoconvex functions, it is then possible to approximate a strongly pseudonvex domain by strongly pseudonvex domains from the outside that contain the closure of the original one. Thus, a bridge from the Cousin I Problem to the Pseudoconvexity Problem is provided; these are the reason why $\sigma(t,z)$ and Σ_t are termed the Levi-Oka polynomial and the Oka hypersurfaces, respectively.

Around that time, P. Lelong introduced the same notion from the potential theoretic viewpoint.

The present proof of Bochner's Theorem 4.3.43 is due to the author.[10] Equation (4.3.40) was inspired by an exercise in Fritzsche–Grauert [18] p. 87. While the existence of an envelope of holomorphy is used in the present proof, in [18] that notion is introduced later, so that the points of consideration seem to be different. Theorem 4.3.39 is due to Makoto Abe [1]. It is nice to see that these are obtained as applications of Oka's Theorem of Boundary Distance Functions 4.3.1. Cf. Exercises 8 and 9 below.

S. Bochner proved Theorem 4.3.43 first for bounded holomorphic functions in [6] (1937), and in [7] (1938) he removed the boundedness condition. The method of the proof was very technical, by means of Legendre polynomial developments. Around that time, K. Stein ([56] 1937) proved the result independently by making use of ellipses in the case of $n = 2$. Since the theorem has a special importance in the theories of homogeneous spaces, hyperfunctions, and partial differential equations, it has had a various proofs since then.

Hörmander [30] §2.5 gives a proof by making use of ellipses, close to Stein's proof in n-dimensional case; it is rather involved. In Hörmander [30] p. 43 Example, some relation with the theory of partial differential equations is mentioned, and in [31] H. Komatsu proved a localized form of Bochner's Tube Theorem obtained due to M. Kashiwara who used it in a fundamental part of M. Sato's hyperfunctions.

Furthermore, Hartogs' phenomenon has an application to the Edge of the Wedge Theorem due to N. Bogoliubov in theoretical physics (cf. Rudin [55] and the references there).

---

[10] J. Noguchi, A brief proof of Bochner's tube theorem and a generalized tube, arXiv, 2020.

Exercises

1. Show that an upper semi-continuous function on a compact set K of $\mathbf{C}^n$ attains the maximum on K.
2. Prove (4.1.2).
3. Prove that a function $\varphi : U \to [-\infty, \infty)$ on an open set $U \subset \mathbf{C}^n$ is upper semi-continuous if and only if $\overline{\lim_{z \to a}}\, \varphi(z) \le \varphi(a)$ at every point $a \in U$.
4. Show that an upper semi-continuous function on an open set of $\mathbf{C}^n$ is the limit of a decreasing sequence of continuous functions on U.
5. Let φ be a subharmonic function on a domain $U \subset \mathbf{C}$, and assume $\varphi \not\equiv -\infty$. Show that φ is integrable on the boundary circle $C(a;r)$ of any disk $\Delta(a;r) \Subset U$.
6. (Hadamard's three circle theorem) Let $R_2 > R_1 > 0$ and let $\varphi(z)$ be a subharmonic function in a neighborhood of the closed ring domain $\{z \in \mathbf{C} : R_1 \le |z| \le R_2\}$. For $R_1 \le r \le R_2$ we set $M_\varphi(r) = \max_{\{|z|=r\}} \varphi(z)$. Then, show that

$$M_\varphi(r) \le \frac{(\log R_2 - \log r) M_\varphi(R_1) + (\log r - \log R_1) M_\varphi(R_2)}{\log R_2 - \log R_1}.$$

   Therefore, $M_\varphi(r)$ is a continuous convex function in $\log r$.
   Hint: $\psi(z) = \varphi(z) - \alpha \log |z|$ $(\alpha \in \mathbf{R})$ is subharmonic. Take $\alpha \in \mathbf{R}$ so that $M_\psi(R_1) = M_\psi(R_2)$. By the Maximum Principle, $M_\psi(r) = M_\varphi(r) - \alpha \log r \le M_\psi(R_1)$.
7. With real numbers $\alpha < \beta$ we define a 1-dimensional tube domain $T = \{z + iy \in \mathbf{C} : \alpha < x < \beta, y \in \mathbf{R}\}$. Let $\varphi(z)$ be a continuous function on T such that $\varphi(z) = \varphi(z + iy)$ $(\forall y \in R)$. Show that $\varphi(z)$ is subharmonic if and only if $\varphi(x)$ $(\alpha < x < \beta)$ is a convex function in x.
   Hint: Use the smoothing.
8. Let $\Omega \subset \mathbf{C}^n$ be a Reinhardt domain (cf. Chap. Exercise 5). We define the logarithmic image of it by $\log \Omega = \{(\log |z_j|) \in \mathbf{R}^n : (z_j) \in \Omega, \forall z_j \ne 0\}$. On the other hand, for a real domain $R \subset \mathbf{R}^n$ we define

$$\exp R = \{(z_j) \in \mathbf{C}^n : |z_j| < e^{\rho_j}, 1 \le j \le n, \exists (\rho_j). \in R\}.$$

   Assume that Ω contains the origin.
   Show that the envelope of holomorphy of Ω is $\exp \mathrm{ch}(\log \Omega)$.
   Hint: Use Theorem 4.3.43.
9. Let $f \in \mathcal{O}(\mathbf{C}^n)$ and set $Q = \{x = (x_j) \in \mathbf{R}^n : f(x + iy) \ne 0, \forall y \in \mathbf{R}^n\}$. Prove that every connected component of Q is convex.
   Remark: This is non-trivial even if $f(z)$ is polynomial.
10. With two real domains $R, Q \subset \mathbf{R}^n$ $(n \ge 2)$ we consider a domain $\Omega = R + iQ \subset \mathbf{C}^n$ of real–imaginary product type. Show that Ω is a domain of holomorphy if and only if both R and Q are (affine) convex.

Hint: Use that in a small neighborhood of $\partial\Omega \supset (\partial R + iQ) \cup (R + i\partial Q)$, $-\log \delta_{\mathrm{P}\Delta}(z, \partial\Omega)$ is plurisubharmonic.

11. Let $\mathfrak{D}/\mathbf{C}^n$ be a domain, and let $v \in \mathbf{C}^n \setminus \{0\}$ be a directional vector. Set

$$R_v(z) = \sup\{\rho > 0 : z + \zeta v \in \mathfrak{D}, \ \forall |\zeta| < \rho\}$$

for $z \in \mathfrak{D}$. Show that if $\mathfrak{D}$ is a domain of holomorphy, then $-\log R_v(z)$ is plurisubharmonic.
The function $R_v(z)$ is called the *Hartogs radius* with respect to v.
Hint: After a linear change of coordinates we may assume $v = (1, 0, \ldots, 0)$. With a small $t > 0$ we take a polydisk $\mathrm{P}\Delta_t = \{(z_1, z_2, \ldots, z_n) : |z_1| < 1, |z_j| < t, 2 \le j \le n\}$. As $t \searrow 0$, $\delta_{\mathrm{P}\Delta_t}(z, \partial\mathfrak{D})$ converges increasingly to $R_v(z)$.

12. Let $M \in \mathbf{R}$ and let $D \subset \mathbf{C}$ be a domain. Let $f(z, w)$ be a holomorphic function in $\Omega = \{(z, w) \in \mathbf{C}^2 : z \in D, \Re w > M\}$. For a point $z_0 \in D$ we denote by $\sigma(z_0)$ the infimum of $s \in \mathbf{R}$ such that $f(z, w)$ is analytically continued over a neighborhood of $\{z = z_0, \ \Re w > s\}$.
Show that $\sigma(z)$ is subharmonic in $z \in D$.
Hint: Set a family $\mathscr{F} = \{f(z, w + ib) : b \in \mathbf{R}\}$. Let $\tilde{\Omega}$ be the $\mathscr{F}$-envelope of Ω. Then, $\tilde{\Omega} = \{(z, w) : z \in D, \Re w > \tau(z)\}$ with some real-valued function $\tau(z)$, which is a domain of holomorphy. Let $R_v(z, w)$ $(= R_v(z, w + ib), \ b \in \mathbf{R})$ be the Hartogs radius with respect to $v = (0, 1)$. Then, $-\log R_v(z, w)$ is plurisubharmonic. For $T > M$, $\tau(z) = T - R_v(z, T)$, and $\log(T - \tau(z))^{-1}$ is subharmonic in $z \in D$. The function $\log(1 - \tau(z)/T)^{-T}$ is subharmonic in z, and monotone decreasing in $T > M$; as $T \nearrow \infty$, the limit is $\tau(z)$.[11]

13. Let $\mathfrak{D}/\mathbf{C}^n$ be a domain. Let $\Omega_{\mathrm{H}} \subset \mathbf{C}^2$ be any Hartogs domain (see §1.1 (b)). Let $\Psi : \Omega_{\mathrm{H}} \to \mathfrak{D}$ be an arbitrary holomorphic map. If Ψ is always analytically continued to a holomorphic map $\hat{\Psi} : \hat{\Omega}_{\mathrm{H}} \to \mathfrak{D}$ from the envelope holomorphy $\hat{\Omega}_{\mathrm{H}}$ (a bi-disk) of Ω_{H} into $\mathfrak{D}$, $\mathfrak{D}$ is said to be *Hartogs pseudoconvex*.
Prove that if $\mathfrak{D}/\mathbf{C}^n$ is Hartogs pseudoconvex, then

$$-\log \delta_{\mathrm{P}\Delta}(p, \partial\mathfrak{D}) \in \mathscr{P}^0(\mathfrak{D}).$$

Hint: Follow the proof of Oka's Theorem 4.3.11, in particular, the arguments after (4.3.14).

14. Let $\mathfrak{D}/\mathbf{C}^n$ be a domain, and let $\mathfrak{D}_\nu \subset \mathfrak{D}$ $(\nu = 1, 2, \ldots)$ be a sequence of subdomains of $\mathfrak{D}$. Assume that every $\mathfrak{D}_\nu$ is pseudoconvex, monotone increasing $\mathfrak{D}_\nu \subset \mathfrak{D}_{\nu+1}$, and $\mathfrak{D} = \bigcup_{\nu=1}^{\infty} \mathfrak{D}_\nu$. Then, show that $\mathfrak{D}$ is pseudoconvex.

15. Show that the domain defined by $|z_1| < |z_2| < \cdots < |z_n|$ in $\mathbf{C}^n$ with coordinates $(z_j)_{1 \le j \le n}$ is a domain of holomorphy.

16. In Example 4.3.72, write down the Oka hypersurfaces Σ_t in terms of coordinates at the boundary point $p_0 = (p_{01}, \ldots, p_{0n}) \in \partial \mathrm{B}(1)$.

[11] This is a bit difficult and due to Bochner–Martin, Several Complex Variables, Princeton Univ. Press, 1948, Chap. 7 §8.

Chapter 5
Pseudoconvex Domains II — Solution

In the previous chapter the Pseudoconvexity Problem is reduced to the problem of asking if a pseudoconvex domain is Stein. In this chapter we solve it affirmatively. It is the high point to prove that a bounded domain with strongly pseudoconvex boundary is Stein (Levi's Problem). We shall give two proofs to it; the first is K. Oka's original one due to an unpublished paper of 1943 by means of the Fredholm integral equation of the second kind type combined with the Joku-Iko Principle, and the second is due to H. Grauert (1958) through L. Schwartz's Fredholm Theorem for compact operators and the bumping method. The comparison is interesting. Each proof has its own advantage.

5.1 The Oka Extension with Estimate

5.1.1 Preparation from Topological Vector Spaces

In the present book all vector spaces are assumed to be defined over complex numbers. Let E be a vector space. In general, E is called a *topological vector space* if E is endowed with a topology such that the algebraic operations, including those with $\mathbf{C}$, are continuous.

There is a method to introduce such a topology on E by a system of *semi-norms* on E as follows. A semi-norm $\|x\|$ $(x \in E)$ is a real-valued function satisfying the following conditions:

(i) $\|x\| \geq 0, \quad x \in E.$
(ii) $\|\lambda x\| = |\lambda| \cdot \|x\|, \quad \lambda \in \mathbf{C}, x \in E.$
(iii) $\|x+y\| \leq \|x\| + \|y\|, \quad x, y \in E.$

If $\|x\| = 0$ implies $x = 0$, $\|x\|$ is called a norm, and defines naturally a metric topology, with which E is a topological vector space.

Assume that E is given a system of semi-norms $\|x\|_j$ $(j = 1, 2, \ldots)$. For a finite set $\Gamma \subset \mathbf{N}$ and positive numbers $\varepsilon_j > 0$ $(j \in \Gamma)$ we consider a subset defined by

J. Noguchi, *Basic Oka Theory in Several Complex Variables*, Universitext,
https://doi.org/10.1007/978-981-97-2056-9_5

(5.1.1) $$U(\Gamma, \{\varepsilon_j\}_{j\in\Gamma}) = \{x \in E : \|x\|_j < \varepsilon_j,\ \forall j \in \Gamma\}.$$

Then it is easy to see that the family of all such $U(\Gamma, \{\varepsilon_j\}_{j\in\Gamma})$ forms a fundamental system of neighborhoods of 0. For a general point $a \in E$, the family of $a + U(\Gamma, \{\varepsilon_j\}_{j\in\Gamma})$ forms a fundamental system of neighborhoods of a, and so E gives rise to a topological vector space. In this case, E is a Hausdorff space if and only if for every $x \in E \setminus \{0\}$ there is a $j \in \mathbf{N}$ with $\|x\|_j \neq 0$. We assume all through the present book that:

5.1.2. Topological vector spaces are Hausdorff.

For two points $x, y \in E$ we set

(5.1.3) $$d(x,y) = \sum_{j=1}^{\infty} \frac{1}{2^j} \cdot \frac{\|x-y\|_j}{1+\|x-y\|_j}.$$

The function $t/(1+t)$ is increasing in $t \geq 0$, and by an easy computation

$$\frac{t+s}{1+t+s} - \frac{t}{1+t} - \frac{s}{1+s} \leq 0, \quad t, s \geq 0.$$

Thus, $d(x,y)$ satisfies the axioms of a metric (distance). Note that $d(x,y)$ has the following invariant property:

(5.1.4) $$d(x+v, w+v) = d(x,w), \quad d(-x,0) = d(x,0).$$

Lemma 5.1.5. *The topology of E is homeomorphic to the metric topology by $d(x,y)$.*

Proof. Since the two topologies are invariant with respect to translations of E, it is sufficient to compare them at $0 \in E$. For $r > 0$ we set $U(r) = \{x \in E : d(x,0) < r\}$. Noting that

$$\sum_{j=N}^{\infty} \frac{1}{2^j} \cdot \frac{\|x-y\|_j}{1+\|x-y\|_j} < \sum_{j=N}^{\infty} \frac{1}{2^j} = \frac{1}{2^{N-1}},$$

we see the following:

5.1.6. (i) *For any neighborhood $U(\Gamma, \{\varepsilon_j\}_{j\in\Gamma})$ of 0, there is some $U(r)$ with $U(r) \subset U(\Gamma, \{\varepsilon_j\}_{j\in\Gamma})$.*
(ii) *For any $U(r)$, there is some $U(\Gamma, \{\varepsilon_j\}_{j\in\Gamma})$ such that $U(\Gamma, \{\varepsilon_j\}_{j\in\Gamma}) \subset U(r)$.*

Therefore the two topologies are homeomorphic to each other. □

Definition 5.1.7. (i) When the above-defined metric $d(x,y)$ is complete, E is called a *Fréchet space*.
(ii) A topological vector space is called a *Baire vector space* if it is Baire as a topological space (cf. Definition 4.2.1).

Proposition 5.1.8. *A Fréchet space is Baire.*

Proof. By Baire Theorem 4.2.3. □

Theorem 5.1.9 (Banach's Open Map Theorem). *Let E be a Fréchet space and let F be a Baire vector space. If $A : E \to F$ is a continuous surjective homomorphism, then A is an open map.*

Proof. By the assumption, E carries a complete metric $d(x,w)$ defined by (5.1.3). Note that besides (5.1.4), $d(x,w)$ has the following property:

$$d(x+w,0) \le d(x+w,w)+d(w,0) = d(x,0)+d(w,0). \tag{5.1.10}$$

We put $U(\varepsilon) = \{x \in E : d(x,0) < \varepsilon\}$, $\varepsilon > 0$.

It suffices to show that for every $\varepsilon > 0$, $A(U(\varepsilon))$ contains $0 \in F$ as an interior point. We first show:

Claim 5.1.11. *The closure $\overline{A(U(\varepsilon))}$ contains $0 \in E$ as an interior point.*

$\because$) From the continuity of the operation $(x,y) \in E \times E \to x-y \in E$ it follows that there is a neighborhood W of $0 \in E$ with $W - W \subset U(\varepsilon)$. Since $E = \bigcup_{\nu=1}^{\infty} \nu W$, $F = \bigcup_{\nu=1}^{\infty} \nu \overline{A(W)}$. Since $\nu\overline{A(W)}$ is a closed subset, by the assumption there is some $\nu_0 \in \mathbf{N}$ such that $\nu_0\overline{A(W)}$ contains an interior point. Therefore $\overline{A(W)}$ contains an interior point x_0, and so $0 \in F$ is an interior point of $\overline{A(W)} - x_0$. Since

$$0 \in \overline{A(W)} - x_0 \subset \overline{A(W)} - \overline{A(W)} = \overline{A(W-W)} \subset \overline{A(U(\varepsilon))},$$

the zero vector $0 \in F$ is an interior point of $\overline{A(U(\varepsilon))}$. △

It follows from Claim 5.1.11 that there is a neighborhood V of $0 \in F$ with $V \subset \overline{A(U(\varepsilon))}$. We set $U_\nu = U\left(\frac{\varepsilon}{2^{\nu+1}}\right)$, $\nu = 1,2,\ldots$. For each $\overline{A(U_\nu)}$ there is a neighborhood V_ν of $0 \in F$ such that $V_\nu \subset \overline{A(U_\nu)}$, $V_\nu \supset V_{\nu+1}$ and $\bigcap_{\nu=1}^{\infty} V_\nu = \{0\}$. The following finishes the proof:

Claim 5.1.12. $A(U(\varepsilon)) \supset V_1$.

$\because$) Take arbitrarily a point $y = y_1 \in V_1$. Because of $y_1 \in \overline{A(U_1)}$, $(y_1 - V_2) \cap A(U_1) \ne \emptyset$; therefore, there are $y_2 \in V_2$ and $x_1 \in U_1$ with $y_1 - y_2 = A(x_1)$. Since $y_2 \in \overline{A(U_2)}$, $(y_2 - V_3) \cap A(U_2) \ne \emptyset$; similarly, there are $y_3 \in V_3$ and $x_2 \in U_2$ with $y_2 - y_3 = A(x_2)$. Thus inductively we take $x_\nu \in U_\nu$ and $y_\nu \in V_\nu$ so that

$$y_\nu - y_{\nu+1} = A(x_\nu), \quad \nu = 1,2,\ldots.$$

Because of the choices of $\{V_\nu\}_\nu$ and $\{y_\nu\}_\nu$, $\lim_{\nu\to\infty} y_\nu = 0$, and

$$\begin{aligned} y = y_1 &= A(x_1) + y_2 = A(x_1) + A(x_2) + y_3 \\ &= \cdots = \sum_{j=1}^{\nu} A(x_j) + y_{\nu+1} = A\left(\sum_{j=1}^{\nu} x_j\right) + y_{\nu+1}. \end{aligned} \tag{5.1.13}$$

We check the convergence of $\sum_{\nu=1}^{\infty} x_\nu$. Since $x_\nu \in U_\nu = U(\frac{\varepsilon}{2^{\nu+1}})$, a computation with (5.1.10) implies that for every $\nu, \mu \in \mathbf{N}$,

$$d\left(\sum_{j=1}^{\nu} x_j, \sum_{j=1}^{\nu+\mu} x_j\right) = d\left(0, \sum_{j=\nu+1}^{\nu+\mu} x_j\right) \le \sum_{j=\nu+1}^{\nu+\mu} d(0, x_j)$$
$$< \sum_{j=\nu+1}^{\nu+\mu} \frac{\varepsilon}{2^{j+1}} < \frac{\varepsilon}{2^{\nu+1}}.$$

Therefore $\sum_{\nu=1}^{\infty} x_\nu$ satisfies the Cauchy condition, and hence converges. We set the limit, $w = \sum_{\nu=1}^{\infty} x_\nu$; by (5.1.13), $y = A(w)$, and also

$$d(0, w) \le \sum_{\nu=1}^{\infty} d(0, x_\nu) \le \sum_{\nu=1}^{\infty} \frac{\varepsilon}{2^{\nu+1}} = \frac{1}{2}\varepsilon < \varepsilon.$$

Thus, the inclusion $A(U(\varepsilon)) \supset V_1$ follows. □

Example 5.1.14. A finite-dimensional vector space, isomorphic to some $\mathbf{C}^N$, with the natural topology is a Fréchet space. While there are a number of examples of Fréchet spaces (cf. Exercise 3 at the end of the chapter), the following is of importance in complex analysis.

(1) Let $\Omega \subset \mathbf{C}^n$ be a domain. We take an increasing open covering $\{\Omega_j\}_{j=1}^{\infty}$ such that

$$(5.1.15) \qquad \emptyset \ne \Omega_j \Subset \Omega_{j+1}, \qquad \Omega = \bigcup_{j=1}^{\infty} \Omega_j.$$

We define a system of semi-norms on $\mathcal{O}(\Omega)$ by

$$\|f\|_j := \|f\|_{\Omega_j} = \sup_{\Omega_j} |f| \; (= \max_{\bar{\Omega}_j} |f|), \quad j = 1, 2, \ldots, \quad f \in \mathcal{O}(\Omega).$$

With the system of semi-norms $\|f\|_{\bar{\Omega}_j}$ ($j \in \mathbf{N}$), $\mathcal{O}(\Omega)$ gives rise to a Fréchet space: The topology of $\mathcal{O}(\Omega)$ is independent of the choice of such an open covering, and equivalent to the topology of locally uniform convergence.

(2) In the case of an unramified domain $\pi : \mathfrak{D} \to \mathbf{C}^n$, we use $\mathfrak{D}_\nu$ of Proposition 3.6.18 to define a system of semi-norms

$$\|f\|_\nu := \|f\|_{\mathfrak{D}_\nu} = \sup_{\mathfrak{D}_\nu} |f| \; (= \max_{\bar{\mathfrak{D}}_\nu} |f|), \quad \nu = 1, 2, \ldots, \quad f \in \mathcal{O}(\mathfrak{D}).$$

Then $\mathcal{O}(\mathfrak{D})$ is a Fréchet space.

5.1.2 The Oka Extension with Estimate

We prove the extension of the Joku-Iko Principle with estimate, which will play an important role in the proof of the Pseudoconvexity Problem (Levi's Problem):

Lemma 5.1.16 (Oka Extension with Estimate). *Let* $\mathrm{P}\Delta$ *be a polydisk, let* $\Sigma \subset \mathrm{P}\Delta$ *be a complex submanifold, and let* $L \Subset \mathrm{P}\Delta$ *be a compact subset. Then for a bounded holomorphic function* $f \in \mathscr{O}(\Sigma)$*, there is some* $F \in \mathscr{O}(\mathrm{P}\Delta)$ *such that*

$$F|_\Sigma = f, \qquad \|F\|_L \le C\|f\|_\Sigma.$$

Here, $C > 0$ *is a positive constant independent from* f.

Proof. Note that $\mathscr{O}(\mathrm{P}\Delta)$ and $\mathscr{O}(\Sigma)$ are Fréchet spaces; in particular, $\mathscr{O}(\Sigma)$ is Baire. By the Oka Extension Theorem 2.5.14 the restriction map

$$A : F \in \mathscr{O}(\mathrm{P}\Delta) \longrightarrow F|_\Sigma \in \mathscr{O}(\Sigma)$$

is a continuous surjective homomorphism. Note that $U = \{F \in \mathscr{O}(\mathrm{P}\Delta) : \|F\|_L < 1\}$ is a neighborhood of 0 in $\mathscr{O}(\mathrm{P}\Delta)$. By Banach's Open Map Theorem 5.1.9, $A(U)$ contains a neighborhood of 0 in $\mathscr{O}(\Sigma)$. Therefore, there is a compact subset $K \Subset \Sigma$ and $\varepsilon > 0$ such that

$$A(U) \supset \{f \in \mathscr{O}(\Sigma) : \|f\|_K < \varepsilon\}.$$

It suffices to choose $C = 1/\varepsilon$. □

5.2 Strongly Pseudoconvex Domains

The aim of this section is to solve Pseudoconvexity Problem III 4.3.54. Let $\pi : \mathfrak{G} \to \mathbf{C}^n$ be a domain. Let $\mathfrak{D} \Subset \mathfrak{G}$ be a bounded subdomain.

Lemma 5.2.1. *If* $\mathfrak{D}$ *is strongly pseudoconvex, then* $\mathfrak{D}$ *is Stein.*

N.B. The original Levi Problem 4.3.65 is solved by this lemma.

Now we first give the proof of Lemma 5.2.1 based on Oka's unpublished papers of 1943.

5.2.1 Oka's Method

(a) Oka's Heftungslemma. For the sake of simplicity we say in general that the closure $\bar{\mathfrak{D}}$ of a domain $\mathfrak{D}$ is *Stein* if $\bar{\mathfrak{D}}$ satisfies the following conditions:

(i) $\partial\mathfrak{D}$ is the boundary of the set of exterior points of $\mathfrak{D}$.
(ii) $\bar{\mathfrak{D}}$ has a fundamental system of Stein open neighborhoods.

In this case, $\mathfrak{D}$ is itself Stein by Corollary 3.7.4.

Let $\mathfrak{D} \Subset \mathfrak{G}$ be a strongly pseudoconvex domain. Then $\partial\mathfrak{D}$ satisfies (i) above, so that we have:

Lemma 5.2.2. *If* $\bar{\mathfrak{D}}$ *is Stein, then* $\mathfrak{D}$ *is Stein.*

This will be used implicitly and frequently.

Suppose that the strongly pseudoconvex domain $\mathfrak{D}$ ($\Subset \mathfrak{G}$) $\xrightarrow{\pi} \mathbf{C}^n$ with the first coordinate $z_1 = x_1 + iy_1$ spreads over an open interval containing $x_1 = 0$. Taking $a_2 < 0 < a_1$ in the interval, we set

$$\text{(5.2.3)} \qquad \mathfrak{D}_1 = \mathfrak{D} \cap \{x_1 < a_1\}, \quad \mathfrak{D}_2 = \mathfrak{D} \cap \{x_1 > a_2\}, \quad \mathfrak{D}_3 = \mathfrak{D}_1 \cap \mathfrak{D}_2 \neq \emptyset.$$

N.B. In general, $\mathfrak{D}_\nu$ ($\nu = 1, 2, 3$) may have several connected components.

Lemma 5.2.4 (Oka's Heftungslemma). *If the closures* $\bar{\mathfrak{D}}_\nu$ ($\nu = 1, 2$) *are Stein, so is* $\bar{\mathfrak{D}}$, *too.*

Remark 5.2.5. By Corollary 3.7.4 (also by Theorem 4.3.74), $\bar{\mathfrak{D}}_3$ is Stein.

The idea here is to take a sufficiently fine partition of $\bar{\mathfrak{D}}$ in Lemma 5.2.1 by real hyperplanes parallel to real and imaginary axes of the complex coordinates, so that every piece of such partition of $\bar{\mathfrak{D}}$ is contained in a univalent Stein domain. Then we sew them together horizontally and vertically with keeping the Steinness, and hence finally get the Steinness of the whole $\bar{\mathfrak{D}}$.

We use the Cousin I Problem to show the Steinness of $\bar{\mathfrak{D}}$ in Lemma 5.2.4.

Lemma 5.2.6. *Assume that* $\bar{\mathfrak{D}}_\nu$ ($\nu = 1, 2$) *are Stein. Then for* $f \in \mathscr{O}(\bar{\mathfrak{D}}_3)$ *there are* $f_\nu \in \mathscr{O}(\bar{\mathfrak{D}}_\nu)$ ($\nu = 1, 2$) *such that*

$$f_1(z) - f_2(z) = f(z), \quad z \in \mathfrak{D}_3.$$

(b) Integral equation of Fredholm type. We prepare for the proof of Lemma 5.2.6. We keep the notation above. With sufficiently large $r > r_0 > 0$, we take the double neighborhoods of the closure of $\pi(\mathfrak{D})$ as follows:

$$\pi(\mathfrak{D}) \Subset \mathrm{P}\Delta_0 = \{(z_j) : |z_j| < r_0\} \Subset \mathrm{P}\Delta = \{(z_j) : |z_j| < r\}.$$

In the z_1-plane we take a line segment, which will be used later for Cousin's integral,

$$\ell_0 = \{z_1 = it : -r_0 \leq t \leq r_0\},$$

where the orientation is the one as t increases.

With $\delta > 0$ such that $a_2 < -\delta < 0 < \delta < a_1$, we set

$$\mathrm{P}\Delta_\delta = \mathrm{P}\Delta \cap \{|x_1| < \delta\}.$$

We choose $\varphi_k \in \mathscr{O}(\mathfrak{D}_3), 1 \leq k \leq m$, satisfying the following (cf. Fig. 5.1).

5.2.7 (Heftungscondition). (i) With $\mathfrak{D}'_3 := \{z \in \mathfrak{D}_3 : |\varphi_k(z)| < 1,\ 1 \leq k \leq m\}$, we have

$$(\partial \mathfrak{D}'_3) \setminus \partial \mathfrak{D}_3 \Subset \{a_2 < x_1 < a_1\},$$

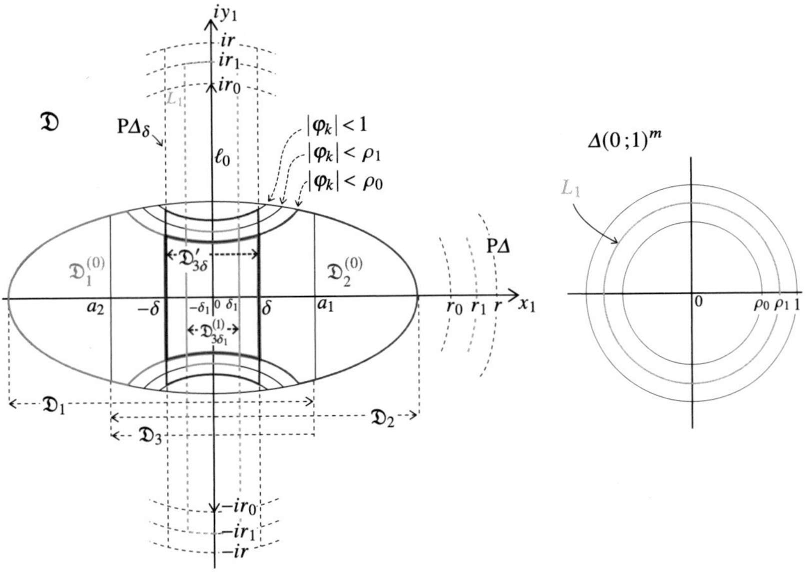

Fig. 5.1 Oka Heftungs–strongly pseudoconvex domains.

$$(\partial\mathfrak{D}_3)\setminus\partial\mathfrak{D}'_3 \Supset (\partial\mathfrak{D}_3)\cap\{|x_1|\le\delta\}.$$

(ii) With $\mathfrak{D}'_{3\delta}:=\mathfrak{D}'_3\cap\{|x_1|<\delta\}\Subset\mathfrak{D}_3$ and $\varphi(z)=(\varphi_1(z),\ldots,\varphi_m(z))$ $(z\in\mathfrak{D}_3)$, the Oka map

$$\mathfrak{D}'_{3\delta}\ni z\longmapsto(z,\varphi(z))\in\widetilde{\mathrm{P}\Delta_\delta}:=\mathrm{P}\Delta_\delta\times\Delta(1)^m \tag{5.2.8}$$

is a proper embedding.

Lemma 5.2.9. *For every compact subset $K\Subset\mathfrak{D}$ the above $\{\varphi_k\}_{k=1}^m\subset\mathscr{O}(\mathfrak{D}_3)$ can be chosen so that*

$$K\cap\mathfrak{D}_3\subset\mathfrak{D}'_3$$

and Heftungscondition 5.2.7 is satisfied.

Proof. For each point $q\in(\partial\mathfrak{D}_3)\cap\{|x_1|\le\delta\}$ we apply Lemma 4.3.70 (iii) to obtain a holomorphic function $\varphi(p)$ on $\bar{\mathfrak{D}}_3\setminus\{q\}$ with polar locus on the Oka hypersurface Σ_0 $(\ni q)$ of $\partial\mathfrak{D}_3$ at q (hence, $|\varphi(q)|=\infty$[1]). Multiplying a small constant $(\neq 0)$ to $\varphi(p)$, we have

$$\|\varphi\|_K<1,\qquad \{|\varphi|\ge 1\}\cap\{x_1=a_1,a_2\}\cap\bar{\mathfrak{D}}_3=\emptyset.$$

[1] This stands for $\overline{\lim}_{p\to q}|\varphi(p)|=\infty$; the same in the sequel.

Since $q \in \{|\varphi| > 1\} \cap (\partial \mathfrak{D}_3) \cap \{|x_1| \le \delta\}$ and $(\partial \mathfrak{D}_3) \cap \{|x_1| \le \delta\}$ is compact, there are finitely many such φ_k satisfying

$$\left(\bigcup_{k=1}^{m} \{|\varphi_k| > 1\}\right) \cap (\partial \mathfrak{D}_3) \cap \{|x_1| \le \delta\} \supset (\partial \mathfrak{D}_3) \cap \{|x_1| \le \delta\}.$$

Since $\bar{\mathfrak{D}}_3$ is Stein (cf. Remark 5.2.5), we may increase the members of $\{\varphi_k\}$ so that (5.2.8) is injective, and the Heftungscondition (ii) is satisfied. □

We take ρ_0, ρ_1 $(0 < \rho_0 < \rho_1 < 1)$, both close to 1, r_1 $(r_0 < r_1 < r)$ close to r, and δ_1 $(0 < \delta_1 < \delta)$ close to δ; we fix them. Set

(5.2.10)
$$\mathfrak{D}_1^{(0)} := (\mathfrak{D}_1 \setminus \mathfrak{D}_3) \cup \{z \in \mathfrak{D}_3 : x_1 < \delta,\ |\varphi_k(z)| < \rho_0,\ 1 \le k \le m\},$$
$$\mathfrak{D}_2^{(0)} := (\mathfrak{D}_2 \setminus \mathfrak{D}_3) \cup \{z \in \mathfrak{D}_3 : x_1 > -\delta,\ |\varphi_k(z)| < \rho_0,\ 1 \le k \le m\}$$

(cf. Fig. 5.1). We first show Lemma 5.2.6 for a little bit smaller $\mathfrak{D}_1^{(0)}, \mathfrak{D}_2^{(0)}$: Here is the main part of Oka's method and the proof gets a bit long.

Lemma 5.2.11. *With the notation above, there exist* $h_\nu \in \mathscr{O}\left(\mathfrak{D}_\nu^{(0)}\right)$ $(\nu = 1, 2)$ *such that*

(5.2.12)
$$h_1(z) - h_2(z) = f(z), \quad z \in \mathfrak{D}_1^{(0)} \cap \mathfrak{D}_2^{(0)}.$$

Proof. With $w = (w_k)_{1 \le k \le m} \in \mathbf{C}^m$ we set

$$L_1 = \{(z, w) \in \mathbf{C}^n \times \mathbf{C}^m : |x_1| \le \delta_1,\ \forall\, |z_j| \le r_1,\ \forall\, |w_k| \le \rho_1\} \ (\Subset \widetilde{\mathrm{P}\Delta_\delta}).$$

We consider a holomorphic function $g \in \mathscr{O}(\mathfrak{D}'_{3\delta})$ such that $\|g\|_{\mathfrak{D}'_{3\delta}} \le M\ (< \infty)$. By virtue of the Oka map (5.2.8) and Lemma 5.1.16 (the Oka Extension with Estimate) there exists $G(z, w) \in \mathscr{O}(\widetilde{\mathrm{P}\Delta_\delta})$ such that

(5.2.13)
$$G(z, \varphi(z)) = g(z)\ (z \in \mathfrak{D}'_{3\delta}), \quad \|G\|_{L_1} \le C\|g\|_{\mathfrak{D}'_{3\delta}} \le CM,$$

where C is a positive constant independent from g. We take Cousin's integral (decomposition) of $G(z, w)$ along ℓ_0:

$$G_\nu(z, w) = \frac{1}{2\pi i} \int_{\ell_0} \frac{G(t, z', w)}{t - z_1}\, dt, \quad \nu = 1, 2,$$
$$|z_1| < r_0,\ z' = (z_2, \ldots, z_n),\ |z_j| < r,\ 2 \le j \le n,$$
$$w = (w_k),\ |w_k| < 1,\ 1 \le k \le m.$$

Here G_1 (resp. G_2) is the branch of the integral defined in the left (resp. right)-hand side of ℓ_0, and analytically continued across ℓ_0 up to $\{x_1 < \delta\}$ (resp. $\{x_1 > -\delta\}$). We then have

$$G_1(z,w) - G_2(z,w) = G(z,w), \quad |x_1| < \delta,\ |z_1| < r_0,$$
$$|z_j| < r,\ 2 \le j \le n, \quad |w_k| < 1,\ 1 \le k \le m.$$

Now, substituting $w = \varphi(z)$ $(z \in \mathfrak{D}'_{3\delta})$, we set $g_\nu(z) = G_\nu(z, \varphi(z)) \in \mathcal{O}(\mathfrak{D}'_{3\delta})$ $(\nu = 1, 2)$, which satisfy

$$g_\nu(z_1, z') = \frac{1}{2\pi i} \int_{\ell_0} \frac{G(t, z', \varphi(z_1, z'))}{t - z_1} dt, \quad \nu = 1, 2,$$
$$g_1(z) - g_2(z) = g(z), \quad z \in \mathfrak{D}'_{3\delta}.$$

(Roughly speaking, we would like to extend the defining domain of each g_ν to $\mathfrak{D}_\nu^{(0)}$.)

Setting $\Gamma_0 = \prod_{k=1}^m \{|w_k| = \rho_0\}$, we have by Cauchy's integral formula

$$G(z,w) = \frac{1}{(2\pi i)^m} \int_{\Gamma_0} \frac{G(z, u_1, \ldots, u_m)}{(u_1 - w_1) \cdots (u_m - w_m)} du_1 \cdots du_m,$$
$$|w_k| < \rho_0,\ 1 \le k \le m.$$

We write $u_0 = t$, $u = (u_1, \ldots, u_m)$, $\tilde{u} = (u_0, u)$, $d\tilde{u} = du_0 du_1 \cdots du_m$. (Pay attention to the variable z_1 in what follows.) For $\nu = 1, 2$ we set

$$G_\nu(z_1, z', w) = \frac{1}{(2\pi i)^{m+1}} \int_{\ell_0 \times \Gamma_0} \frac{G(u_0, z', u)}{(u_0 - z_1)(u_1 - w_1) \cdots (u_m - w_m)} d\tilde{u},$$
$$\chi(\tilde{u}, z_1, z') = \frac{1}{(2\pi i)^{m+1}(u_0 - z_1)(u_1 - \varphi_1(z_1, z')) \cdots (u_m - \varphi_m(z_1, z'))},$$
$$g_\nu(z_1, z') = \int_{\ell_0 \times \Gamma_0} \chi(\tilde{u}, z_1, z') G(u_0, z', u) d\tilde{u}, \ z \in \mathfrak{D}'_{3\delta} \cap \bigcap_{k=1}^m \{|\varphi_k(z)| < \rho_0\}.$$

With a sufficiently small cylinder neighborhood V (Stein) of $\ell_0 \times \Gamma_0 \subset \mathbf{C}^{m+1}$, we are going to define Cousin I data on $V \times \mathfrak{D}_1$ covered by $V \times \mathfrak{D}_3$ and $V \times (\mathfrak{D}_1 \cap \{x_1 < a'_2\})$ with $a'_2 > a_2$ chosen sufficiently close to a_2 below. We firstly take

$$\chi(\tilde{u}, z_1, z') \in \mathcal{M}(V \times \mathfrak{D}_3).$$

Note that its polar set is a closed subset of $V \times \mathfrak{D}_3$, and has no intersection with the real hyperplanes $x_1 = a_\nu$ $(\nu = 1, 2)$, and hence with $x_1 = a'_2 > a_2$ close to a_2. We secondly take 0 on $V \times (\mathfrak{D}_1 \cap \{x_1 < a'_2\})$, so that we obtain Cousin I data on $V \times \mathfrak{D}_1$. Since $V \times \mathfrak{D}_1$ is Stein, the Cousin I Problem for the data has a solution by Theorem 3.7.9:

$$\exists \chi_1(\tilde{u}, z_1, z') \in \mathcal{M}(V \times \mathfrak{D}_1), \quad \chi - \chi_1|_{V \times \mathfrak{D}_3} \in \mathcal{O}(V \times \mathfrak{D}_3).$$

Similarly, we have

$$\exists \chi_2(\tilde{u}, z_1, z') \in \mathcal{M}(V \times \mathfrak{D}_2), \quad \chi - \chi_2|_{V \times \mathfrak{D}_3} \in \mathcal{O}(V \times \mathfrak{D}_3).$$

It is noticed that the defining domain of the variable z of each $\chi_\nu(\tilde{u},z)$ $(\nu=1,2)$ is extended from $\mathfrak{D}_3$ to $\mathfrak{D}_\nu$, keeping the same poles as those of $\chi(\tilde{u},z)$. We extended the defining domain of χ by virtue of Cousin I, but the substitution of χ by χ_ν causes necessarily an error; to make the error small, we modify χ_ν.

Note that $V\times\mathfrak{D}_3$ is $\mathscr{O}(V\times\mathfrak{D}_\nu)$-convex $(\nu=1,2)$ and that $(\ell_0\times\Gamma_0)\times\bar{\mathfrak{D}}'_{3\delta}\Subset V\times\mathfrak{D}_3$. By the Oka–Weil Approximation Theorem 3.7.7 we have for every $\varepsilon>0$:

$$\exists\gamma_\nu\in\mathscr{O}(V\times\mathfrak{D}_\nu),\quad \|\chi-\chi_\nu-\gamma_\nu\|_{(\ell_0\times\Gamma_0)\times\bar{\mathfrak{D}}'_{3\delta}}<\frac{\varepsilon}{2},\quad \nu=1,2,$$
$$K_\nu:=\chi-\chi_\nu-\gamma_\nu\in\mathscr{O}(V\times\mathfrak{D}_\nu),\quad \|K_\nu\|_{(\ell_0\times\Gamma_0)\times\bar{\mathfrak{D}}'_{3\delta}}<\frac{\varepsilon}{2}.$$

Set

(5.2.14)
$$\begin{aligned}I_\nu(G)(z_1,z')&=\int_{\ell_0\times\Gamma_0}(\chi(\tilde{u},z_1,z')-K_\nu(\tilde{u},z_1,z'))G(u_0,z',u)d\tilde{u}\\&=\int_{\ell_0\times\Gamma_0}(\chi_\nu(\tilde{u},z_1,z')+\gamma_\nu(\tilde{u},z_1,z'))G(u_0,z',u)d\tilde{u},\\&\qquad\nu=1,2,\\I_\nu(G)(z)&\in\mathscr{O}\left(\mathfrak{D}_\nu^{(0)}\right),\quad \nu=1,2.\end{aligned}$$

Then we set

$$g_0(z)=I_1(G)(z)-I_2(G)(z)=g(z)-\int_{\ell_0\times\Gamma_0}(K_1-K_2)Gd\tilde{u},$$
$$K=K_1-K_2\in\mathscr{O}(V\times\mathfrak{D}_3),\quad \|K\|_{(\ell_0\times\Gamma_0)\times\bar{\mathfrak{D}}'_{3\delta}}<\varepsilon,$$
$$K(G)(z)=\int_{\ell_0\times\Gamma_0}K(\tilde{u},z)G(u_0,z',u)d\tilde{u}\in\mathscr{O}(\mathfrak{D}_3),$$
$$\|K(G)\|_{\bar{\mathfrak{D}}'_{3\delta}}\le(2r_0)(2\pi\rho_0)^m\varepsilon\|G\|_{L_1}\le\varepsilon C_1\|G\|_{L_1}\le\varepsilon C_1CM,$$

where $C_1:=(2r_0)(2\pi\rho_0)^m$. Thus the following integral equation holds:

$$g(z)=K(G)(z)+g_0(z),\quad z\in\mathfrak{D}'_{3\delta}.$$

- While g_0 is deduced from g, g_0 satisfies

$$g_0=g-K(G)=I_1(G)-I_2(G),$$

so that as for g_0 the Cousin I Problem is solved.

- We consider this in the opposite way: For the given $f\in\mathscr{O}(\mathfrak{D}'_{3\delta})$ we consider the following *Fredholm integral equation of the second kind type* (see Remark 5.2.20) with inhomogeneous term $g_0=f$:

(5.2.15)
$$g(z)=K(G)(z)+f(z),\quad z\in\mathfrak{D}'_{3\delta}.$$

If we find a solution $g \in \mathscr{O}(\mathfrak{D}'_{3\delta})$ of the above equation, we use it to obtain

$$f\,(= g_0) = I_1(G) - I_2(G),$$

which solves the Cousin I Problem for f.

In what follows, we solve (5.2.15) by means of the iterative approximation method:

(i) Put $f_0 := f \in \mathscr{O}(\mathfrak{D}_3)$. Then, $\|f_0\|_{\mathfrak{D}'_{3\delta}} \le M\ (< \infty)$.

(ii) By Lemma 5.1.16 (Oka Extension with Estimate), $f_0 \rightsquigarrow F_0 \in \mathscr{O}(\widetilde{\mathrm{P}\Delta_\delta})$, $F_0|_{\mathfrak{D}'_{3\delta}} = f_0$, $\|F_0\|_{L_1} \le CM$.

(iii) $f_1 := K(F_0) \in \mathscr{O}(\mathfrak{D}_3)$, $\|f_1\|_{\mathfrak{D}'_{3\delta}} \le \varepsilon C_1 CM$.

(iv) By Lemma 5.1.16 (Oka Extension with Estimate), $f_1 \rightsquigarrow F_1 \in \mathscr{O}(\widetilde{\mathrm{P}\Delta_\delta})$, $F_1|_{\mathfrak{D}'_{3\delta}} = f_1$, $\|F_1\|_{L_1} \le \varepsilon C_1 C^2 M$.

(v) $f_2 := K(F_1) \in \mathscr{O}(\mathfrak{D}_3)$, $\|f_2\|_{\mathfrak{D}'_{3\delta}} \le (\varepsilon C_1 C)^2 M$.

$\vdots$

(vi) $f_\mu := K(F_{\mu-1}) \in \mathscr{O}(\mathfrak{D}_3)$, $\|f_\mu\|_{\mathfrak{D}'_{3\delta}} \le (\varepsilon C_1 C)^\mu M$.

(vii) Take $\varepsilon > 0$ so that $\theta := \varepsilon C_1 C < 1$. Then,

(viii) $\|f_\mu\|_{\mathfrak{D}'_{3\delta}} \le \theta^\mu M$.

(ix) By Lemma 5.1.16 (Oka Extension with Estimate), $f_\mu \rightsquigarrow F_\mu \in \mathscr{O}(\widetilde{\mathrm{P}\Delta_\delta})$, $F_\mu|_{\mathfrak{D}'_{3\delta}} = f_\mu$, and $\|F_\mu\|_{L_1} \le \theta^\mu CM$.

(x) $g := f_0 + f_1 + f_2 + \cdots \in \mathscr{O}(\mathfrak{D}'_{3\delta})$: This converges uniformly on $\mathfrak{D}'_{3\delta}$ by a majorant convergent series, and is bounded.

(xi) $G := F_0 + F_1 + F_2 + \cdots \in \mathscr{O}(L_1^\circ)$ with the interior L_1° of L_1 converges uniformly on L_1 by a majorant convergent series.

(xii) $G(z, \varphi(z)) = g(z)$, $z \in \mathfrak{D}^{(1)}_{3\delta_1} := \{z \in \mathfrak{D}'_{3\delta} : (z, \varphi(z)) \in L_1^\circ\} = \{z \in \mathfrak{D}'_{3\delta} : |x_1| < \delta_1, |\varphi_k(z)| < \rho_1, 1 \le k \le m\}$ ($\Subset \mathfrak{D}'_{3\delta}$); that is, G is an extension of g through the Oka map $\mathfrak{D}^{(1)}_{3\delta_1} \ni z \mapsto (z, \varphi(z)) \in L_1^\circ$, and

$$g = f_0 + K(F_0) + K(F_1) + \cdots = K(G) + f_0 \quad (\text{on } \mathfrak{D}^{(1)}_{3\delta_1}).$$

Claim 5.2.16. *With the preparation above, we define* $I_\nu(G)(z) \in \mathscr{O}\left(\mathfrak{D}^{(0)}_\nu\right)$ $(\nu = 1, 2)$ *by* (5.2.14). *Then,*

$$I_1(G)(z) - I_2(G)(z) = g(z) - K(G)(z) = f(z), \quad z \in \mathfrak{D}^{(0)}_1 \cap \mathfrak{D}^{(0)}_2 \cap \mathfrak{D}^{(1)}_{3\delta_1}.$$

In fact, with f_μ and F_μ $(\mu = 0, 1, 2, \ldots)$ constructed above inductively, we get

$$I_1(F_\mu) - I_2(F_\mu) = f_\mu - K(F_\mu), \quad \mu \ge 0.$$

Summing up each side with respect to $\mu = 0, 1, \ldots, N$, we deduce from $f_0 = f$ and $f_{\mu+1} = K(F_\mu)$ that

$$I_1\left(\sum_{\mu=0}^{N} F_\mu\right) - I_2\left(\sum_{\mu=0}^{N} F_\mu\right) = \sum_{\mu=0}^{N} f_\mu - K\left(\sum_{\mu=0}^{N} F_\mu\right) = f - K(F_N).$$

Since $K(F_N)(z) \to 0$ ($z \in \mathfrak{D}^{(1)}_{3\delta_1}$) as $N \to \infty$, it follows that

$$(5.2.17)\ \ I_1(G)(z) - I_2(G)(z) = g(z) - K(G)(z) = f(z), \quad z \in \mathfrak{D}^{(0)}_1 \cap \mathfrak{D}^{(0)}_2 \cap \mathfrak{D}^{(1)}_{3\delta_1}.$$

Thus integral equation (5.2.15) is solved in $z \in \mathfrak{D}^{(0)}_1 \cap \mathfrak{D}^{(0)}_2 \cap \mathfrak{D}^{(1)}_{3\delta_1}$.

Here, since $I_\nu(G)$ ($\nu = 1,2$) are holomorphic in $\mathfrak{D}^{(0)}_\nu$, and δ_1 can be arbitrarily close to δ, it follows from the Identity Theorem 1.1.46 that (5.2.17) holds in $\mathfrak{D}^{(0)}_1 \cap \mathfrak{D}^{(0)}_2$. Hence we obtain (5.2.12) with $h_\nu = I_\nu(G)$ ($\nu = 1,2$). □

Proof of Lemma 5.2.6. By the assumption we take a little bit larger strongly pseudoconvex domains $\tilde{\mathfrak{D}} \Supset \mathfrak{D}$, and define $\tilde{\mathfrak{D}}_\nu$ ($\nu = 1,2,3$) as in (5.2.3), so that the closures $\overline{\tilde{\mathfrak{D}}}_\nu$ ($\nu = 1,2$) are Stein. Take an element $f \in \mathscr{O}\left(\overline{\tilde{\mathfrak{D}}_3}\right)$. We then apply the above arguments for the domains $\tilde{\mathfrak{D}}_\nu$ ($\nu = 1,2,3$). By Lemma 5.2.9 we may assume that $\tilde{\mathfrak{D}}^{(0)}_\nu \supset \mathfrak{D}_\nu$ ($\nu = 1,2$). Therefore, by restricting the obtained $I_\nu(G) \in \mathscr{O}\left(\tilde{\mathfrak{D}}^{(0)}_\nu\right)$ ($\nu = 1,2$) to $\mathfrak{D}_\nu$, we get the required solutions f_ν. □

Proof of Oka's Heftungslemma 5.2.4. (i) Similarly to the above we take double strongly pseudoconvex domains as follows:

$$\tilde{\mathfrak{D}} = \tilde{\mathfrak{D}}_1 \cup \tilde{\mathfrak{D}}_2 \Supset \tilde{\mathfrak{D}}' = \tilde{\mathfrak{D}}'_1 \cup \tilde{\mathfrak{D}}'_2 \Supset \mathfrak{D} = \mathfrak{D}_1 \cup \mathfrak{D}_2,$$

where $\tilde{\mathfrak{D}}_\nu$ and $\tilde{\mathfrak{D}}'_\nu$ ($\nu = 1,2$) are Stein. We would like to show $\tilde{\mathfrak{D}}'$ to be Stein. We are going to construct a meromorphic function with a pole at $q \in \partial\tilde{\mathfrak{D}}'$, and holomorphic on $\overline{\tilde{\mathfrak{D}}'} \setminus \{q\}$. Assume $q \in \partial\tilde{\mathfrak{D}}'_1$. By a parallel transition we may assume $q \notin \overline{\tilde{\mathfrak{D}}'_2}$. By Lemma 4.3.70 (iii) there is a meromorphic function $f \in \mathscr{M}(\tilde{\mathfrak{D}}'_1)$ with polar locus on the Oka hypersurface Σ_0 ($\ni q$) of $\partial\tilde{\mathfrak{D}}'$ at q ($|f(q)| = \infty$) and holomorphic on $\tilde{\mathfrak{D}}'_1 \setminus \{q\}$. By Lemma 5.2.6 there are functions $f_\nu \in \mathscr{O}(\tilde{\mathfrak{D}}'_\nu)$ ($\nu = 1,2$) satisfying

$$f_2(z) - f_1(z) = f(z), \quad z \in \tilde{\mathfrak{D}}'_1 \cap \tilde{\mathfrak{D}}'_2.$$

Then $F = f + f_1 = f_2$ is a meromorphic function on $\tilde{\mathfrak{D}}'$ with a pole at $q \in \partial\tilde{\mathfrak{D}}'$ ($|F(q)| = \infty$) and holomorphic on $\overline{\tilde{\mathfrak{D}}'} \setminus \{q\}$. Therefore $\tilde{\mathfrak{D}}'$ is holomorphically convex.

(ii) To show the holomorphic separation we assume that $\tilde{\mathfrak{D}}' \xrightarrow{\pi} \mathbf{C}^n$ is not univalent. Let $p_1, p_2 \in \tilde{\mathfrak{D}}'$ be distinct two points with $z_0 := \pi(p_1) = \pi(p_2)$. We consider a half line ℓ in $\mathbf{C}^n$ with initial point z_0 and with any direction. Let ℓ_ν ($\nu = 1,2$) be the connected component of $\pi^{-1}\ell$ containing p_ν, respectively. Since $\tilde{\mathfrak{D}}'$ is bounded, ℓ_ν must intersect the boundary $\partial\tilde{\mathfrak{D}}'$. We consider a moving point P on ℓ starting from z_0. Correspondingly, a moving point Q_ν with $\pi(Q_\nu) = P$ on ℓ_ν is obtained for each ν. Whichever will be fine, we suppose that Q_1 reaches to $\partial\tilde{\mathfrak{D}}'$ on the same time or earlier than Q_2. We denote the point $q_1 \in \ell_1 \cap \partial\tilde{\mathfrak{D}}'$, and the corresponding point on ℓ_2 by $q_2 \in \ell_2 \cap \overline{\tilde{\mathfrak{D}}'}$ with $\pi(q_1) = \pi(q_2)$. Then, $q_1 \neq q_2$. By the same arguments as above F was obtained, we obtain a meromorphic function g with a pole at $q_1 \in \partial\tilde{\mathfrak{D}}'$ ($|g(q_1)| = \infty$), holomorphic on $\overline{\tilde{\mathfrak{D}}'} \setminus \{q_1\}$. By the uniqueness of analytic continuation, we have as convergent power series locally in $z - z_0$ that

$$\underline{g}_{p_1} \neq \underline{g}_{p_2}. \tag{5.2.18}$$

Therefore it follows from Proposition 3.6.9 that $\tilde{\mathfrak{D}}' \to \mathbf{C}^n$ is holomorphically separable.[2]

Now, $\tilde{\mathfrak{D}}'$ is Stein. Since $\tilde{\mathfrak{D}}'$ is a Stein neighborhood of $\bar{\mathfrak{D}}$, arbitrarily close to $\bar{\mathfrak{D}}$, and $\partial\mathfrak{D}$ is a strongly pseudoconvex boundary, the Steinness of $\mathfrak{D}$ is deduced from Corollary 3.7.4. □

(c) *Proof of Lemma 5.2.1.* (c1) We first partition $\bar{\mathfrak{D}}$ by real hyperplanes parallel to the real and imaginary axes to small closed domains, each piece of which is a closed Stein domain.

Let λ be the defining strongly pseudoconvex function of the boundary $\partial\mathfrak{D}$. For a point $p_0 \in \partial\mathfrak{D}$ we take a Stein neighborhood $V(\ni p)$ contained in the open set where λ is defined, so that $V \cap \mathfrak{D}$ is Stein by Theorem 4.3.74.

Similarly, we have that $\bar{V}$ and $\bar{V} \cap \bar{\mathfrak{D}}$ are Stein.

(c2) With complex coordinates $z_j = x_{2j-1} + i x_{2j}$ $(1 \leq j \leq n)$ in real and imaginary parts and a sufficiently large $N \in \mathbf{N}$ we have

$$\pi(\bar{\mathfrak{D}}) \Subset \{(x_k) : -N < x_k < N,\ 1 \leq k \leq 2n\}.$$

We set a closed cuboid $E_0 = \{(x_k) : -N \leq x_k \leq N,\ 1 \leq k \leq 2n\}$. For axes x_k we take partitions

$$-N = c_{k0} < c_{k1} < \cdots < c_{kL} = N \quad (1 \leq k \leq 2n) \tag{5.2.19}$$

and set $E_{h_1 h_2 \ldots h_{2n}} = \{(x_k) : c_{k h_k - 1} \leq x_k \leq c_{k h_k},\ 1 \leq k \leq 2n\}$ $(1 \leq h_1, \ldots, h_{2n} \leq L)$. Taking the partitions (5.2.19) sufficiently fine, we see by the arguments of (c1) that every non-empty connected component $\bar{\mathfrak{D}}^{(l)}_{h_1 h_2 \ldots h_{2n}}$ of $(\pi^{-1} E_{h_1 h_2 \ldots h_{2n}}) \cap \bar{\mathfrak{D}}$ with finitely many l is univalent and Stein. Firstly with fixed $h_2, h_3, \ldots, h_{2n}$ arbitrarily, we run h_1 from 1 to $L-1$. We consider two adjacent $\bar{\mathfrak{D}}^{(l)}_{1 h_2 \ldots h_{2n}}$ and $\bar{\mathfrak{D}}^{(l')}_{2 h_2 \ldots h_{2n}}$ sharing the edge $x_1 = c_{11}$. By the definition there is a pseudoconvex neighborhood U of the union $\bar{\mathfrak{D}}^{(l)}_{1 h_2 \ldots h_{2n}} \cup \bar{\mathfrak{D}}^{(l')}_{2 h_2 \ldots h_{2n}}$. It follows from Proposition 4.3.53 that there is a neighborhood V with strongly pseudoconvex boundary such that $\bar{\mathfrak{D}}^{(l)}_{1 h_2 \ldots h_{2n}} \cup \bar{\mathfrak{D}}^{(l')}_{2 h_2 \ldots h_{2n}} \Subset V \Subset U$.

[2] The present proof is taken from Oka's unpublished paper XI of 1943. In the published paper Oka IX (1953) it is shown by another proof. In Gunning–Rossi [26] the proof relies on the coherence of a direct image sheaf by a finite morphism, which is rather sophisticated. The present proof of Oka XI of 1943 is not found in the literature to the best of the author's knowledge, and hence is original even now.

We now apply Oka's Heftungslemma 5.2.4 to deduce that V is Stein, and hence $\bar{\mathfrak{D}}^{(l)}_{1h_2\dots h_{2n}} \cup \bar{\mathfrak{D}}^{(l')}_{2h_2\dots h_{2n}}$ is Stein. Next, we consider $\bar{\mathfrak{D}}^{(l)}_{1h_2\dots h_{2n}} \cup \bar{\mathfrak{D}}^{(l')}_{2h_2\dots h_{2n}}$ and the adjacent $\bar{\mathfrak{D}}^{(l'')}_{3h_2\dots h_{2n}}$ sharing the edge $x_1 = c_{12}$. In the same way as above by applying Oka's Heftungslemma 5.2.4, we deduce that the merged closed domain $\bar{\mathfrak{D}}^{(l)}_{1h_2\dots h_{2n}} \cup \bar{\mathfrak{D}}^{(l')}_{2h_2\dots h_{2n}} \cup \bar{\mathfrak{D}}^{(l'')}_{3h_2\dots h_{2n}}$ is Stein; we proceed with this up to $h_1 = L-1$, and for all connected components $\bar{\mathfrak{D}}^{(l)}_{1h_2\dots h_{2n}}$. Denote the merged connected components by $\bar{\mathfrak{D}}^{(l)}_{h_2\dots h_{2n}}$. (Note that this is no longer univalent in general.) We do the same for the next axis x_2, and then up to x_{2n}. Thus we deduce that $\bar{\mathfrak{D}}$ is Stein. □

Remark 5.2.20. (i) In general, an integral equation of the type

$$g(z) = \int K(z,\zeta)g(\zeta)d\zeta + f(z)$$

is called a *Fredholm integral equation of the second kind*, where $K(z,w)$ (so-called kernal function) and $f(z)$ are given and $g(z)$ is unknown. In (5.2.15) the integral part is given by $K(G)$ with the Oka extension $G(z)$ of $g(z)$. So we call (5.2.15) a *Fredholm integral equation of the second kind type.*
More generally, one may consider

$$g(z) = \lambda \int K(z,\zeta)g(\zeta)d\zeta + f(z)$$

with parameter $\lambda(\in \mathbf{C})$ (spectrum); the associated Fredholm integral equation of the first kind is given by

$$g(z) = \lambda \int K(z,\zeta)g(\zeta)d\zeta.$$

Therefore one may consider:

$$g(z) = \lambda K(G)(z) + f(z),$$
$$g(z) = \lambda K(G)(z).$$

Here, we deal with only the case $\lambda = 1$.

(ii) This idea was first described in K. Oka, Proc. Imperial Acad. Tokyo (1941) and then in Oka VI, Tôhoku Math. J. **49** (1942), in which he solved the Pseudoconvexity Problem for 2-dimensional univalent domains by making use of Weil's integral formula. This idea of thinking in the reverse direction is non-trivial. The way of writing the present part is due to Oka's unpublished paper XI (1943); in the published Oka IX (1953) the contents shown there are the same, but integral equation (5.2.15) is not written explicitly as an equation, probably because it was the third time for him to write down the part.

N.B. With the preparation above we are now close to the proof of Oka's pseudoconvexity Theorem 5.3.1. For those readers who are learning the present subject for the first time, it is recommended to move to Theorem 5.3.1 and to finish its proof or to read §5.3, and then proceed to the next section after grasping the whole figure.

5.2.2 Grauert's Method

(1) Topological Vector Spaces. In Grauert's method it is necessary to prepare one step more in topological linear spaces.

Let E be a topological vector space with semi-norms $\|\cdot\|_j$ $(j \in \mathbf{N})$ as in §5.1.1. We begin with the following elementary property.

Proposition 5.2.21. *Let E be Baire and let $F \subset E$ be a closed vector subspace. Then the quotient space E/F is Baire.*

Proof. Let $q : E \to E/F$ be the quotient homomorphism. By definition q is an open map. Therefore, if $G \subset E/F$ contains no interior point, so does $q^{-1}G$. Therefore, if $G_\nu \subset E/F$ $(\nu \in \mathbf{N})$ are countably many closed subsets without interior point, the union $\bigcup_{\nu \in \mathbf{N}} G_\nu$ contains no interior point. □

Proposition 5.2.22. *A finite-dimensional vector subspace of E is closed.*

Proof. Let F be a finite-dimensional vector subspace of E with basis $\{v_j\}_{j=1}^{q}$. We set $v_0 = 0$ and denote by F_h $(\subset F)$ the vector subspace spanned by $v_0, v_1, \ldots, v_h$ $(0 \le h \le q)$. We use the induction on h. By the Hausdorff assumption F_0 is closed. Suppose that F_{h-1} $(1 \le h \le q)$ is closed.

Taking an arbitrary element $x \in F_h \setminus F_{h-1}$, we set

$$x = y + \alpha v_h, \quad y \in F_{h-1},\ \alpha \in \mathbf{C} \setminus \{0\},$$

where α is uniquely determined. Since $v_h \notin F_{h-1}$, it follows from the induction hypothesis that there are $N \in \mathbf{N}$ and $\delta > 0$ satisfying

$$(v_h + \{u \in E : \|u\|_j < \delta, 1 \le j \le N\}) \cap F_{h-1} = \emptyset.$$

Since $v_h - (1/\alpha)x = -(1/\alpha)y \in F_{h-1}$, the above implies

$$(1/\alpha)x \notin \{u \in E : \|u\|_j < \delta, 1 \le j \le N\},$$
$$\max_{1 \le j \le N} \|(1/\alpha)x\|_j \ge \delta,$$
$$|\alpha| \le \frac{1}{\delta} \max_{1 \le j \le N} \|x\|_j, \quad x \in F_h. \tag{5.2.23}$$

Note that the last equation holds trivially for $\alpha = 0$.

Suppose that a sequence $\{x_\nu\}_{\nu=1}^{\infty}$ of points of F_h converges to $x_0 \in E$. Write

$$x_\nu = y_\nu + \alpha_\nu v_h, \quad y_\nu \in F_{h-1}, \quad \nu = 1, 2, \ldots.$$

For every $\varepsilon > 0$ there is a number $M \in \mathbf{N}$ such that

$$\max_{1 \le j \le N} \|x_\nu - x_\mu\| < \varepsilon\delta, \quad \nu, \mu \ge M.$$

From this and (5.2.23) it follows that

$$|\alpha_\nu - \alpha_\mu| < \varepsilon, \quad \nu, \mu \ge M.$$

Therefore $\{\alpha_\nu\}$ forms a Cauchy sequence, and hence converges. With $\lim \alpha_\nu = \alpha_0$ we have that $\lim y_\nu = y_0 = x_0 - \alpha_0 v_h$; since F_{h-1} is closed, $y_0 \in F_{h-1}$. Thus $x_0 = y_0 + \alpha_0 v_h \in F_h$, and so F_h is closed. □

Definition 5.2.24. A linear continuous map $\psi : E \to F$ between topological vector spaces is said to be *completely continuous* or a *compact operator* if there exists a neighborhood U of the origin of E with $\psi(U) \Subset F$ (i.e., the closure $\overline{\psi(U)}$ is compact in F).

Now we prove *L. Schwartz's Fredholm Theorem*:[3]

Theorem 5.2.25 (L. Schwartz). *Let E be a Fréchet space and let F be a Baire vector space. Let $\phi : E \to F$ be a continuous linear surjection, and let $\psi : E \to F$ be a completely continuous linear map. Then the image $(\phi + \psi)(E)$ is closed and the co-kernel* $\mathrm{Coker}(\phi + \psi) := F/(\phi + \psi)(E)$ *is finite dimensional.*

Proof.[4] (a) Because of the length of the proof (2.5 pp.) we present a rough plan of it. We will imitate the proof of "*a locally compact topological vector space being finite dimensional*". First by the assumption we take a neighborhood $U = -U$ of $0 \in E$ with compact $K := \overline{\psi(U)}$. By the Open Map Theorem 5.1.9 $V = \phi(U)$ is an open neighborhood of $0 \in F$. Since K is compact, there are finitely many points $b_j \in K$, $1 \le j \le l$, such that $K \subset \bigcup_{j=1}^{l} \left(b_j + \frac{1}{2}V\right)$. Let $G \subset F$ denote the linear subspace spanned by $\{b_j\}_{1 \le j \le l}$. Set $F \to H := F/\big((\phi + \psi)(E) + G\big)$ (assuming the quotient is defined). Then V has an image W, an open neighborhood of $0 \in H$. It follows from the construction that $W \subset \frac{1}{2}W$. Hence, $W \subset \left(\frac{1}{2}\right)^\nu W$ ($\forall \nu \in \mathbf{N}$), and so $W = \{0\}$; $F/(\phi + \psi)(E) \cong G/(G \cap (\phi + \psi)(E))$ is finite dimensional.

But the closedness of $(\phi + \psi)(E) + G$ is not proved, and hence the argument is incomplete. In fact, we will prove them simultaneously.

---

[3] This naming is due to A. Huckleberry, Jahresber. Dtsch. Math. Ver. **115** (2013), 21–45). According to him, it is due to A. Andreotti. Since this counters the Fredholm integral equation of the second kind type (5.2.15) which Oka used, it is a quite appropriate term (see p. ix footnote 3)).

[4] The idea of the following proof, largely shortened and simplified, is due to J.-P. Demailly (cf. [39] §7.3.4). The former proofs are rather long, twenty–thirty pages (cf., e.g., Bers [5], Grauert–Remmert [24] VI), and so it was often left out or relied on other sources on a key point (e.g., Hitotsumatsu [29] Chap. 12 §2, Gunning–Rossi [26] Appendix B, Fritzsche–Grauert [18] §5.3, etc.).

(b) Set $\phi_0 = \phi + \psi : E \to F$. It suffices to show the existence of a finite dimensional subspace $S \subset F$ such that the composite $\check{\phi}_0$ of ϕ_0 and the quotient map $F \to F/S$ is surjective. For with the direct decomposition $S = S' \oplus (S \cap \phi_0(E))$ $F = \phi_0(E) \oplus S'$ algebraically. Note that S' is finite dimensional and hence Fréchet, and that $E/\ker\phi_0$ is Hausdorff. We consider the following continuous linear surjection and injective surjection:

$$\widetilde{\phi_0} : x \oplus y \in E \oplus S' \to \phi_0(x) + y \in F,$$
$$\widehat{\phi_0} : [x] \oplus y \in \widehat{E} := (E/\mathrm{Ker}\,\phi_0) \oplus S' \to \phi_0(x) + y \in F.$$

By Theorem 5.1.9 $\tilde{\phi}_0$ is an open map, and so is $\widehat{\phi_0}$. Therefore $\widehat{\phi_0}$ is a linear homeomorphism. Since $(E/\mathrm{Ker}\,\phi_0) \oplus \{0\}$ $(\subset \widehat{E})$ is closed, so is $\widehat{\phi_0}((E/\mathrm{Ker}\,\phi_0) \oplus \{0\}) = \phi_0(E)$. Because $\mathrm{Coker}\,\phi_0 = F/\phi_0(E) \cong S'$, $\mathrm{Coker}\,\phi_0$ is finite dimensional.

We show the existence of S above: By the assumption there is a convex neighborhood U_0 of $0 \in E$ such that $-U_0 = U_0$ and $K := \overline{\psi(U_0)}$ is compact. Since ϕ is surjective, $V_0 := \phi(U_0)$ is open by Theorem 5.1.9. We consider an open covering $K \subset \bigcup_{b \in K} \left(b + \frac{1}{2}V_0\right)$. Since K is compact, there are finitely many points $b_j \in K$, $1 \le j \le l$, such that $K \subset \bigcup_{j=1}^{l} \left(b_j + \frac{1}{2}V_0\right)$. Let $S = \langle b_1, \ldots, b_l\rangle$ denote the vector subspace of finite dimension spanned by $b_j, 1 \le j \le l$. Then S is closed by Proposition 5.2.22, so that the quotient space F/S is Hausdorff and Baire by Proposition 5.2.21. Let $\pi : F \to F/S$ be the quotient map, and set $\tilde{V}_0 = \pi(V_0)$. Note that $\tilde{K} := \pi(K)$ is compact and $\tilde{K} \subset \frac{1}{2}\tilde{V}_0$. Replacing F with F/S we may assume from the beginning that

$$K \subset \frac{1}{2}V_0;$$

then we are going to show the surjectivity of ϕ_0.

(c) Since $\phi_0(E)$ is a linear subspace of F, the following claim implies $\phi_0(E) = F$.

Claim 5.2.26. $\phi_0(E) \supset V_0$.

$\because$) Take an arbitrary point $y_0 \in V_0$. There is a point $x_0 \in U_0$ with $\phi(x_0) = y_0$. Since

$$y_1 = y_0 - \phi_0(x_0) = -\psi(x_0) \in K \subset \frac{1}{2}V_0 = \phi\left(\frac{1}{2}U_0\right),$$

there is a point $x_1 \in \frac{1}{2}U_0$ with $\phi(x_1) = y_1$, and then

$$\begin{aligned} y_2 &:= y_1 - \phi_0(x_1) = -\psi(x_1) \in \psi\left(\frac{1}{2}U_0\right) = \frac{1}{2}\psi(U_0) \\ &\subset \frac{1}{2}K \subset \frac{1}{2^2}V_0 = \phi\left(\frac{1}{2^2}U_0\right). \end{aligned}$$

Hence there is a point $x_2 \in \frac{1}{2^2}U_0$ with $y_2 = \phi(x_2)$. Inductively we choose $x_\nu \in \frac{1}{2^\nu}U_0$, $y_\nu = \phi(x_\nu)$, $\nu = 1, 2, \ldots$, so that

$$y_{\nu+1} = y_\nu - \phi_0(x_\nu) \in \frac{1}{2^\nu} K \subset \phi\left(\frac{1}{2^{\nu+1}} U_0\right).$$

For any seminorm $\|\cdot\|_j$ of F there is an $M_j > 0$ with $\|y\|_j \le M_j$ $(\forall\, y \in K)$. It follows that

$$\|y_\nu\| \le \frac{M_j}{2^{\nu-1}} \to 0, \qquad \nu \to \infty.$$

Thus, $\lim_{\nu\to\infty} y_\nu = 0$ and

(5.2.27)
$$\begin{aligned} y_{\nu+1} &= y_\nu - \phi_0(x_\nu) = y_{\nu-1} - \phi_0(x_{\nu-1}) - \phi_0(x_\nu) \\ &= \cdots = y_0 - \phi_0\left(\sum_{j=0}^{\nu} x_j\right). \end{aligned}$$

We would like to re-choose them so that $\sum_{j=0}^\infty x_j$ converges. Let $d_E(\cdot,\cdot)$ be the complete distance defined by (5.1.3) on E. Set $U(r) = \{x \in E : d(x,0) < r\}$ $(r > 0)$. We take a fundamental neighborhood system $\{U_p\}_{p=0}^\infty$ of $0 \in E$ as follows:

(i) U_0 is already taken, and may be assumed to satisfy $U_0 \subset U(1)$. Moreover, $U_p \subset U(2^{-p})$, $p = 1,2,\ldots$.
(ii) Every U_p is convex and symmetric; i.e., $-U_p = U_p$.
(iii) $U_{p+1} \subset \frac{1}{2} U_p$, $p = 0,1,\ldots$.

We consider an open covering of K,

$$K \subset \phi\left(\left(\bigcup_{\mu=1}^\infty 2^\mu U_p\right) \cap \frac{1}{2} U_0\right) = \bigcup_{\mu=1}^\infty \phi\left((2^\mu U_p) \cap \frac{1}{2} U_0\right).$$

Then there is a number $N(p) (\ge 1)$ such that

(5.2.28)
$$K \subset \phi\left(\left(2^{N(p)} U_p\right) \cap \frac{1}{2} U_0\right).$$

We may assume that $N(p) < N(p+1)$ $(p = 1,2,\ldots)$. For $0 \le \nu \le N(1)$ we take x_ν chosen above, and set

$$\tilde{x}_0 = x_0 + \cdots + x_{N(1)}.$$

For $N(p) < \nu \le N(p+1)$ $(p = 1,2,\ldots)$ we have by (5.2.28)

$$\frac{1}{2^{\nu-1}} K \subset \phi\left(\left(2^{N(p)-\nu+1} U_p\right) \cap \frac{1}{2^\nu} U_0\right).$$

Since $y_\nu \in \frac{1}{2^{\nu-1}} K$, we take $x_\nu \in \left(2^{N(p)-\nu+1} U_p\right) \cap \frac{1}{2^\nu} U_0$ with $\phi(x_\nu) = y_\nu$. Then,

(5.2.29)
$$\tilde{x}_p := x_{N(p)+1} + \cdots + x_{N(p+1)} \in \left(1 + \frac{1}{2} + \cdots + \frac{1}{2^{N(p+1)-N(p)-1}}\right) U_p$$

$$\subset 2U_p \subset U_{p-1} \subset U\left(\frac{1}{2^{p-1}}\right);$$

(5.2.30) $$d_E(\tilde{x}_p, 0) < \frac{1}{2^{p-1}}.$$

For every $p > q > q_0$ we have by (5.1.10) and (5.2.30)

$$\begin{aligned} d_E\left(\sum_{\nu=0}^{p} \tilde{x}_\nu, \sum_{\nu=0}^{q} \tilde{x}_\nu\right) &\le \sum_{\nu=q+1}^{p} d_E(\tilde{x}_\nu, 0) < \sum_{\nu=q+1}^{p} \frac{1}{2^{\nu-1}} \\ &< \frac{1}{2^{q-1}} \le \frac{1}{2^{q_0}} \to 0 \quad (q_0 \to \infty). \end{aligned}$$

Therefore $\sum_{\nu=0}^{\infty} \tilde{x}_\nu$ is a Cauchy series. Since d_E is complete, there is a limit $w = \sum_{\nu=0}^{\infty} \tilde{x}_\nu \in E$. By (5.2.27)

$$y_{N(p+1)+1} = y_0 - \phi_0\left(\sum_{\nu=0}^{p} \tilde{x}_\nu\right).$$

By letting $p \to \infty$, we get $y_0 = \phi_0(w)$. Hence, $\phi_0(E) \supset V_0$. □

Example 5.2.31. Let $\mathfrak{G}/\mathbf{C}^n$ be a domain and let $\mathfrak{D} \Subset \mathfrak{G}$ be a bounded subdomain. Then the restriction

$$\psi : f \in \mathcal{O}(\mathfrak{G}) \longrightarrow f|_{\mathfrak{D}} \in \mathcal{O}(\mathfrak{D})$$

is a completely continuous linear map from a Fréchet space to a Baire vector space.

(2) The 1st Cohomology. Let $\mathscr{U} = \{U_\alpha\}_{\alpha \in \Gamma}$ be an open covering of a domain $\mathfrak{D}$ $(/\mathbf{C}^n)$. For any open covering of $\mathfrak{D}$ there is a locally finite and at most countable refinement (see Proposition 3.4.15). As in §3.4 an open covering is always assumed to be locally finite and at most countable.

We put

(5.2.32) $$\begin{aligned} C^0(\mathscr{U}, \mathcal{O}_{\mathfrak{D}}) &= \{(f_\alpha)_{\alpha \in \Gamma} : f_\alpha \in \mathcal{O}(U_\alpha),\ \alpha \in \Gamma\}, \\ C^1(\mathscr{U}, \mathcal{O}_{\mathfrak{D}}) &= \left\{(f_{\alpha\beta}) : f_{\alpha\beta} \in \mathcal{O}(U_\alpha \cap U_\beta),\ \ \alpha, \beta \in \Gamma\right\}. \end{aligned}$$

Here, for $U_\alpha \cap U_\beta = \emptyset$, we associate $f_{\alpha\beta} = 0$ as convention. The sets $C^\nu(\mathscr{U}, \mathcal{O}_{\mathfrak{D}})$ $(\nu = 0, 1)$ naturally form complex linear spaces, of which element is called a ν-*cochain* (of holomorphic functions). A 1-cochain $(f_{\alpha\beta})$ satisfying the following, so-called, *cocycle conditions* is called a 1-*cocycle* (of holomorphic functions):

(5.2.33) $$\begin{aligned} &f_{\alpha\beta} + f_{\beta\alpha} = 0 && (\text{on } U_\alpha \cap U_\beta), \\ &f_{\alpha\beta} + f_{\beta\gamma} + f_{\gamma\alpha} = 0 && (\text{on } U_\alpha \cap U_\beta \cap U_\gamma). \end{aligned}$$

Here if $U_\alpha \cap U_\beta = \emptyset$ or $U_\alpha \cap U_\beta \cap U_\gamma = \emptyset$, we consider that those conditions are satisfied. The set of all 1-cocycles is denoted by $Z^1(\mathscr{U}, \mathcal{O}_{\mathfrak{D}})$; $Z^1(\mathscr{U}, \mathcal{O}_{\mathfrak{D}})$ is a linear subspace of $C^1(\mathscr{U}, \mathcal{O}_{\mathfrak{D}})$. The linear map

(5.2.34) $$\delta : (f_\alpha) \in C^0(\mathscr{U}, \mathscr{O}_{\mathfrak{D}}) \to (f_{\alpha\beta}) := (f_\beta - f_\alpha) \in C^1(\mathscr{U}, \mathscr{O}_{\mathfrak{D}})$$

is called a *coboundary operator*, and the image $\delta(f_\alpha)$ is called a 1-*coboundary*. It is immediate that $B^1(\mathscr{U}, \mathscr{O}_{\mathfrak{D}}) := \delta(C^0(\mathscr{U}, \mathscr{O}_{\mathfrak{D}}))$ is a linear subspace of $Z^1(\mathscr{U}, \mathscr{O}_{\mathfrak{D}})$. The quotient space

(5.2.35) $$H^1(\mathscr{U}, \mathscr{O}_{\mathfrak{D}}) = Z^1(\mathscr{U}, \mathscr{O}_{\mathfrak{D}})/B^1(\mathscr{U}, \mathscr{O}_{\mathfrak{D}})$$

is called the *1st covering cohomology* (of holomorphic functions on $\mathfrak{D}$), of which element is called a *1st covering cohomology class*.

Example 5.2.36. We consider Cousin I data $\{(U_\alpha, f_\alpha)\}_{\alpha\in\Gamma}$ ($f_\alpha \in \mathscr{M}(U_\alpha)$) on $\mathfrak{D}$. Set

(5.2.37) $$f_{\alpha\beta} = f_\alpha - f_\beta \in \mathscr{O}(U_\alpha \cap U_\beta) \quad (U_\alpha \cap U_\beta \neq \emptyset).$$

Then $(f_{\alpha\beta}) \in Z^1(\mathscr{U}, \mathscr{O}_{\mathfrak{D}})$. It is called the 1-cocycle induced from the Cousin I data $\{(U_\alpha, f_\alpha)\}_{\alpha\in\Gamma}$. The 1st covering cohomology class $[(f_{\alpha\beta})] \in H^1(\mathscr{U}, \mathscr{O}_{\mathfrak{D}})$ being 0 means that there is an element $(g_\alpha) \in C^0(\mathscr{U}, \mathscr{O}_{\mathfrak{D}})$ with $(f_{\alpha\beta}) = \delta(g_\alpha)$; that is, $f_{\alpha\beta} = g_\beta - g_\alpha$. Hence

$$f_\alpha + g_\alpha = f_\beta + g_\beta, \quad \text{on } U_\alpha \cap U_\beta.$$

Since g_α are holomorphic, the Cousin I Problem is solved.

The converse is easily checked to hold, too.

Let $\mathscr{V} = \{V_\lambda\}_{\lambda\in\Lambda}$ be a refinement of $\mathscr{U} = \{U_\alpha\}_{\alpha\in\Gamma}$ with $\phi : \Lambda \to \Gamma$. Through the restrictions of ν-cochain ($\nu = 0, 1$) of $\mathscr{U}$ we have the following linear homomorphisms:

$$\phi^* : (f_\alpha) \in C^0(\mathscr{U}, \mathscr{O}_{\mathfrak{D}}) \to (f_{\phi(\lambda)}|_{V_\lambda}) \in C^0(\mathscr{V}, \mathscr{O}_{\mathfrak{D}}),$$
$$\phi^* : (f_{\alpha\beta}) \in C^1(\mathscr{U}, \mathscr{O}_{\mathfrak{D}}) \to (f_{\phi(\lambda)\phi(\beta)}|_{V_\lambda \cap V_\lambda}) \in C^1(\mathscr{V}, \mathscr{O}_{\mathfrak{D}}).$$

By definition

$$\phi^*(Z^1(\mathscr{U}, \mathscr{O}_{\mathfrak{D}})) \subset Z^1(\mathscr{V}, \mathscr{O}_{\mathfrak{D}}), \quad \phi^*(B^1(\mathscr{U}, \mathscr{O}_{\mathfrak{D}})) \subset B^1(\mathscr{V}, \mathscr{O}_{\mathfrak{D}}).$$

Therefore they induce a linear homomorphism

(5.2.38) $$\phi^* : H^1(\mathscr{U}, \mathscr{O}_{\mathfrak{D}}) \to H^1(\mathscr{V}, \mathscr{O}_{\mathfrak{D}}).$$

If $\mathscr{V}$ is a refinement of $\mathscr{U}$ through another $\psi : \Lambda \to \Gamma$, we set

$$\theta : f = (f_{\alpha\beta}) \in Z^1(\mathscr{U}, \mathscr{O}_{\mathfrak{D}}) \longrightarrow \theta(f) := (f_{\phi(\lambda)\psi(\lambda)}|_{V_\lambda}) \in C^0(\mathscr{V}, \mathscr{O}_{\mathfrak{D}}).$$

Then, by a computation we see that

(5.2.39) $$\delta \circ \theta(f) = \psi^* f - \phi^* f,$$

and that

$$\psi^* = \phi^* : H^1(\mathscr{U}, \mathscr{O}_{\mathfrak{D}}) \to H^1(\mathscr{V}, \mathscr{O}_{\mathfrak{D}}). \tag{5.2.40}$$

For any two coverings of $\mathfrak{D}$ there is a common refinement of them, so that the whole family of open coverings of $\mathfrak{D}$ forms an ordered (or directed) family by the relation $\mathscr{U} < \mathscr{V}$ $(\mu : \mathscr{V} \to \mathscr{U})$. For $\mathscr{U} < \mathscr{V} < \mathscr{W}$ we have the following commutative diagram:

$$\begin{array}{ccc} H^1(\mathscr{U}, \mathscr{O}_{\mathfrak{D}}) & \longrightarrow & H^1(\mathscr{V}, \mathscr{O}_{\mathfrak{D}}) \\ & \circlearrowright \searrow & \downarrow \\ & & H^1(\mathscr{W}, \mathscr{O}_{\mathfrak{D}}). \end{array}$$

We first take a disjoint union

$$\Xi^1(\mathscr{O}_{\mathfrak{D}}) = \bigsqcup_{\mathscr{U}} H^1(\mathscr{U}, \mathscr{O}_{\mathfrak{D}}).$$

Two elements $f \in H^1(\mathscr{U}, \mathscr{O}_{\mathfrak{D}}) \subset \Xi^1(\mathscr{O}_{\mathfrak{D}})$ and $g \in H^1(\mathscr{V}, \mathscr{O}_{\mathfrak{D}}) \subset \Xi^1(\mathscr{O}_{\mathfrak{D}})$ are defined to be equivalent, $f \sim g$, if there is a common refinement $\mathscr{W}$ of $\mathscr{U}$ and $\mathscr{V}$

$$\mu : \mathscr{W} \to \mathscr{U}, \qquad \nu : \mathscr{W} \to \mathscr{V},$$

satisfying

$$\mu^*(f) = \nu^*(g) \in H^1(\mathscr{W}, \mathscr{O}_{\mathfrak{D}}).$$

As a quotient space by this equivalence relation the *inductive limit* (also called a *direct limit*) is defined by

$$H^1(\mathfrak{D}, \mathscr{O}_{\mathfrak{D}}) = \varinjlim_{\mathscr{U}} H^1(\mathscr{U}, \mathscr{O}_{\mathfrak{D}}) = \Xi^1(\mathscr{O}_{\mathfrak{D}})/\sim, \tag{5.2.41}$$

which is called the *1st (Čech) cohomology* of holomorphic functions on $\mathfrak{D}$. An element of $H^1(\mathfrak{D}, \mathscr{O}_{\mathfrak{D}})$ is called a *1st (Čech) cohomology class* of $\mathscr{O}_{\mathfrak{D}}$.

Proposition 5.2.42. *The natural homomorphism*

$$H^1(\mathscr{U}, \mathscr{O}_{\mathfrak{D}}) \longrightarrow H^1(\mathfrak{D}, \mathscr{O}_{\mathfrak{D}})$$

is injective.

Proof. Let $f = (f_{\alpha\beta}) \in Z^1(\mathscr{U}, \mathscr{O}_{\mathfrak{D}})$ be a 1-cocycle. Since $\delta f = 0$, we get

$$f_{\beta\gamma} - f_{\alpha\gamma} + f_{\alpha\beta} = 0 \quad \text{on} \quad U_\alpha \cap U_\beta \cap U_\gamma. \tag{5.2.43}$$

Suppose that the image $[(f_{\alpha\beta})]$ of this $(f_{\alpha\beta})$ in $H^1(\mathfrak{D}, \mathscr{O}_{\mathfrak{D}})$ is 0. Then there are a refinement $\mathscr{V} = \{V_\lambda\}_{\lambda \in \Lambda} > \mathscr{U} = \{U_\alpha\}_{\alpha \in \Phi}$ of $\mathscr{U}$ $(\varphi : \Lambda \to \Phi)$ and $(g_\lambda) \in C^0(\mathscr{V}, \mathscr{O}_{\mathfrak{D}})$ such that $\delta(g_\lambda) = (f_{\varphi(\lambda)\varphi(\mu)})$; i.e.,

(5.2.44) $$f_{\varphi(\lambda)\varphi(\mu)} = g_\mu - g_\lambda \quad \text{on } V_\lambda \cap V_\mu.$$

Noting that $U_\alpha = \bigcup_\lambda (V_\lambda \cap U_\alpha)$, we set on each $V_\lambda \cap U_\alpha$

$$h_{\alpha\lambda} = g_\lambda + f_{\varphi(\lambda)\alpha}.$$

It follows from (5.2.43) and (5.2.44) that on $V_\lambda \cap V_\mu \cap U_\alpha$,

$$\begin{aligned} h_{\alpha\lambda} - h_{\alpha\mu} &= g_\lambda - g_\mu + f_{\varphi(\lambda)\alpha} - f_{\varphi(\mu)\alpha} \\ &= f_{\varphi(\mu)\varphi(\lambda)} + f_{\varphi(\lambda)\alpha} - f_{\varphi(\mu)\alpha} = 0. \end{aligned}$$

Therefore, $(h_{\alpha\lambda})$ determines $h_\alpha \in \mathscr{O}(U_\alpha)$ so that $h_\alpha|_{U_\alpha \cap V_\lambda} = h_{\alpha\lambda}$. Taking any $x \in U_\alpha \cap U_\beta$, we take $V_\lambda \ni x$, and then have that

$$\begin{aligned} h_\beta(x) - h_\alpha(x) &= h_{\beta\lambda}(x) - h_{\alpha\lambda}(x) \\ &= g_\lambda(x) + f_{\varphi(\lambda)\beta}(x) - g_\lambda(x) - f_{\varphi(\lambda)\alpha}(x) \\ &= f_{\varphi(\lambda)\beta}(x) - f_{\varphi(\lambda)\alpha}(x) \\ &= f_{\alpha\beta}(x). \end{aligned}$$

This means that $(h_\alpha) \in C^0(\mathscr{U}, \mathscr{O}_{\mathfrak{D}})$ and $\delta(h_\alpha) = (f_{\alpha\beta})$, so that $[(f_{\alpha\beta})] = 0$ in $H^1(\mathscr{U}, \mathscr{O}_{\mathfrak{D}})$. □

Remark 5.2.45. With the notation in Example 5.2.36, Cousin I data $\{(U_\alpha, f_\alpha)\}_{\alpha\in\Gamma}$ on $\mathfrak{D}$ induces a 1-cocycle f and then a 1st cohomology class $[f] \in H^1(\mathfrak{D}, \mathscr{O}_{\mathfrak{D}})$. The original Cousin I data is solvable if and only is $[f] = 0$. In particular, if $H^1(\mathfrak{D}, \mathscr{O}_{\mathfrak{D}}) = 0$, any Cousin I data on $\mathfrak{D}$ has a solution.

Theorem 5.2.46. *Every continuous Cousin data on a domain $\mathfrak{D}$ $(/\mathbf{C}^n)$ has a solution if and only if $H^1(\mathfrak{D}, \mathscr{O}_{\mathfrak{D}}) = 0$.*

Proof. We begin with the proof of "if" part. Let $\{(U_\alpha, f_\alpha)\}_{\alpha\in\Gamma}$ be continuous Cousin I data on $\mathfrak{D}$. Then f_α is a continuous function on U_α and with

$$f_{\alpha\beta} := f_\alpha - f_\beta \in \mathscr{O}(U_\alpha \cap U_\beta).$$

$f = (f_{\alpha\beta})$ is a 1-cocycle. If $H^1(\mathfrak{D}, \mathscr{O}_{\mathfrak{D}}) = 0$, Proposition 5.2.42 implies that there is a 0-cochain $g = (g_\alpha) \in C^0(\mathscr{U}, \mathscr{O}_{\mathfrak{D}})$ with $\delta(g) = (f)$; i.e.,

$$f_\alpha + g_\alpha = f_\beta + g_\beta \quad \text{on} \quad U_\alpha \cap U_\beta.$$

This gives a solution of the continuous Cousin data $\{(U_\alpha, f_\alpha)\}$.

We show the converse. It suffices to prove that $H^1(\mathscr{U}, \mathscr{O}_{\mathfrak{D}}) = 0$ for an arbitrary open covering $\mathscr{U} = \{U_\alpha\}_{\alpha\in\Gamma}$ of $\mathfrak{D}$. It is noted that $\mathscr{U}$ is locally finite and at most countable. Similarly to Proposition 3.4.16, we take a continuous partition of unity $\{\chi_\alpha\}_{\alpha\in\Gamma}$ associated with $\mathscr{U}$:

$$0 \le \chi_\alpha \in \mathscr{C}^0(U_\alpha), \quad \operatorname{Supp} \chi_\alpha \subset U_\alpha, \quad \sum_{\alpha \in \Gamma} \chi_\alpha = 1.$$

Take an element $f = (f_{\alpha\beta}) \in Z^1(\mathscr{U}, \mathscr{O}_{\mathfrak{D}})$ arbitrarily. We extend $\chi_\gamma f_{\alpha\gamma}$ to be 0 on $U_\alpha \setminus U_\gamma$ as a continuous function on U_α. Set

$$g_\alpha = \sum_\gamma \chi_\gamma f_{\alpha\gamma}. \tag{5.2.47}$$

Then g_α is a continuous function on U_α. It follows from an easy computation that $g_\alpha - g_\beta = f_{\alpha\beta} \in \mathscr{O}(U_\alpha \cap U_\beta)$. Hence, $\{(U_\alpha, g_\alpha)\}$ is continuous Cousin data. By the assumption there is a continuous function G on $\mathfrak{D}$ such that

$$h_\alpha := G - g_\alpha \in \mathscr{O}(U_\alpha), \quad \forall U_\alpha.$$

We see that $h = (h_\alpha) \in C^0(\mathscr{U}, \mathscr{O}_{\mathfrak{D}})$ and $\delta(h) = f$; hence $[f] = 0$. □

This together with Oka's Theorem 3.7.9 implies:

Theorem 5.2.48 (Oka). *If $\mathfrak{D}$ $(/\mathbf{C}^n)$ is a Stein domain, then $H^1(\mathfrak{D}, \mathscr{O}_{\mathfrak{D}}) = 0$.*

An open covering $\{U_\alpha\}$ of $\mathfrak{D}$ is called a *Stein covering* if all U_α are Stein.

Lemma 5.2.49. *Let $\mathscr{U} = \{U_\alpha\}_{\alpha \in \Gamma}$ be a Stein covering of $\mathfrak{D}$. Then the natural homomorphism*

$$H^1(\mathscr{U}, \mathscr{O}_{\mathfrak{D}}) \to H^1(\mathfrak{D}, \mathscr{O}_{\mathfrak{D}})$$

is an isomorphism.

Proof. By Proposition 5.2.42 the homomorphism is injective, and so it suffices to show the surjectivity. Take an element $[f] \in H^1(\mathfrak{D}, \mathscr{O}_{\mathfrak{D}})$. Then there is an open covering $\mathscr{V} = \{V_\lambda\}_{\lambda \in \Lambda}$ such that $f = (f_{\lambda\mu}) \in Z^1(\mathscr{V}, \mathscr{O}_{\mathfrak{D}})$. Without loss of generality $\mathscr{V}$ is assumed to be a refinement of $\mathscr{U}$ with $\phi : \Lambda \to \Gamma$. By making use of a partition of unity associated to $\mathscr{V}$ we have continuous Cousin data $\{(V_\lambda, g_\lambda)\}_{\lambda \in \Lambda}$ with respect to $\mathscr{V}$ as in (5.2.47):

$$g_\lambda - g_\mu = f_{\lambda\mu} \in \mathscr{O}(V_\lambda \cap V_\mu).$$

For each U_α we consider an open covering $U_\alpha = \bigcup_\lambda (U_\alpha \cap V_\lambda)$. Since U_α is Stein, by Theorem 5.2.48 there are continuous functions G_α on U_α such that

$$G_\alpha - g_\lambda \in \mathscr{O}(U_\alpha \cap V_\lambda), \quad \forall V_\lambda.$$

Setting $h_{\alpha\beta} := G_\alpha - G_\beta$, we have

$$h_{\alpha\beta} = G_\alpha - g_\lambda - (G_\beta - g_\lambda) \in \mathscr{O}(U_\alpha \cap U_\beta \cap V_\lambda)$$

for every V_λ. It follows that $h_{\alpha\beta} \in \mathscr{O}(U_\alpha \cap U_\beta)$ and $(h_{\alpha\beta}) \in Z^1(\mathscr{U}, \mathscr{O}_{\mathfrak{D}})$. We see that

$$f_{\lambda\mu} - h_{\phi(\lambda)\phi(\mu)} = (g_\lambda - G_{\phi(\lambda)}) - (g_\mu - G_{\phi(\mu)});$$

the first term in parentheses is holomorphic in V_λ, and the second is holomorphic in V_μ. Thus

$$(f_{\lambda\mu}) - \phi^*(h_{\alpha\beta}) \in B^1(\mathscr{V}, \mathscr{O}_{\mathfrak{D}}), \quad [f] \in \phi^* H^1(\mathscr{U}, \mathscr{O}_{\mathfrak{D}}). \qquad \square$$

(3) Grauert's Theorem. We prove a finite-dimensionality theorem due to H. Grauert [22]:

Theorem 5.2.50 (Grauert). *Let $\mathfrak{G}$ $(/\mathbf{C}^n)$ be a domain, and let $\mathfrak{D} \Subset \mathfrak{G}$ be a strongly pseudoconvex domain. Then*

$$\dim_{\mathbf{C}} H^1(\mathfrak{D}, \mathscr{O}_{\mathfrak{D}}) < \infty.$$

Remark 5.2.51. Roughly speaking, we proved by Oka's method that $H^1(\mathfrak{D}, \mathscr{O}_{\mathfrak{D}}) = 0$ for a strongly pseudoconvex domain $\mathfrak{D}$.

Proof. We recall the assumption:

5.2.52. There is a strongly plurisubharmonic function φ defined in a neighborhood T of $\partial\mathfrak{D}$ and $c > 0$ such that

$$\mathfrak{D} \cap T = \{\varphi < 0\}, \quad \partial\mathfrak{D} = \{\varphi = 0\} \subset \{-c < \varphi < c\} \Subset T.$$

We extend φ to be $-c$ on $\mathfrak{D} \setminus \{-c < \varphi < 0\}$ as a continuous plurisubharmonic function on $\tilde{\mathfrak{D}} := \mathfrak{D} \cup T$. We set $T_c = \{\varphi < c\}$.

Step 1. For each point $a \in \partial\mathfrak{D}$ we take a univalent open ball neighborhood $U = \mathrm{B}(a;\delta) \Subset T_c$. By Theorem 4.3.74, $U \cap \mathfrak{D}$ is Stein. We take the double neighborhoods $V = \mathrm{B}(a;\delta/2) \Subset U$ of a; $V \cap \mathfrak{D}$ is also Stein. Since $\partial\mathfrak{D}$ is compact, we can cover it with finitely many such $V_i \Subset U_i$:

(5.2.53)
$$\partial\mathfrak{D} \subset \bigcup_{i=1}^{l} V_i \Subset \bigcup_{i=1}^{l} U_i \quad (l < \infty).$$

Since $\mathfrak{D} \setminus \bigcup_{i=1}^{l} V_i$ is compact, it is covered by finitely many double open ball neighborhoods:

(5.2.54)
$$V_i = B(a_i;\delta_i/2) \Subset U_i = B(a_i;\delta_i) \Subset \mathfrak{D}, \qquad i = l+1, \ldots, L.$$

Since $\mathscr{V} = \{\mathfrak{D} \cap V_i\}$ and $\mathscr{U} = \{\mathfrak{D} \cap U_i\}$ are Stein coverings of $\mathfrak{D}$, it follows from Lemma 5.2.49 that

(5.2.55)
$$H^1(\mathfrak{D}, \mathscr{O}_{\mathfrak{D}}) \cong H^1(\mathscr{V}, \mathscr{O}_{\mathfrak{D}}) \cong H^1(\mathscr{U}, \mathscr{O}_{\mathfrak{D}}).$$

Step 2. Take a C^∞ function $c_1(z) \ge 0$ such that

$$\operatorname{Supp} c_1 \subset U_1, \qquad c_1|_{V_1} = 1.$$

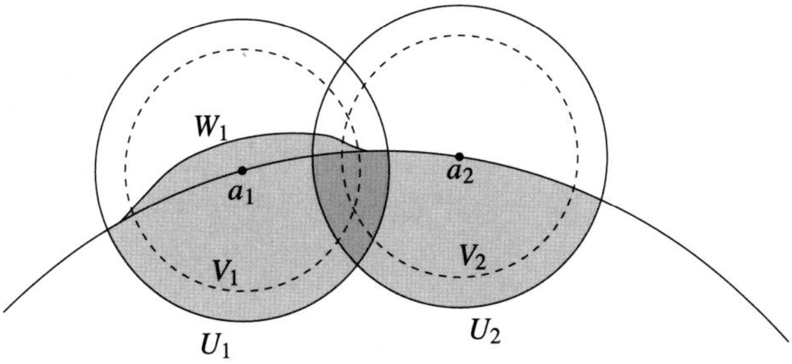

Fig. 5.2 Boundary bumping method 1.

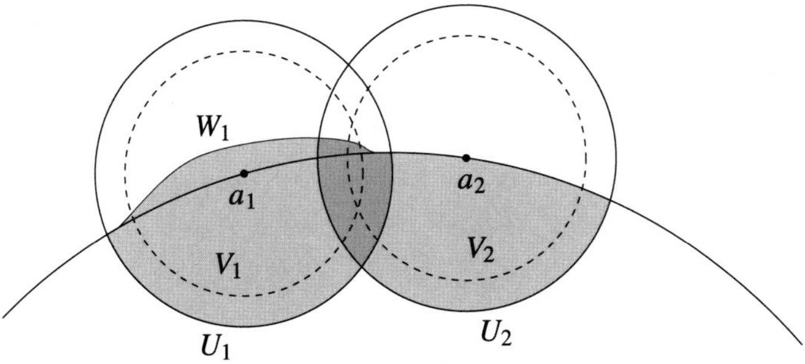

Fig. 5.3 Boundary bumping method 2.

With a sufficiently small $\varepsilon > 0$, $\varphi_\varepsilon(z) := \varphi(z) - \varepsilon c_1(z)$ is strongly plurisubharmonic in T_c. We set

(5.2.56) $$W_1 = U_1 \cap \{\varphi_\varepsilon < 0\}$$

(cf. Fig. 5.2). By Theorem 4.3.74 W_1 is Stein. With $\varepsilon > 0$ smaller if necessary we have

$$\overline{V_1 \cap \mathfrak{D}} \Subset W_1.$$

Moreover, for other U_j having non-empty intersection with U_1, $\{z \in U_j : \varphi_\varepsilon(z) < 0\}$ is Stein (cf. Fig. 5.3). We set

$$U_1^{(1)} = W_1, \qquad U_j^{(1)} = U_j \cap \mathfrak{D}, \quad j \geq 2,$$
$$\mathscr{U}^{(1)} = \{U_j^{(1)}\}_{j=1}^L, \ \mathfrak{D}^{(1)} = \bigcup_{j=1}^L U_j^{(1)}.$$

It follows that $\mathscr{U}^{(1)}$ is a Stein covering of $\mathfrak{D}^{(1)}$, and that for U_{j_0}, U_{j_1} $(j_0 \neq j_1)$ and $U_{j_0}^{(1)}, U_{j_1}^{(1)}$ with the same pair of indices (j_0, j_1), $U_{j_0} \cap U_{j_1} \cap \mathfrak{D} = U_{j_0}^{(1)} \cap U_{j_1}^{(1)}$. Therefore we obtain the following equality and the surjection induced from the

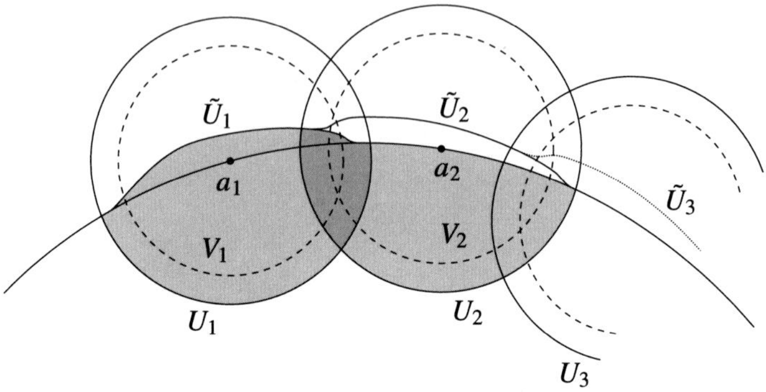

Fig. 5.4 Boundary bumping method 3.

restrictions:

$$Z^1(\mathscr{U},\mathscr{O}_{\mathfrak{D}}) = Z^1(\mathscr{U}^{(1)},\mathscr{O}_{\mathfrak{D}^{(1)}}),\tag{5.2.57}$$
$$H^1(\mathfrak{D}^{(1)},\mathscr{O}_{\mathfrak{D}^{(1)}}) \cong H^1(\mathscr{U}^{(1)},\mathscr{O}_{\mathfrak{D}^{(1)}}) \to H^1(\mathscr{U},\mathscr{O}_{\mathfrak{D}}) \cong H^1(\mathfrak{D},\mathscr{O}_{\mathfrak{D}}) \to 0.$$

Step 3. We change the covering of $\mathfrak{D}^{(1)}$ as follows. Now, W_1 is already given in (5.2.56). Set

$$W_j = \mathfrak{D}^{(1)} \cap U_j, \qquad j \geq 2,$$
$$\mathscr{W} = \{W_j\}_{j=1}^{L}.$$

Since all W_j are Stein, we get

$$H^1(\mathfrak{D}^{(1)},\mathscr{O}_{\mathfrak{D}^{(1)}}) \cong H^1(\mathscr{W},\mathscr{O}_{\mathfrak{D}^{(1)}}).\tag{5.2.58}$$

Step 4. For $\mathfrak{D}^{(1)} = \bigcup_j W_j$ and W_2 we practice the procedures of Steps 2 and 3. Repeating this procedure l-times, we enlarge outward all $U_i \cap \partial\mathfrak{D}$, $i = 1,2,\ldots,l$, and denote the resulting covering of $\partial\mathfrak{D}$ by

$$\tilde{U}_1, \tilde{U}_2,, \ldots, \tilde{U}_l$$

(cf. Fig. 5.4), and after the $(l+1)$th we put without change,

$$\tilde{U}_i = U_i, \qquad l+1 \leq i \leq L.$$

Now we set

$$\tilde{\mathscr{U}} = \{\tilde{U}_i\}_{i=1}^{L}, \qquad \tilde{\mathfrak{D}} = \bigcup_{i=1}^{L} \tilde{U}_i.$$

From the construction and (5.2.57) we infer that

(5.2.59)
$$V_i \cap \mathfrak{D} \Subset \tilde{U}_i, \qquad 1 \le i \le L,$$
$$\tilde{\rho} : H^1(\tilde{\mathscr{U}}, \mathscr{O}_{\tilde{\mathfrak{D}}}) \to H^1(\mathscr{V}, \mathscr{O}_{\mathfrak{D}}) \to 0.$$

Here, $\tilde{\rho}$ is the homomorphism naturally induced from the restrictions. Therefore we obtain the following surjective homomorphism:

(5.2.60)
$$\Psi : \xi \oplus \eta \in Z^1(\tilde{\mathscr{U}}, \mathscr{O}_{\tilde{\mathfrak{D}}}) \oplus C^0(\mathscr{V}, \mathscr{O}_{\mathfrak{D}}) \to \rho(\xi) + \delta\eta \in Z^1(\mathscr{V}, \mathscr{O}_{\mathfrak{D}}) \to 0,$$

where $\rho : Z^1(\tilde{\mathscr{U}}, \mathscr{O}_{\tilde{\mathfrak{D}}}) \to Z^1(\mathscr{V}, \mathscr{O}_{\mathfrak{D}})$ denotes the homomorphism induced from the restrictions from $\tilde{U}_\alpha \cap \tilde{U}_\beta$ to $V_\alpha \cap V_\beta \cap \mathfrak{D}$, and by definition

$$H^1(\mathscr{V}, \mathscr{O}_{\tilde{\mathfrak{D}}}) = Z^1(\mathscr{V}, \mathscr{O}_{\tilde{\mathfrak{D}}}) / \delta C^0(\mathscr{V}, \mathscr{O}_{\tilde{\mathfrak{D}}}).$$

Since $V_\alpha \cap V_\beta \cap \mathfrak{D} \Subset \tilde{U}_\alpha \cap \tilde{U}_\beta$, ρ is completely continuous (see Example 5.2.31). It follows from Schwartz's Fredholm Theorem 5.2.25 that

$$\operatorname{Coker}(\Psi - \rho) = Z^1(\mathscr{V}, \mathscr{O}_{\mathfrak{D}}) / \delta C^0(\mathscr{V}, \mathscr{O}_{\mathfrak{D}}) = H^1(\mathscr{V}, \mathscr{O}_{\mathfrak{D}})$$

is finite dimensional. We see by (5.2.55) that $\dim H^1(\mathfrak{D}, \mathscr{O}_{\mathfrak{D}}) < \infty$. □

The above method of the proof is called the *bumping method* of Grauert, and is effective even for so-called complex spaces with singularities (cf. [39] Chap. 8).

(4) Proof of Lemma 5.2.1. (a) (Holomorphic convexity) Let φ be a plurisubharmonic function taken in 5.2.52 such that $\mathfrak{D} = \{\varphi < 0\}$ and φ is strongly plurisubharmonic in a neighborhood of $\partial\mathfrak{D}$. For a boundary point $p_0 \in \partial\mathfrak{D}$ we take the Levi polynomial $P(z)$ of φ at p_0, defined by (4.3.67). Restricting (4.3.69) to $t = 0$, we have for a sufficiently small $r > 0$ Cousin I data with $U' = \mathfrak{D}_r = \{\varphi < r\}$ for U' in the argument of (4.3.67) and $g = \frac{1}{P(z)}$ in a neighborhood of p_0 such that $|g(p_0)| = \infty$. With $g_k = \frac{1}{P(z)^k}$ $(k = 1, 2, \ldots)$ in the neighborhood of p_0 and 0 on $\mathfrak{D}_r \setminus \{P = 0\}$ we obtain Cousin I data on $\mathfrak{D}_r$, which induce 1st cohomology classes $f_k \in H^1(\mathfrak{D}_r, \mathscr{O}_{\mathfrak{D}_r})$ $(k = 1, 2, \ldots)$. Since the vector space $H^1(\mathfrak{D}_r, \mathscr{O}_{\mathfrak{D}_r})$ is finite dimensional by Grauert's Theorem 5.2.50, there are finitely many $c_1, c_2, \ldots, c_k \in \mathbf{C}$ $(c_k \ne 0)$ such that

(5.2.61)
$$c_1 f_1 + c_2 f_2 + \cdots + c_k f_k = 0 \in H^1(\mathfrak{D}_r, \mathscr{O}_{\mathfrak{D}_r}).$$

This means that there is a meromorphic function $F(z)$ in $\mathfrak{D}_r$ such that

(5.2.62)
$$F(z) - \sum_{j=1}^{k} \frac{c_j}{P(z)^j}$$

is holomorphic in a neighborhood of p_0 and $F(z)$ has no other poles. Since $c_k \ne 0$, $|F(p_0)| = \infty$ and $F \in \mathscr{O}(\bar{\mathfrak{D}} \setminus \{p_0\})$. There is such a function $F(z)$ for every point of $\partial\mathfrak{D}$, and hence $\mathfrak{D}$ is holomorphically convex.

(b) (Holomorphic separation) Let $p_1, p_2 \in \mathfrak{D}$ be points such that $p_1 \neq p_2$ and $\pi(p_1) = \pi(p_2)$. We use arguments similar to the proof (b) of Oka's Heftungslemma 5.2.4, and use the same notation as there. As in (5.2.61) and (5.2.62), we define Cousin I data such that in a neighborhood of $q_1 \in \partial\mathfrak{D}$ poles are given by

$$\sum_{j=1}^{k} \frac{c_j}{P(z)^j}, \quad c_k \neq 0,$$

and in a neighborhood of another q_2 it has no pole (holomorphic). Solving the Cousin I Problem, we have a meromorphic function g on $\mathfrak{D}_r$. It follows that $g \in \mathscr{O}(\mathfrak{D})$ and as in (5.2.18), $\underline{g}_{p_1} \neq \underline{g}_{p_2}$ by the uniqueness of analytic continuation; Proposition 3.6.9 implies that $\mathfrak{D}$ is holomorphically separated.

Thus, $\mathfrak{D}$ is Stein. □

5.3 Oka's Pseudoconvexity Theorem

By making use of Lemma 5.2.1 proved by two methods due to Oka and Grauert we describe the solution of the final goal of the Pseudoconvexity Problem.

Theorem 5.3.1 (Oka 1943/'53**).** *A pseudoconvex domain over* $\mathbf{C}^n$ *is Stein.*

Proof. Let $\pi : \mathfrak{D} \to \mathbf{C}^n$ be a pseudoconvex domain with a pseudoconvex exhaustion function $\varphi : \mathfrak{D} \to [-\infty, \infty)$. We fix a point $p_0 \in \mathfrak{D}$. Taking $\varphi(p_0) < c_1 < c_2 < c_3$ ($\in \mathbf{R}$) arbitrarily, we denote by $\mathfrak{D}_\nu$ the connected components of $\{\varphi < c_\nu\}$ ($\Subset \mathfrak{D}$) containing p_0 ($\nu = 1, 2, 3$). With a sufficiently small $\varepsilon > 0$ we define the smoothing $\varphi_\varepsilon(p)$ of φ defined on $\bar{\mathfrak{D}}_3$. Since $\|\pi(p)\|^2$ is strongly plurisubharmonic,

$$\tilde{\varphi}_\varepsilon(p) = \varphi_\varepsilon(p) + \varepsilon\|\pi(p)\|^2, \quad p \in \bar{\mathfrak{D}}_3$$

is strongly plurisubharmonic. Here c_3 is chosen larger if necessary, so that $\bar{\mathfrak{D}}_2 \Subset \mathfrak{D}_3$. For any neighborhood $W \supset \partial\mathfrak{D}_3$ there is a sufficiently small $\varepsilon > 0$ such that

$$\mathfrak{D}_2 \Subset \mathfrak{D}'_3 := \{p \in \mathfrak{D}_3 : \tilde{\varphi}_\varepsilon(p) < c_3\}.$$

By Lemma 5.2.1, $\mathfrak{D}'_3$ is Stein. Since φ is a plurisubharmonic function on $\mathfrak{D}'_3$ and $\mathfrak{D}_\nu$ ($\nu = 1, 2$) are connected components of the sublevel sets, Theorem 4.3.74 implies that they are Stein and moreover $\mathfrak{D}_1$ is $\mathscr{O}(\mathfrak{D}_2)$-convex. Therefore, $(\mathfrak{D}_1, \mathfrak{D}_2)$ is a Runge pair.

Taking a divergent sequence $c_\nu \nearrow \infty$ ($\nu = 1, 2, \ldots,\ c_1 > \varphi(p_0)$), we set $\mathfrak{D}_\nu$ to be the connected components of $\{\varphi < c_\nu\}$ containing p_0. Then $\mathfrak{D} = \bigcup_{\nu=1}^{\infty} \mathfrak{D}_\nu$ and it is inferred from Theorem 3.7.8 that $\mathfrak{D}$ is Stein.

Thus the proof of Oka's Pseudoconvexity Theorem is completed. □

We summarize the results obtained above:

Theorem 5.3.2. *Let* $\mathfrak{D}$ $(/\mathbf{C}^n)$ *be a domain and let* $\delta_{\mathrm{P}\Delta}(p, \partial\mathfrak{D})$ *be the boundary distance function. Then the following seven conditions are equivalent:*

(i) $\mathfrak{D}$ *is Stein.*
(ii) $\mathfrak{D}$ *is holomorphically convex.*
(iii) $\mathfrak{D}$ *is* C^0*-pseudoconvex.*
(iv) $\mathfrak{D}$ *is pseudoconvex.*
(v) $-\log \delta_{\mathrm{P}\Delta}(p, \partial\mathfrak{D})$ *is plurisubharmonic.*
(vi) $\mathfrak{D}$ *is Hartogs pseudoconvex.* [5]
(vii) $\mathfrak{D}$ *is a domain of holomorphy.*

The proof is a corollary of what has already been proved, but it is stated since the content has its own interest; the proof is left to the readers (see Exercise 8 at the end of the chapter).

Remark 5.3.3. The plurisubharmonicity of $-\log \delta_{\mathrm{P}\Delta}(p, \partial\mathfrak{D})$ of condition (i) of Theorem 5.3.2 is sufficient only "near the boundary of $\mathfrak{D}$"; e.g., it is sufficient to assume that $-\log \delta_{\mathrm{P}\Delta}(p, \partial\mathfrak{D})$ is plurisubharmonic in $\{p \in \mathfrak{D} : \delta_{\mathrm{P}\Delta}(p, \partial\mathfrak{D}) < c\}$ with a small $c > 0$.

Remark 5.3.4 (Singular spaces). It is an interesting and significant problem to ask what happens for domains with singularities. Because of the introductory nature of the present book we are not going into the details of the case of singular spaces, but just give an overview how the results obtained above are extended to singular spaces and leave the details to the references.

A so-called complex space with allowing singularities is defined (cf., e.g., Noguchi [39] §6.9, Chap. 8). It is possible to define plurisubharmonic functions and strongly plurisubharmonic functions on a complex space (cf., e.g., Vâjâitu [58] and the references there). There are the generalizations of Oka's method to singular complex spaces by Nishino [36], [37] Chap. 9, and Andreotti–Narasimhan [3]; as for Grauert's method, cf. Narasimhan [34]. By V. Vâjâitu [58] Lemma 1 we have the following general result:

Theorem 5.3.5. *If a complex space X satisfies the following two conditions, then X is Stein.*

(i) *X is pseudoconvex.*
(ii) *Every relatively open subset of X admits a strongly plurisubharmonic function.*

Remark 5.3.6 (Generalizations of $\pi : \mathfrak{D} \to \mathbf{C}^n$). Docquire–Grauert [14] in 1960 generalized Theorem 5.3.1 to an abstract unramified domain $\varpi : X \to Y$ such that X and Y are complex manifolds and ϖ is a locally biholomorphic map. They proved that *if every point* $p \in Y$ *has a Stein neighborhood* V *with the Stein inverse* $\varpi^{-1}V$ (locally Stein) *and if* Y *is Stein, then* X *is Stein.* The result was further generalized to the case of singular complex spaces by Vâjâitu [59] in 2008 (cf. also the references there).

[5] See Chap. 4, Exercise 13.

There is a generalization of Theorem 5.3.1 to another direction by R. Fujita [19] in 1963 and by A. Takeuchi [57] in 1964 for the compactification of $\mathbf{C}^n$ by n-dimensional complex projective space $\mathbf{P}^n(\mathbf{C})$:

Theorem 5.3.7 (Fujita, Takeuchi). *If $\varpi : X \to \mathbf{P}^n(\mathbf{C})$ is a locally biholomorphic map and locally Stein (as above), then X is Stein.*

Remark 5.3.8 (Ramified domains over $\mathbf{C}^n$). We say that a complex space X is *weakly holomorphically separable*[6] if the following condition is satisfied: For every point $a \in X$ there are finitely many holomorphic functions f_k $(1 \le k \le l)$ on X vanishing at a such that a is an isolated point of the analytic subset $\{x \in X : f_k(x) = 0, 1 \le k \le l\}$. Andreotti–Narasimhan [3] p. 356 proved that:

Lemma 5.3.9. *A weakly holomorphically separable complex space X satisfies condition (ii) of Theorem 5.3.5.*

We consider the case where $\pi : X \to \mathbf{C}^n$ is a ramified domain. (cf. Remark 3.6.3); naturally, such X is weakly holomorphically separable. Therefore in this case, the Pseudoconvexity Problem is valid if there is a pseudoconvex exhaustion function on X (cf. [3] Theorem 3 for C^0-pseudoconvex case):

Theorem 5.3.10. *If there is a pseudoconvex exhaustion function on a ramified domain $\pi : X \to \mathbf{C}^n$, then X is Stein.*

Now the Pseudoconvexity Problem for ramified domains is stated as follows:

Problem 5.3.11 (Pseudoconvexity Problem for Ramified Domains)**.** Let X be a complex space and let $\pi : X \to \mathbf{C}^n$ be a ramified domain. If every point $z \in \mathbf{C}^n$ carries a neighborhood U with the Stein inverse $\pi^{-1}U$, then is X Stein?

The Pseudoconvexity Problem for Ramified Domains 5.3.11 is, however, countered by an example of J.E. Fornæss [16] with a two-sheeted ramified domain over $\mathbf{C}^2$, so that the problem cannot be true in its original form (cf. Noguchi [41] for an affirmative result). Subsequently, the problem is reduced to asking:

Problem 5.3.12. What separates the validity of Theorem 5.3.10 and the non-validity of Problem 5.3.11?

Note. The Pseudoconvexity Problem is often referred to as *Levi's Problem* in a large sense; e.g., cf. I. Lieb [33].[7] As written in Oka IX (Jpn. J. Math. **23** (1953), p. 138; [48], p. 211) Oka IX was published as an intermediate report and was a modified version of of the unpublished papers VII–XI of 1943 (in Japanese, [50] Posthumous Papers Vol. 1) which gave the first complete solution of the Pseudoconvexity Problem for unramified domains over $\mathbf{C}^n$ with $n \ge 2$ general: The papers

[6] Cf. "Weak Sepration Axiom" in Grauert–Remmert [24] Introduction.

[7] This is a fine-detailed survey on Levi's Problem, beginning with the origin of the problem, by which we understand the state of the art of the study in Europe then. It devotes much space in detail to the works of K. Oka; however, in the journal names of the referred papers, the use of the abbreviation "Jap." should be avoided and replaced with "Jpn." or the full-spelling "Japanese".

were modified by means of "coherence" or "idéal de domaines indéterminés", which were invented in order to solve the problem for domains with *ramifications* (Oka VII written 1948, VIII 1951). He titled a section at p. 138 of Oka IX 1953 as

B. Problème inverse de Hartogs.—Point de départ.

The present book began with Hartogs' phenomenon in Chap. 1. Theorem 5.3.2 (vi) is just a statement, returning there. In Levi's Problem 4.3.65 a regularity (differentiability) of the boundary is assumed, but there is no such assumption in Theorem 5.3.2: Here one sees that K. Oka investigated the problem by going back to the origin. While the Pseudoconvexity Problem was clearly grasped as a boundary problem by assuming the boundary regularity, it is interesting to learn that the problem was solved by introducing a new class of pseudoconvex or plurisubharmonic functions without regularity.

Oka's method directly proves the solvability of the Continuous Cousin Problem on a strongly pseudoconvex domain, that is, $H^1(\mathfrak{D}, \mathscr{O}_{\mathfrak{D}}) = 0$. The key Lemma 5.2.4 is called "Oka's Heftungslemma" (German), which is due to Andreotti-Narasimhan [3]; this paper is in English and Oka's in French. One wonders why a German term was used.

Grauert's Theorem 5.2.50 asserts the finite dimensionality, relaxing the claim. Although the claim is relaxed, it holds as far as the boundary is strongly pseudoconvex even in a complex space, and hence has a number of applications; for instance, Kodaira's Embedding Theorem in a generalized form for singular complex spaces is deuced from it (cf., e.g, [39] Chap. 8).

On the other hand, Oka's method is to solve an integral equation for given Cousin I data by means of iterative approximations; it is direct and constructive. It is also possible to generalize the proof for singular complex spaces (cf. Nishino [36], [37] Chap. 9, Andreotti-Narasimhan [3]).

H. Grauert wrote the following comments in his "Selected Papers" (Vol. I pp. 155–156, Springer 1994) together with C.L. Siegel's comments:

> Oka's methods are very complicated. At first he proved (rather simply) that in any unbranched pseudoconvex domain X there is a continuous strictly plurisubharmonic function $p(x)$ which converges to $+\infty$ as x goes to the (ideal) boundary of X.[8] Then he got the existence of holomorphic functions f from this property. In [19][9] the existence of the f comes from a theorem of L. Schwartz in functional analysis (topological vector spaces, see: H. Cartan, Séminaire E.N.S. 1953/54, Exposés XVI and XVII). The approach is much simpler, but my predecessor in Göttingen C.L. Siegel nevertheless did not like it: Oka's method is constructive and this one is not!

If the proof of L. Schwartz's Theorem 5.2.25[10] is included in Grauert's method, it might be not so convincing to say "much simpler". It is rather interesting to see a

[8] Theorem 4.3.15 in the present book.

[9] [22] at the end of the present book.

[10] Now it is not so long as seen in the present book, but it was originally more than twenty pages or so.

strong similarity in the arguments of the two methods; they are the part that solves the Fredholm integral equation of the second kind type (5.2.15) in Oka's method, and in Grauert's, the part that obtains a convergent series in the proof of L. Schwartz's Fredholm Theorem 5.2.25.[11] The present author does not know how they are really related mathematically.

Exercises

1. Let X be a Hausdorff topological space, and let G_1, G_2 $(\subset X)$ be closed sets without interior points. Show that $G_1 \cup G_2$ does not contain an interior point.
2. Show 5.1.6 (i), (ii).
3. Let the notation be as in Example 5.1.14 (1). For $f \in \mathscr{C}^0(\mathfrak{D})$ we define semi-norms $\|f\|_{\bar{\mathfrak{D}}_j} = \max_{\bar{\mathfrak{D}}_j} |f|$ $(j \in \mathbf{N})$, and endow $\mathscr{C}^0(\mathfrak{D})$ with the topology defined by $\|f\|_j$ $(j \in \mathbf{N})$. Show that $\mathscr{C}^0(\mathfrak{D})$ is Fréchet.
4. Let $\mathscr{U} = \{U_\alpha\}_{\alpha \in \Gamma}$ be an open covering of a domain $\mathfrak{D}(/\mathbf{C}^n)$. Show that for every compact subset $K \Subset \mathfrak{D}$ the cardinality of $\{\alpha \in \Gamma : U_\alpha \cap K \neq \emptyset\}$ is finite if and only if $\mathscr{U}$ is locally finite.
5. Let $\delta : (f_\alpha) \in C^0(\mathscr{U}, \mathscr{O}) \to \delta(f_\alpha) \in C^1(\mathscr{U}, \mathscr{O})$ be defined as (5.2.34). Show that $\delta(f_\alpha) = 0$ if and only if there exists an element $f \in \mathscr{O}(\mathfrak{D})$ with $f_\alpha = f|_{U_\alpha}$ $(\forall \alpha \in \Gamma)$.
6. Prove (5.2.39).
7. Let $\mathfrak{D}(/\mathbf{C}^n)$ be a domain and let $\varphi : \mathfrak{D} \to \mathbf{R}$ be an upper semi-continuous exhaustion function. Fix a point $p_0 \in \mathfrak{D}$. With an arbitrarily given monotone increasing sequence, $\varphi(p_0) < c_1 < c_\nu \nearrow \infty$ $(\nu = 2, 3, \ldots)$, we denote by $\mathfrak{D}_\nu$ the connected component of $\{\varphi < c_\nu\}$ containing p_0. Prove $\mathfrak{D} = \bigcup_{\nu=1}^{\infty} \mathfrak{D}_\nu$.
8. Give a proof of Theorem 5.3.2.
9. (Behnke–Stein Theorem) Let $\mathfrak{D}/\mathbf{C}^n$ be a domain and let $\mathfrak{D}_\nu \subset \mathfrak{D}$ $(\nu = 1, 2, \ldots)$ be a sequence of subdomains such that all $\mathfrak{D}_\nu$ are Stein, $\mathfrak{D}_\nu \subset \mathfrak{D}_{\nu+1}$, and $\bigcup_\nu \mathfrak{D}_\nu = \mathfrak{D}$. Then, prove that $\mathfrak{D}$ is Stein.
   Hint: Use Oka's Theorem of Boundary Distance Functions.
10. Let $\mathfrak{D}/\mathbf{C}^n$ be a Stein domain, and let $\mathscr{F} \subset \mathscr{O}(\mathfrak{D})$ be an infinite family. We define that $z \in \mathfrak{D}$ belongs $\mathfrak{D}(\mathscr{F})$ if there are a neighborhood $U \ni z$ and $M > 0$ such that
$$|f(z)| \leq M, \quad \forall z \in U, \forall f \in \mathscr{F}.$$
    Suppose that $\mathfrak{D}(\mathscr{F}) \neq \emptyset$.
    Then, prove that $\mathfrak{D}(\mathscr{F})$ is Stein as follows:
    a. For $z \in \mathfrak{D}$ we set $\varphi(z) = \sup_{f \in \mathscr{F}} |f(z)| \leq \infty$, and furthermore
$$\tilde{\varphi}(z) = \overline{\lim_{p \to z}}\, \varphi(p), \quad \Omega = \{z \in \mathfrak{D} : \tilde{\varphi}(z) < \infty\}.$$

[11] For the wording, see p. ix, Footnote 3).

Show that $\Omega = \mathfrak{D}(\mathscr{F})$.

b. Show that $\tilde{\varphi} : \mathfrak{D}(\mathscr{F}) \to \mathbf{R}$ is a plurisubharmonic function.

c. Let $\psi : \mathfrak{D} \to [-\infty, \infty)$ be a pseudoconvex exhaustion function, and set

$$\Phi(z) = \max\{\tilde{\varphi}(z), \psi(z)\} \in \mathbf{R}, \quad z \in \mathfrak{D}(\mathscr{F}).$$

Show that $\Phi(z)$ is an exhaustion function on $\mathfrak{D}(\mathscr{F})$.[12]

11. A family $\mathscr{F}$ of holomorphic functions (or maps into $\mathbf{C}^n$) on a domain Ω is called a *normal family* if every sequence of $\mathscr{F}$ admits a locally uniformly convergent subsequence.

Let $f : \mathbf{C}^n \to \mathbf{C}^n$ be a holomorphic map, and let $f^\nu := \overbrace{f \circ \cdots \circ f}^{\nu}$ denote the νth iterates. Set $\mathscr{F} = \{f^\nu : \nu \in \mathbf{N}\}$. $\mathscr{F}$ is called a complex dynamical system on $\mathbf{C}^n$. We denote by $F(f)$, called a Fatou set, the set of all points $z \in \mathbf{C}^n$ having a neighborhood U such that the restriction $\mathscr{F}|_U = \{f^\nu|_U : \nu \in \mathbf{N}\}$ is a normal family.
Then, prove that $F(f)$ is Stein.

12. Extend Corollary 4.3.81 for unramified domains over $\mathbf{C}^n$.

[12] In general, $\tilde{\varphi}(z)$ is upper semi-continuous, but is not assumed to be continuous, so that if the pseudoconvex exhaustion function is restricted to being continuous, $\mathfrak{D}(\mathscr{F})$ being Stein does not follow.

Afterword — Historical Comments

The Pseudoconvexity Problem has an origin in a study of the shape of singularities of holomorphic or meromorphic functions in domains of several variables. According to K. Oka's lecture [51], K. Weierstrass thought that the shape of singularities is arbitrary, and hence the study in that direction had been delayed considerably. Later, however, into the 1900s, due to the studies of E. Fabry, W.G. Osgood, F. Hartogs, E.E. Levi, etc., it was found that the singularities have a specific feature; it is the notion of "pseudoconvexity". Behnke and Thullen published a monograph [4] in 1934 that summarized the current results and problems. K. Oka read it, was attracted to the Three Big Problems described there, and began to concentrate on the study of them, discontinuing his research on the composition problem of holomorphic functions and some works related to complex dynamics. These Three Big Problems were then not thought to be solvable. The chronicle of the affirmative solutions of the problems is roughly as follows (cf. [42]):

(i) The 1st Problem (Approximation) and the 2nd Problem (Cousin I, II): The case of univalent domains of general dimension by Oka I (1936) — III (1939) ([48], [49]).

(ii) The 3rd Problem (Pseudoconvexity): The case of univalent domains of dimension $n = 2$ by Oka [47] (1941) (announcement), Oka VI (1942) ([48], [49]).

(iii) The 3rd Problem (Pseudoconvexity) together with the 1st and 2nd Problems: The case of unramified domains over $\mathbf{C}^n$ of general dimension $n \geq 2$ by Oka's unpublished papers VII–XI of 1943 ([50], [44]).
In VII (1943) Oka formulated and proved a kind of "primitive coherence" which is sufficient for the proofs.

(iv) The 3rd Problem (Pseudoconvexity): The case of univalent domains of general dimension $n \geq 2$ by S. Hitotsumatsu [28] (1949).[1]

(v) The 3rd Problem (Pseudoconvexity): The case of unramified domains over $\mathbf{C}^n$ of general dimension $n \geq 2$ by Oka IX (1953) ([48], [49]).

[1] Probably because the paper is written in Japanese, it is rarely referred to, unfortunately.

J. Noguchi, *Basic Oka Theory in Several Complex Variables*, Universitext,
https://doi.org/10.1007/978-981-97-2056-9

(vi) The 3rd Problem (Pseudoconvexity): The case of univalent domains of general dimension $n \geq 2$ by Bremermann [8] (1954) and by Norguet [45] (1954), independently. Their method relies on Weil's integral formula, which was developed in (ii) above and is the same as Hitotsumatsu's in (iv) above.

(vii) The 3rd Problem (Pseudoconvexity): Another proof of (v) above by Grauert [22] (1958).

Just after Oka wrote VII (1948/50), the cohomology theory due to Cartan–Serre for coherent sheaves was initiated, and then after Oka IX (1953) the theory of $\bar{\partial}$-equation as a part of elliptic partial differential equations due to Morrey, Kohn, Hörmander, ... , developed; new theories and new analytic methods were introduced (cf., e.g., Hörmander [30], Henkin–Leiter [27], Range [53], Laurent-Thiébaut [32], Ohsawa [46], etc.).

After solving affirmatively the Three Big Problems in (iii) above without publishing them, Oka continued the study in order to establish them for ramified domains over $\mathbf{C}^n$ which may have singularities, and invented the notion of "coherence" or "idéal de domaines indéterminés" in his words (*Oka's Three Coherence Theorems* proved by Oka VII (1948/50) and VIII (1951); as for the 2nd Coherence Theorem, H. Cartan (1950) gave his own proof; cf. [39] Chap. 9). From the viewpoint of analytic continuation of analytic functions ramification points are naturally introduced, and so there is a definite necessity to deal with ramifications and singularities in domains. Oka's solution of the Pseudoconvexity Problem in 1943 was published ten years later as Oka IX (1953), where the proofs were modified in a form to use the 1st and 2nd Coherence Theorems of Oka VII and VIII. Oka himself wrote that Oka IX is an intermediate report towards the final goal. Oka's Three Coherence Theorems developed into the theory of Oka–Cartan–Serre–Grauert– $\cdots$, giving a broad influence in mathematics, and changing the description style of complex analysis and related fields after the 1950s; the influence has reached even to theoretical physics as mentioned in M. Gonokami [2] [21].

As mentioned after Problem 5.3.11, the Pseudoconvexity Problem for ramified domains was countered by J.E. Fornæss [16]. This created uncertainty in how to formulate the problem (*Oka's Dream*, Oka VII Introduction). While there are some affirmative results (cf. Narasimhan [35], Noguchi [41]), it is expected to be explored more.

---

[2] He was the President of the University of Tokyo. His research subjects are quantum electronics and quantum photonics.

References

1. Makoto Abe, Tube domains over $\mathbf{C}^n$, Memoirs Fac. Sci., Kyushu Univ. Ser. A **39** (2) (1985), 253–259.
2. Y. Aihara and J. Noguchi, Complex Analysis in One and Several Variables (in Japanese), Shokabo, Tokyo, 2024.
3. A. Andreotti and R. Narasimhan, Oka's Heftungslemma and the Levi problem, Trans. Amer. Math. Soc. **111** (1964), 345–366.
4. H. Behnke and P. Thullen, Theorie der Funktionen mehrerer komplexer Veränderlichen, Ergeb. Math. Grenzgeb. Bd. 3, Springer-Verlag, Heidelberg, 1934.
5. L. Bers, Introduction to Several Complex Variables, Lecture Notes, Courant Inst. Math. Sci., New York University, 1964.
6. S. Bochner, Bounded analytic functions in several variables and multiple Laplace integrals, Amer. J. Math. **59** (1937), 732–738.
7. ———, A theorem of analytic continuation of functions in several variables, Ann. Math. **39** no. 1 (1938), 14–19.
8. H.J. Bremermann, Über die Äquivalenz der pseudokonvexen Gebiete und der Holomorphiegebiete im Raum von n komplexen Veränderlichen, Math. Ann. **128** (1954), 63–91.
9. H. Cartan, Sur les matrices holomorphes de n variables complexes, J. Math. pure appl. **19** (1940), 1–26.
10. H. Cartan, Idéaux et modules de fonctions analytiques de variables complexes Bull. Soc. Math. France **78** (1950), 29–64.
11. H. Cartan und P. Thullen, Regularitäts- und Konvergenzbereiche, Math. Ann. **106** (1932), 617–647.
12. P. Cousin, Sur les fonctions de n variables complexes, Acta Math. **19** (1895), 1–61.
13. A. Czarnecki, M. Kulczycki and W. Lubawski, On the connectedness of boundary and complement for domains, Ann. Polon. Math. **103** (2) (2012), 189–191.
14. F. Docquier und H. Grauert, Levischen problem und Rungescher Satz für Teilgebiete Steinscher Mannigfaltigkeiten, Math. Ann. **140** (1960), 94–123.
15. E. Fabry, Sur les rayons de convergence d'une série double, C. R. Acad. Sci., Paris **134** (1902), 1190–1192.
16. J.E. Fornæss, A counterexample for the Levi problem for branched Riemann domains over $\mathbf{C}^n$, Math. Ann. **234** (1978), 275–277.
17. F. Forstnerič, Stein Manifold and Holomorphic Mappings, Ergeb. Math. Grenzgeb. 3, Vol. 56, Springer-Verlag, Berlin–Heidelberg, 2011: 2nd Edition, 2017.
18. K. Fritzsche and H. Grauert, From Holomorphic Functions to Complex Manifolds, G.T.M. 213, Springer-Verlag, New York, 2002.
19. R. Fujita, Domaines sans point critique intérieur sur l'espace projectif complexe, J. Math. Soc. Jpn. **15** (1963), 443–473.

J. Noguchi, *Basic Oka Theory in Several Complex Variables*, Universitext, https://doi.org/10.1007/978-981-97-2056-9

20. R. Godement, Topologie Algébrique et Théorie des Faisceaux, Hermann, Paris, 1964.
21. M. Gonokami, Address by the President of the University of Tokyo at the 2020 Autumn Semester Diploma Presentation and Commencement Ceremony, 18 Sep. 2020; https://www.u-tokyo.ac.jp/ja/about/president/b_message02_07.html.
22. H. Grauert, On Levi's problem and the imbedding of real-analytic manifolds, Ann. Math. **68** (1958), 460–472.
23. ———, Über Modifikationen und exzeptionelle analytische Mengen, Math. Ann. **146** (1962), 331–368.
24. H. Grauert and R. Remmert, Theorie der Steinschen Räume, Springer-Verlag, 1977. Translated into English by Alan Huckleberry, Theory of Stein Spaces, Springer-Verlag, 1979. Translated into Japanese by K. Miyajima, Stein Kukan Ron, Springer, Tokyo, 2009.
25. ———, Coherent Analytic Sheaves, Grundl. der Math. Wissen. vol. 265, Springer-Verlag, Berlin, 1984.
26. R.C. Gunning and H. Rossi, Analytic Functions of Several Complex Variables, Prentice-Hall; AMS Chelsea Publ., Amer. Math. Soc., Providence Rhode Island, 1965.
27. G. Henkin and J. Leiterer, Theory of Functions on Complex Manifolds, Beikhäuser Verlag, Basel-Bostaon-Stuttgart, 1984.
28. S. Hitotsumatsu, On Oka's Heftungs Theorem (in Japanese), Sugaku **1** (4) (1949), 304–307, Math. Soc. Jpn.
29. ———, Analytic Function Theory of Several Variables, Baifukan, Tokyo, 1960.
30. L. Hörmander, Introduction to Complex Analysis in Several Variables, 3rd Edition, 1990; 1st Edition, 1966, North-Holland.
31. H. Komatsu, A local version of Bochner's tube theorem, J. Fac. Sci. Univ. Tokyo, Ser. IA **19** (1972), 201–214.
32. C. Laurent-Thiébaut, Holomorphic Function Theory in Several Variables, Springer-Verlag, London, 2011.
33. I. Lieb, Le problème de Levi, Gazette Math., Soc. Math. France **115** (2008), 9–34.
34. R. Narasimhan, The Levi problem for complex spaces, Math. Ann. **142** (1961), 355-365; ibid. II, Math. Ann. **146** (1962), 195–216.
35. ———, The Levi Problem and pseudo-convex domains, l'Enseign. Math. **24** (1978), 161–172.
36. T. Nishino, Sur les espaces analytiques holomorphiquement complets, J. Math. Kyoto Univ. **1** (1962), 247–254.
37. ——, Function Theory in Several Complex Variables, transl. by N. Levenberg and H. Yamaguchi, Amer. Math. Soc. Providence, R.I., 2001; Jpn. Edition, The University Tokyo Press, Tokyo, 1996.
38. J. Noguchi, Introduction to Complex Analysis, MMONO **168**, Amer. Math. Soc. Rhode Island, 1997; Jpn. First Edition, Shokabo, Tokyo, 1993.
39. ———, Analytic Function Theory of Several Variables—Elements of Oka's Coherence, Springer, Singapore, 2016, corrected version 2023; Jpn. 1st Edition 2013, 2nd Edition 2019, Asakura Shoten, Tokyo.
40. ———, A remark to a division algorithm in the proof of Oka's First Coherence Theorem, Internat. J. Math. **26** No. 4 (2015); DOI: 10.1142/S0129167X15400054.
41. ———, Inverse of Abelian integrals and ramified Riemann domains, Math. Ann. **367** (2017), 229–249; DOI: 10.1007/s00208-016-1384-3.
42. ———, A brief chronicle of the Levi (Hartogs' Inverse) Problem, Coherence and an open problem, ICCM Notices **7** No. 2 (2019), 19–24: DOI: https://dx.doi.org/10.4310/ICCM.2019.v7.n2.a2 ; arXiv:1807.08246.
43. ———, A New Introduction to Oka Theory—Basics of Several Complex Variables (in Japanese), Shokabo, Tokyo, 2021.
44. ———, On Kiyoshi Oka's unpublished papers in 1943, ICCM Notices **10** no. 1 (2022), 44–70; arXiv:2103.07647.
45. F. Norguet, Sur les domains d'holomorphie des fonctions uniformes de plusieurs variables complexes (Passage du local au global), Bull. Soc. Math. France **82** (1954), 137–159.

46. T. Ohsawa, L^2 Approaches in Several Complex Variables, Springer Mono. Math., Springer, Tokyo, 2015.
47. K. Oka, Sur les domaines pseudoconvexes, Proc. Imperial Acad. Tokyo (1941), 7–10.
48. ———, Sur les fonctions analytiques de plusieurs variables, Iwanami Shoten, Tokyo, 1961.
49. ———, Collected Works, Translated by R. Narasimhan, Ed. R. Remmert, Springer-Verlag, Berlin–Heidelberg–New York–Tokyo, 1984.
50. K. Oka Digital Archives, Library of Nara Women's University, URL https://www.nara-wu.ac.jp/aic/gdb/nwugdb/oka/.
51. ———, On analytic functions of several variables, Lecture at Yukawa Institute for Theoretical Physics, Kyoto University, 1964. https://www.nara-wu.ac.jp/aic/gdb/nwugdb/oka/fram/mi.html.
52. H. Poincaré, Les fonctions analytiques de deux variables et la représentation conforme, Rend. Ciro. Math. Palermo **23** (1907), 185–220.
53. R.M. Range, Holomorphic Functions and Integral Representations in Several Complex Variables, GTM 108, Springer-Verlag, New York-Berlin-Heidelberg-Tokyo, 1986.
54. K. Reinhardt, Über Abbildungen durch analytische Funktionen zweier Veränderlicher, Math. Ann. bf 83 (1921), 211–255.
55. W. Rudin, Lectures on the Edge-of-the-wedge Theorem, CBMS No. 6, Amer. Math. Soc,, Providence R.I., 1971.
56. K. Stein, Zur Theorie der Funktionen mehrerer komplexen Veränderlichen. Die Regularitätshüllen niederdimensionaler Mannigfaltigkeit, Math. Ann. **114** (1937), 543–569.
57. A. Takeuchi, Domaines pseudoconvexes infinis et la métrique riemannienne dans un espace projectif, J. Math. Soc. Jpn. **16** (1964), 159–181.
58. V. Vâjâitu, Stein spaces with plurisubharmonic bounded exhaustion functions, Math. Ann. **336** (2006), 539–550.
59. ——, Locally Stein domains over holomorphically convex manifolds, J. Math. Kyoto Univ. **48** (1) (2008), 133–148.

Index

A
absolute convergence, 7
additive group (commutative group, abelian group), 31
analytic continuation, 14, 118
analytic function, 4
analytic interpolation problem, 112
analytic local ring, 33
analytic polyhedral domain, 83
analytic polyhedron, 83, 123
analytic (sub)set, 24
analytic sheaf, 34
anti-holomorphic partial differential operator, 3
Approximation Theorem, 18, 85, 89, 123

B
Baire Category Theorem, 140
Baire space, 139
Baire vector space, 176
ball, 1
ball neighborhood, 117
Banach's Open Map Theorem, 177
base of a tube domain, 154
base point, 116
Behnke–Stein Theorem, 206
biholomorphic, 23
bijection, xv
bijective, xv
Bochner's Tube Theorem, 156
Bogolyubov, Nikolay (1909–1992), 172
boundary distance function, 74, 78, 117
bounded domain, 157
bumping method of Grauert, 201

C
Cartan, Henri (1904–2008), vii, 39, 53
Cartan's matrix decomposition, 57
Cartan–Thullen Theorem, 78, 81
Cauchy condition (uniform), 7
Cauchy integral transform, 109
Cauchy kernel, 5
Cauchy's integral formula, 4
Cauchy–Riemann equations, 4
Čech cohomology, 195
C^k-pseudoconvex, 146
closed domain, 2
closed polydisk, 2
closed rectangle, 57
coboundary, 194
coboundary operator, 194
cochain, 193
cocycle, 193
cocycle condition, 193
coherent analytic sheaf, 39
coherent sheaf, 39
cohomology, 195
cohomology class (Čech), 195
commutative ring, 32
compact operator, 190
completely continuous, 190
complex hypersurface, 24
complex Jacobi matrix, 21
complex Jacobian, 21
complex submanifold, 27
continuous, 3
continuous Cousin data, 93
Continuous Cousin Problem, 93
convergent majorant, 7
convergent power series, 8
convex (affine), 2
convex function, 155

J. Noguchi, *Basic Oka Theory in Several Complex Variables*, Universitext,
https://doi.org/10.1007/978-981-97-2056-9

convex hull, 2
Cousin, Pierre (1867–1933), 28
Cousin decomposition, 21, 28
Cousin I data, 91
Cousin I Problem, 91
Cousin II data, 101
Cousin II Problem, 101
Cousin integral, 20
covering cohomology, 194
covering cohomology class, 194
cuboid, 57
cylinder, 2
cylinder domain, 2

D
defining function of $\mathfrak{D}$, 159
$\bar{\partial}$-closed, 109
$\bar{\partial}$-equation, 108
dimension of cuboid, 63
direct limit, 195
direct product sheaf, 34
disk, 1
Dolbeault, Pierre (1924–2015), 111
Dolbeault's Lemma, 110
domain, 2
domain (over $\mathbf{C}^n$), 116
domain of convergence, 9
domain of existence, 75, 119
domain of holomorphy, 75, 119

E
Edge of the Wedge Theorem, 172
entire function, 11
envelope of holomorphy, 119
equivalence relation, xv
euclidean norm, 1
exact sequence, 67
exhaustion function, 146
Extension Lemma, 68

F
$\mathscr{F}$-convex domain, 75
$\mathscr{F}$-convex hull, 75
$\mathscr{F}$-convex set, 75
$\mathscr{F}$-envelope, 119
$\mathscr{F}$-separable, 118
Fabry, Eugène (1856–1944), 9, 28, 209
filter base, 124
finite generator system, 36
finitely sheeted, 117
Fréchet space, 176
Fredholm integral equation of the second kind, vi, 188
Fredholm integral equation of the second kind type, vi, vii, 184, 188
Fujita, R., 204

G
generalized interpolation, 113
geometric ideal sheaf, 38
geometric syzygy, 66
Grauert, Hans (1930–2011), 198

H
Hadamard's three circles theorem, 173
Hartogs, Friedlich (1874–1943), v, 14, 98,209
Hartogs domain, 16
Hartogs Extension, 98
Hartogs pseudoconvex, 174
Hartogs separate analyticity theorem, 142
Hartogs' Inverse Problem, 165
Hartogs' phenomenon, 14
Hartogs' radius, 174
Hartogs' Theorem, 98
Heftungslemma, 179
holomorphic, 4, 117
holomorphic extension, 118
holomorphic function on a submanifold, 28
holomorphic local coordinate system, 27
holomorphic map (or mapping), 22, 118
holomorphic partial differential operator, 3
holomorphic tangent space, 162
holomorphically convex, 75, 122

holomorphically convex domain, 75
holomorphically convex hull, 75
holomorphically convex set, 75
holomorphically isomorphic, 23
holomorphically separable, 118
homogeneous polynomial expansion, 11
homomorphism (ring), 118

I
ideal, 32
ideal boundary, 125
ideal boundary point, 125
idéal de domaines indéterminés, 70
ideal sheaf of A, 37
ideal sheaves of analytic subsets, 38
ideal sheaves of complex submanifolds, 38
Identity Theorem, 12
implicit function theorem, 22
induced continuous Cousin data, 94
induction on cuboid dimension, 67, 68, 94
inductive limit, 195
infinitely sheeted, 117
injection, xv
injective, xv
interpolation problem, 112
inverse function theorem, 22
isomorphism (ring), 118

J
Jensen's formula, 132
Joku-Iko Principle, v, vi, vii, 63, 68

L
Lelong, Pierre (1912–2011), 134, 172
Levi, Eugenio Elia (1883–1917), v, 162, 172, 209
Levi condition, 163
Levi form, 134
Levi polynomial, 166
Levi pseudoconvex, 163
Levi pseudoconvex point, 163
Levi's Problem, 165
Levi–Krzoska condition, 163
Levi–Oka polynomial, 166
Liouville's Theorem, 12
local extension, 28
local pseudoconvexity, 158
locally finite, 36, 92
locally finite generator system, 36
logarithmically convex, 10

M
main part, 100
majorant, 7
Maximum Principle, 13, 135
merged system, 63
meromorphic function, 89
Mittag-Leffler Theorem, 100
module, 32
monotone decreasing, xv
monotone increasing, xv
multi-sheeted, 117
multivalent, 117

N
Nishino, Toshio (1932–2005), 203
non-singular point, 27
normal family, 207

O
Oka, Kiyoshi (1901–1978), v, vii, 47, 172
Oka Extension, 68
Oka Extension with Estimate, 179, 182, 185
Oka hypersurface, 166
Oka map, 83
Oka Principle, 102, 125
Oka Syzygy, 63
Oka's First Coherence Theorem, 39, 47
Oka's Heftungslemma, 179, 180
Oka's Joku-Iko Principle, 63, 68,
Oka's Pseudoconvexity Theorem, 202
Oka's Second Coherence Theorem, 39
Oka's Theorem of Boundary Distance Functions, 83, 145
Oka's Third Coherence Theorem, 40

Oka–Weil Approximation Theorem, 85, 123
open ball, 1
Open Map Theorem, 177
operator norm, 54
order of zero, 24
Osgood, William Fogg (1864–1943), 98, 209

P

partial sum, 7
partition of unity, 93
plurisubharmonic, 134, 139
Poincaré, Henri (1854–1912), 23, 28
point over z, 116
polar set, 90, 91
pole, 90
polydisk, 2, 117
polydisk neighborhood, 117
polynomial polyhedral domain, 83
polynomial polyhedron, 83
polynomial ring, 33
polynomial-like element, 49
polynomial-like germ, 49
polynomial-like section, 49
polynomially convex domain, 75
polynomially convex set, 75
polyradius, 2
power series, 8
proper, xvi
pseudoconvex, 134, 139, 146
pseudoconvex boundary, 158
pseudoconvex boundary point, 158
pseudoconvex hull, 171
Pseudoconvexity Problem, 165
Pseudoconvexity Problem I, 146
Pseudoconvexity Problem II, 147
Pseudoconvexity Problem III, 161
Pseudoconvexity Problem for ramified domains, 204

R

ramified domain (over $\mathbf{C}^n$), 116
refinement, 92
Reinhardt domain, 29, 173
relation sheaf, 39
relative boundary, 125
relative boundary point, 125
relative isomorphism, 118
relative map, 118
relatively compact, xv
relatively compact hull, 73
relatively isomorphic, 118
Remmert, Reinhold (1930–2016), 172
removability of totally real subspaces, 16
restriction of sheaves, 34
Riemann's Extension Theorem, 26
Riemann's Mapping Theorem, 23
ring, 32
Runge Approximation Theorem, 89
Runge pair, 86, 123

S

schlicht domain, 116
Schwartz's Fredholm Theorem, 190
section, 34, 35
section space, 35
semi-norm, 175
separate analyticity, 139
separately analytic, 4
separately holomorphic, 4
sequence of functions, 6
series of functions, 7
sheaf, 33
sheaf of ideals, 34
sheet number, 117
σ-compact, 121
singular point, 27
smooth point, 27
smoothing, 137
solvable (analytically), 101
solvable (topologically), 101
standard coordinate system, 26, 43
standard polydisk, 26, 43
Stein, Karl (1913–2000), 122, 156
Stein covering, 197
Stein domain, 122
strong Levi condition, 163

strong Levi–Krzoska condition, 163
strongly Levi pseudoconvex, 163
strongly Levi pseudoconvex point, 163
strongly plurisubharmonic, 134
strongly pseudoconvex, 134
strongly pseudoconvex boundary, 159
strongly pseudoconvex boundary point, 159
strongly pseudoconvex domain, 159
subharmonic, 133
sublevel set, 146
submodule, 32
sup-norm (supremum norm), 18
support, xv
surjection, xv
surjective, xv
syzygy, 67

T

Takeuchi, A., 203
Three Big Problems, vi
topological solution, 101
topological vector space, 175
topologically solvable, 101
totally real subspace, 16
tube, 154
tube domain, 154
Tube Theorem, 156

U

unit, 32
unit ball, 1
unit disk, 1
unit element, 32
unit polydisk, 2
univalent domain, 116
unramified domain, 116
unramified Riemann domain, 116

W

weakly holomorphicaly separable, 204
Weierstrass decomposition, 47
Weierstrass polynomial, 46
Weierstrass Theorem, 105
Weierstrass' Preparation Theorem, 43
Weil, André (1906–1998), 85
Weil condition, 126

Z

zero set (of meromorphic functions), 90
zero sheaf, 34

Symbols

$A \Subset B$, xv
$\hat{A}_{\mathscr{F}}$, 75
$\hat{A}_{\Omega}$, 75
$\hat{A}_{\mathrm{poly}}$, 75
$|\alpha|$, 3
$\alpha!$, 3
$\mathrm{B}(a;r)$, 1
$\mathrm{B}(p;R)$, 117
$\mathrm{B}(r)$, 1
$\mathscr{C}^*(U)$, 101
$\mathbf{C}[(z_j)]$, 4
$\mathbf{C}[z,\varphi]$, 83
$\mathrm{ch}(B)$, 2
$\mathrm{ch}(R)$, 154
C^k, xv
$\mathscr{C}^k(U)$, xv
$\mathscr{C}^k(W)$, 117
$d(A,B)$, 74
$d(z,A)$, 74
$d(z,\partial\Omega)$, 74
$d\bar{z}_j$, 108
$\partial^*\mathfrak{D}$, 125
∂^α, 4
$\bar{\partial}$, 108
$\Delta(a;r)$, 1
$\Delta(r)$, 1
$\bar{\Delta}(a;r)$, 1
$\bar{\Delta}(r)$, 1
$\delta_{\mathrm{B}(R)}(p,\partial\mathfrak{D})$, 117
$\delta_{\mathrm{B}(R)}(z,\partial\Omega)$, 78
$\delta_{\mathrm{P}\Delta}(A,\partial\Omega)$, 79
$\delta_{\mathrm{P}\Delta}(p,\partial\mathfrak{D})$, 117
$\delta_{\mathrm{P}\Delta}(z,\partial\Omega)$, 78
∂U, xv
dz_j, 108
f_a, 33
$f|_E$, xv
$\mathscr{F}|_U$, 34
$\underline{f}_p$, 118
$\Gamma(E,\mathscr{F})$, 35
$\Gamma(U,\mathscr{F})$, 35
$\mathscr{I}\langle A\rangle$, 37
$\mathfrak{I}z$, xv
$|\varphi(q)| = \infty$, 181
$\cong$, 23
$\widehat{K}_{\mathscr{F}}$, 75
$\widehat{K}_{\mathscr{P}(\mathfrak{D})}$, 171
$K^{\star}_{\Omega}$, 73
$L[\varphi](v)$, 134
$L[\varphi](z;v)$, 134
$M(S,T)$, 56
$\mathscr{M}(\Omega)$, 89
$M_{\varphi}(a;r)$, 132
$N(S,T)$, 56
$\|A\|$, 54
$\|A\|_E$, 54
$\|f\|_A$, 18
$\mathscr{O}(\Omega)$, 4
$\mathscr{O}^*(U)$, 101
$\mathscr{O}_a$, 33
$\mathscr{O}(E)$, 35
$\Omega(f)$, 9
Ω_{H}, 16
$\mathscr{O}_{\Omega}$, 33
$\mathscr{O}^q_{\Omega}$, 34
$o(r^{\alpha})$, xvi
$\mathrm{ord}_a f$, 24
$\mathscr{O}(S)$, 28
$\mathscr{O}(W)$, 117
$\mathscr{P}(\mathfrak{D})$, 139
$\mathscr{P}(U)$, 134
$p(z)$, 117
$\mathrm{P}\Delta(a;(r_j))$, 2
$\mathrm{P}\Delta(p;r)$, 117
$\mathrm{P}\Delta((r_j))$, 2
$\mathscr{P}^k(\mathfrak{D})$, 139
$\mathscr{P}^k(U)$, 134
r^{α}, 8
$\mathfrak{R}z$, xv
$\mathbf{R}_+$, xv
$R[X_1,\ldots,X_N]$, 33
$\mathscr{U} < \mathscr{V}$, 92
Z_0, 90
z^{α}, 8
Z_{∞}, 90
$\|z\|_{\mathrm{P}\Delta}$, 78
$\mathbf{Z}_+$, xv

J. Noguchi, *Basic Oka Theory in Several Complex Variables*, Universitext,
https://doi.org/10.1007/978-981-97-2056-9

Printed in the United States
by Baker & Taylor Publisher Services